PLASTICS
TECHNOLOGY

PLASTICS TECHNOLOGY

ROBERT V. MILBY
Associate Professor
Los Angeles Trade-Technical College
Los Angeles, California

McGRAW-HILL BOOK COMPANY
New York   San Francisco   St. Louis   Düsseldorf   Johannesburg
Kuala Lumpur   London   Mexico   Montreal   New Delhi
Panama   Rio de Janeiro   Singapore   Sydney   Toronto

TO LOIS

This book was set in Univers 55 by Holmes Typography, Inc. The editors were
Robert Buchanan and Eva Marie Strock, the designer was Janet Durey
Bollow, and the production supervisor was Michael A. Ungersma. The
drawings were done by John Foster.

PLASTICS TECHNOLOGY

Printed in the United States of America.

LIBRARY OF CONGRESS CATALOGING IN PUBLICATION DATA

Milby, Robert V
   Plastics technology.

  Bibliography :   p. 566
   1.   Plastics.   2.   Technicians in industry.
I.   Title.
TP1120.M499   668.4   72–7408
ISBN   07–041918–3

 8 9 10 11 12 KPKP 8 9 8 7 6 5 4 3

CONTENTS

PREFACE

Within the plastics industry (as well as in others such as electronics, chemical, and metallurgy), there is a continuing need for a person having a particular type of skill— a *technician*. The position of technician is relatively new in our industrial complex. It was created automatically by the requirements of technological developments in industry for people who have the necessary technical background (short of an engineering degree), plus the required manipulatory and verbal skills that enable them to function on a level between production and engineering.

All technicians do not possess or acquire the *same* skills, nor do they need them. Appendix A sets forth guidelines for the various facets of the plastics industry in which the technician can participate.

The technician's role is prestigious, satisfying, and financially rewarding. It can be a stepping-stone to a management position or to a teaching career.

This book is designed for students seeking direction toward a career. It is also aimed at those already employed in the plastics industry who may find it helpful toward advancement or as a reference. Even though the information is presented in depth, the approach assumes no previous knowledge of the subject or previous training in any process.

The book is based on many years of training technicians, plus broad industrial experience. The author believes in a close correlation between theoretical concepts and the practical approach. For this reason, the student not only is made acquainted with the methods used in industry, but with the reasons behind the methods and the results obtained.

Much emphasis is placed on safety, particularly in handling potentially hazardous chemicals, in operating processing machinery, and in using auxiliary and testing equipment.

This book is divided into three sections:

Sec. I: Thermoset Materials and Processes

Sec. II: Thermoplastic Materials and Processes

Sec. III: Test Procedures

The step-by-step procedures for molding processes, laminating, fabricating, casting, and testing are designed to involve the reader in the actual situation.

The author particularly acknowledges the assistance of William J. Dewar, Works Manager, Furane Plastics, Inc.; Jack Reinhold, President, Reinhold Plastics; and Wayne Scherer, Associate Professor and colleague, Los Angeles Trade-Technical College, for their technical assistance in preparing the manuscript. Ron Sheridan provided invaluable help in setting up and taking many of the photographs in the laboratory. My gratitude is also extended to Cleo Williams for providing her secretarial skills in preparing the manuscript. Finally, I wish to acknowledge the cooperation of the many students who have attended classes at Los Angeles Trade-Technical College and in their careers have proven the worthiness of relevant technical training.

The author is also grateful to the many companies that supplied photographs to supplement the text.

ROBERT V. MILBY

THERMOSET MATERIALS AND PROCESSES

A glance at a dictionary discloses that the word *plastic* means *capable of being formed*. This broad definition can be used to include such commonplace materials as clay, wax, cement, plaster, glass, and metals, for indeed these substances are plastic in that they can undergo change of shape under proper forming conditions and retain that new shape.

Archeological findings have unearthed much evidence that man has consistently used *natural* plastic materials such as mud (clay), glass (obsidian), silicates (cement), and metals (iron, copper, and gold) to fashion his dwellings, his tools, and his weapons, plus artifacts of religion, love, and recreation. Man's evolutionary development has also demonstrated that as he becomes more skilled his use of materials becomes increasingly more sophisticated. Historically, man has been characterized as living in the Stone Age, then progressing through a period when wood was predominant, and more recently as living in the Metals Age.

It is now generally accepted that modern civilization is entering the *Plastics Age*. Perhaps we should call the next historical-industrial period the *Age of Synthetics* or the *Age of Polymers*. There is a vast difference between the plastic materials that provide the basis for our present and already extensive plastics industry (as it is now called) and materials that were predominant during earlier historical times. The plastics materials that form the basis of this text do not occur in nature. They are man-made; in other words, they are synthetically prepared from simple chemicals found everywhere in nature. By today's standards, plastic is accepted as a particular type of material able to stand on its own merits and meet challenges never before presented.

Modern man cannot mine for plastics as he does for coal and minerals. He cannot find plastics materials growing as do plants and trees. Yet nature has provided an abundance of raw materials from which contemporary man has learned to syn-

1

thesize plastics for creating all the objects necessary to meet his basic and aesthetic needs. Technological advances in mining and refining minerals, in extraction and separation of chemicals, and in controlling chemical reactions to obtain pure ingredients make it possible to manufacture products superior in many ways to wood, metal, and natural fibers. Through specific chemical processes, man has learned to extract basic chemicals from petroleum, natural gas, coal, air, water, and agricultural by-products and to recombine them into wholly new substances—namely, plastics.

In the course of this text we shall discover that the foundation of the plastics industry is the chemical industry and that it is worldwide in scope.

HISTORICAL BACKGROUND

Plastics are not new substances. In 1969 the plastics industry began its second century of progress. Until 1863, when a shortage of ivory (used for billiard balls) prompted the firm of Phelen and Collander to offer an award for any satisfactory replacement, only natural-occurring materials were used.

The challenge to find an acceptable material was accepted by two brothers who were not chemists but printers: John and Isaiah Hyatt, of Albany, New York. They reacted nitric acid with cotton to produce nitrated cellulose. This substance had previously been made by others for use as gun cotton. In that form, however, it did not meet the requirement for a plastic material because it could not be formed (shaped) and retain a new shape.

The basis for issuance of the first U.S. patent dealing with plastic material was the Hyatts' incorporation of camphor to make an otherwise intractable substance

I-1 Modern chemical processing plant for production of polystyrene. (*Reprinted by permission of Dow Chemical Co.*)

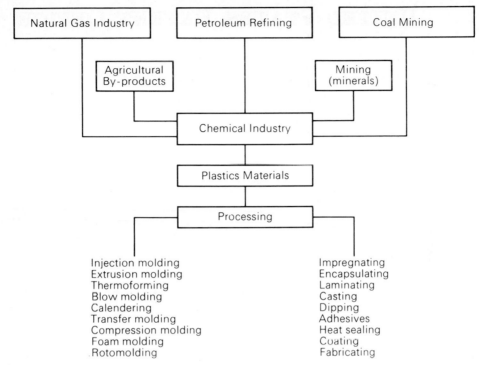

I-2   Organization of the plastics industry.

moldable. Camphor served as a chemical modifier (plasticizer) and was the fore-runner of yet another phase of chemical technology. Over a billion pounds of various plasticizers are now used annually to modify plastics for particular requirements.

The first plastic, trade named Celluloid, was to set the stage for establishing one of the most expansive, dynamic, and comprehensive industries ever to influence our way of life. Starting as one small company founded by the Hyatts to produce combs, brush backs, mirror frames, piano keys, and novelties, the industry a century later included 5,700 separate companies with annual sales of $7 billion and payroll of $1.5 billion.

The growth of the industry since the end of World War II has been 1,300 percent. On an annual basis, in the period 1960 to 1969 expansion developed at a rate of 15 percent/year, which was approximately four times the rate of our Gross National Product.

FUTURE TRENDS

What of the future? In the decade 1970 to 1980, it is expected that the volume of plastics used will expand another 800 percent while the use of steel products will increase only 56 percent for the same period. In terms of jobs, this phenomenal

United States Patent Office.

JOHN W. HYATT, JR., AND ISAIAH S. HYATT, OF ALBANY, NEW YORK.

IMPROVEMENT IN TREATING AND MOLDING PYROXYLINE.

Specification forming part of Letters Patent No. **105,338,** dated July 12, 1870.

We, JOHN W. HYATT, Jr., and ISAIAH S. HYATT, both of Albany, in the county of Albany and State of New York, have invented a new and Improved Process of Dissolving Pyroxyline and of Making Solid Collodion, of which the following is a specification:

Our invention consists, first, of so preparing pyroxyline that pigments and other substances in a powdered condition can be easily and thoroughly mixed therewith before the pyroxyline is subjected to the action of a solvent; secondly, of mixing with the pyroxyline so prepared any desirable pigment, coloring matter, or other material, and also any substance in a powdered state which may be vaporized or liquefied and converted into a solvent of pyroxyline by the application of heat; and, thirdly, of subjecting the compound so made to heavy pressure while heated, so that the least practicable proportion of solvent may be used in the production of solid collodion and its compounds.

The following is a description of our process: First, we prepare the pyroxyline by grinding it in water until it is reduced to a fine pulp by means of a machine similar to those employed in grinding paper-pulp. Second, any suitable white or coloring pigment or dyes, when desired, are then mixed and thoroughly ground with the pyroxyline pulp, or any powdered or granulated material is incorporated that may be adapted to the purpose of the manufacture. While the ground pulp is still wet we mix therewith finely-pulverized gum-camphor in about the proportions of one part (by weight) of the camphor to two parts of the pyroxyline when in a dry state. These proportions may be somewhat varied with good results. The gum-camphor may be comminuted by grinding in water, by pounding, or rolling; or, if preferred, the camphor may be dissolved in alcohol or spirits of wine, and then precipitated by adding water, the alcohol leaving the camphor and uniting with the water, when both the alcohol and the water may be drawn off, leaving the camphor in a very finely-divided state. After the powdered camphor is thoroughly mixed with the wet pyroxyline pulp and the other ingredients, we expel the water as far as possible by straining the mixture and subjecting it to an immense pressure in a perforated vessel. This leaves the mixture in a comparatively solid and dry state, but containing sufficient moisture to prevent the pyroxyline from burning or exploding during the remaining process. Third, the mixture is then placed in a mold of any appropriate form, which is heated by steam or by any convenient method, to from 150° to 300° Fahrenheit, to suit the proportion of camphor and the size of the mass, and is subjected to a heavy pressure in a hydraulic or other press. The heat, according to the degree used, vaporizes or liquefies the camphor, and thus converts it into a solvent of the pyroxyline. By introducing the solvent in the manner here described, and using heat to make the solvent active, and pressure to force it into intimate contact every particle of the pyroxyline, we are able to use a less proportion of this or any solvent which depends upon heat for its activity than has ever been known heretofore. After keeping the mixture under heat and pressure long enough to complete the solvent action throughout the mass it is cooled while still under pressure, and then taken out of the mold. The product is a solid about the consistency of sole-leather, but which subsequently becomes as hard as horn or bone by the evaporation of the camphor. Before the camphor is evaporated the material is easily softened by heat, and may be molded into any desirable form, which neither changes nor appreciably shrinks in hardening.

We are aware that camphor made into a solution with alcohol or other solvents of camphor has been used in a liquid state as a solvent of xyloidine. Such use of camphor as a solvent of pyroxyline we disclaim.

Claims.

We claim as our invention—

1. Grinding pyroxyline into a pulp, as and for the purpose described.

2. The use of finely-comminuted camphor-gum mixed with pyroxyline pulp, and rendered a solvent thereof by the application of heat, substantially as described.

3. In conjunction with such use of camphor-gum, the employment of pressure, and continuing the same until the mold and contents are cooled, substantially as described.

JOHN W. HYATT, JR.
ISAIAH S. HYATT.

Witnesses:
WM. H. SLINGERLAND,
C. M. HYATT.

growth will mean that 200,000 additional workers skilled in the ways of the industry must be found.

Plastics will markedly affect all other major industries, either competing with, or supplementing traditional materials and methods. The major markets for plastic products will be centered in the following areas:

Architectural   Wood, stucco, and brick housing will yield more and more to modular concepts prefabricated from plastics, which can be rapidly erected. Flooring, plumbing, wall covering, and lighting will incorporate more plastics materials.

Packaging   Containers, wrappings, and protective coverings will compete strongly with glass, metal, and paper.

Transportation   Automotive, marine, and allied uses will continue to absorb increasing amounts of plastics per unit.

Electronics   Tape recorders, computers, television, radio, telephone, and communications networks will provide a continually expanding market.

Furniture   Structural foam, replacing wood, will continue to enable designers to use plastics in many large items such as bedroom-suite furniture. Molded or thermoformed seating will remain a major area of application in institutions and stadiums as well as in households. The major share of cushioning and bedding will continue to be made of flexible plastic foams.

Appliances   Trends will continue in the use of plastics as major components of small and major appliances.

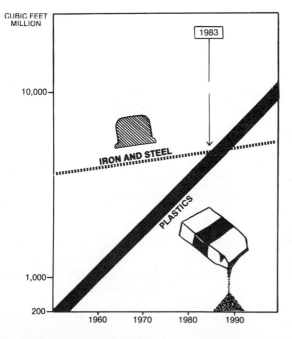

CUBIC FEET MILLION

1983

10,000—

IRON AND STEEL

PLASTICS

1,000—

200—

1960   1970   1980   1990

I-4   World consumption of raw materials by volume. In the 1980s the volume of plastics will exceed that of metals. Plastics is the nation's most spectacular growth industry. Since World War II it has been growing at a rate double that of all manufacturing. In the U.S. its growth is 4 times that of the GNP or about 15%/yr. In 1983 the Plastics Age will arrive. The volume of plastics used for all purposes will exceed the volume of metals. Between now and the 1980s plastics will affect every industry and every type of manufacturing. It will eventually become the basis of the American economy. (*Reprinted by permission of Society of the Plastics Industry, Inc. based on a figure from Reinhold Publishing Co.*)

I-5 The growth of plastics by selected industries. Estimated U.S. consumption of plastics in millions of pounds by selected industry. Plastics is a major industry deeply rooted in almost every type of manufacturing. It has displaced old, established materials such as glass, metals, paper, and wood, but it has also created entirely new markets and new product applications never before possible. (*Reprinted by permission of The Society of the Plastics Industry, Inc. based on a figure from Reinhold Publishing Co.*)

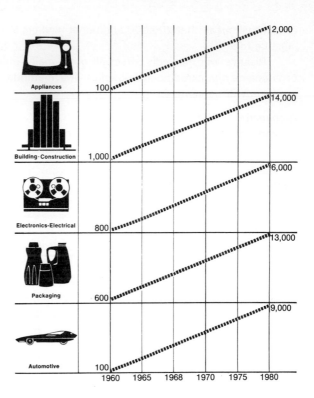

Agriculture   Film and sheeting will increasingly aid growers and ranchers in protecting crops, providing drainage, conserving water, and storing silage. Plastic conduit will be used extensively to convey irrigation water.

Recreation   More leisure time, plus expansion of recreation facilities, will promote greater use of plastics in vehicles, sporting goods, playing surfaces, and arenas.

Aircraft and Aerospace   Helicopters, both commercial and military, will continue to provide a large market for plastics. Fighter and reconnaissance craft, and military transports and cargo planes, are now and will constantly be built, and older planes will be remodeled and repaired.

Production of large commercial aircraft will continue to decline, but private crafts will still increase. Plastics will continue to play a dominant role in the development of improved aircraft as well as all space vehicles.

Although these markets represent areas of great growth potential, significant gains will also be evident in such fields as luggage, toys, apparel, and medical uses.

Chapters 2 to 6 and 15 to 22 describe the important plastics that serve these major markets, the properties that each contributes to a specific application, and the methods by which the product is manufactured and tested. (Lists of manufacturers may be found in Appendix B, grouped by chapter.)

GENERAL NATURE OF SYNTHETIC PLASTICS

1

Plastics literally surround us from cradle to grave. There is scarcely a facet of our lives that does not involve plastics in one form or another, yet few are aware of their scope. In addition to commonplace applications such as toothbrushes, wrapping film, pot handles, television housings, and upholstery, millions of pounds of plastics are converted annually into enamels and lacquers, adhesives, wax, floor tile, wall paper, and roofing materials. Plastics in fiber form provide the basis for synthetic wigs, indoor-outdoor carpet, turf for playing fields, and wearing apparel. Plastic foam is used to seal mine tunnels, and liquid plastic can improve oil-well drilling. Body implants, exterior prosthetics, and artificial dentures are made of plastic.

If the plastics industry and the many applications of its products could be described in one word, that word would be *adaptable*. Plastics take many forms and are capable of undergoing many unique processes to meet widely divergent requirements. It is not unusual to find the same type of plastic (an epoxy, for example) modified for use as a flexible adhesive, altered for encapsulating electric coils, or converted to provide a rigid resin for a molded gear. The epoxy, combined with fibrous glass, can form the structure of a laminated boat or may be mixed with asphalt for a roadway.

Modern life is constantly presenting new demands with which industry must cope. Because no other material is as adaptable as plastic, the natural inclination of researchers, designers, and engineers is to consider plastics as a means of providing solutions to the accelerating problems of housing, transportation, food production, and ecology.

The plastics industry is not dependent upon any single element as is steel; it is broadly based to furnish a variety of materials derived by combining elements that are obtained from many sources.

PROPERTIES OF PLASTIC

Plastics have made and will continue to make an impact in so many aspects of our lives because of the following characteristics.

Light weight

A typical plastic is one-half as heavy as aluminum and one-sixth the weight of steel. This feature provides many benefits: products that utilize plastics are easier to carry and use, and they often bring about a reduction of transportation costs. The aerospace industry, by replacing heavier metal components with plastics, realizes greater pay-loads in aircraft, rockets, and space vehicles.

Electrical insulation

Plastics, unless modified, do not conduct electricity, so they are classified as dielectric (nonconductive). However, incorporating conductive fillers such as metallic salts can render a plastic substance conductive.

The dielectric value of plastics makes them suitable for insulating wires, for wall plugs and switches, and for coating motor windings. Complicated electronic networks found in television, radio, computer, and telephonic circuits would be cumbersome

1-1 Marbon's Cycolac research-racing vehicle thermoformed in two pieces. Vehicle weighs 1,400 lb, with an overall length of 156 in. (*Reprinted by permission of Marbon Div., Borg-Warner Corp.*)

and less dependable without plastics as their primary insulation. Also, plastic housings for appliances and power tools are safeguards against electrical shock.

Pleasant to the touch

Plastics are poor heat conductors compared with metals, glass, or ceramics. As a result, one can come in contact with a plastic substance more safely at elevated or subnormal temperatures without injury or discomfort. This important property alone accounts for the use of plastics in such items as pot handles, steering wheels, electric-iron handles, and housings for appliances. Astronauts returning to earth are protected from extremely high temperatures by plastic heat shields having low thermal conductivity.

Color

Although many wood and metal products are provided with a surface treatment to protect or enhance them after fabrication, many plastics are precolored at the original point of manufacture. Thus the color is embodied throughout the entire structure, precluding the expensive steps required in painting; also, scratches are less evident. Many of the basic plastics are light colored or transparent, making it possible to incorporate any desired tint or degree of opacity. As many as 30,000 colors are cataloged by some plastics manufacturers so that any color match can be obtained.

Ability to be metallized

In addition to being able to undergo standard electroplating methods, plastics can be given a metallized coating to match chromium or other metal surfaces. Thin films

1-2 Switches are molded from a variety of plastics to provide insulation, color, and serviceability. [*Reprinted by permission of Plastics Engineering Co. (PLENCO).*]

1-3 Shower head metallized to match plumbing is produced from acetal resin. (*Reprinted by permission of E. I. du Pont de Nemours & Co.*)

of aluminum, copper, and other metals are deposited on plastic products for appearance or functional requirements.

Transparency

The ability of plastic to permit passage of light, while at the same time providing strength and ease of forming, has catapulted many plastic materials to the forefront in packaging, lighting, glazing, and lens applications.

1-4 Transparent sheets of Acrylite acrylic plastic form the basis for pool enclosure. (*Reprinted by permission of American Cyanamid Co.*)

Water resistance

A major attribute of most plastics is an ability to withstand fresh or salt water. Although wood and metal require protection from wet environments, most plastics are not seriously affected. As a result of inherent water resistance, plastics are used for irrigation and plumbing systems, pump and valve parts, boats, dinnerware, shower curtains, sporting goods, and rainwear.

Chemical resistance

Because plastics are formulated from specific chemicals, they resist the action of other chemicals *on a selective basis*. For example, one plastic may be unaffected by strong acids or alkalies, yet it may be attacked by cleaning fluid or motor oil. Plastics have replaced stainless-steel installations in plants that require fume ducting to carry off harmful chemical vapors. The chemical industry is now able to package and transport dangerous chemicals by using tough, chemical-resisting plastics in place of glass, stainless-steel, or waxed vessels.

1-5   Kettenburg K-41 yacht attests to increasing use of plastics for marine applications. Note high gloss, as exemplified by reflections of bystanders, in the smooth hull. (*Reprinted by permission of Kettenburg Marine Div., Whittaker Corp.*)

Hygienic and nonallergenic

Plastic articles can be manufactured to have smooth, high-gloss surfaces that provide no lodging places for dirt and bacteria. This leads to their selection for medical and surgical devices, hospital equipment, toys, and nursery items. The increasing use of plastics in prosthetic surgery attests to the absence of allergic reaction in the human body.

Mildew and fungus resistance

Unless materials such as canvas, wood, and leather products are chemically treated before or after manufacture, their service life is limited in geographical areas where climate induces mildew or fungus attack. Governmental agencies increasingly purchase plastic products for use by civilian and military personnel assigned to tropical areas.

Can be mass produced

Most manufacturing operations in the plastics industry are geared either to a continuous output of industrial and commercial products or to rapid, automated, or semiautomatic machine cycles that turn out a prodigious number of molded parts per machine per day. It is this capability that brings many products within reach of the average consumer, but it does not reflect the enormous investment necessary for equipment, molds, and trained personnel.

1-6 All-plastic chlorine dioxide bleach tower in paper-mill operation. Tower and elbow sections produced from filament-wound Hetron polyester resin with glass-fiber reinforcement. (*Reprinted by permission of Ershig's, Inc.*)

1-7   Resistance to mildew and fungus is characterized by vertical greenhouse. Completely automated, with plants moving vertically on an elongated ferris wheel, the operation is attended by only one man. Polyester-glass panels transmit 95 percent of available light. (*Reprinted by permission of Filon Div., Vistron Corp.*)

Provide design freedom

Whether a product being proposed is new or simply being updated, plastics generally come under early consideration by the product designer charged with creating more

1-8   Containers mass produced at a rate of more than 100,000 per day on this giant roll-fed thermoformer. (*Reprinted by permission of Brown Machine Co.*)

sales appeal, cost reduction, or greater functionality. Design engineers can incorporate concepts never before possible with wood, metal, or glass. In many cases, the knowledgeable designer can take advantage of unique assembly methods or molding techniques to improve styling and reduce manufacturing cost.

Adaptability

In many instances, only a slight change in chemical formulation will result in a variation of a major property (such as impact strength) to meet a specific end-use requirement. Thus, although 40 distinct plastic materials are produced in significant commercial quantities, some can be modified to provide hundreds, and even thousands, of grades of one basic type.

It should be mentioned that no single plastic possesses the best of all the features under discussion. Instead, the most recent and often spectacular plastic product to appear on the market is a compromise among all the properties, the limitations imposed by practical design, and material and fabrication equipment costs.

It would seem, in comparing plastics with other, more traditional materials such as copper, steel, wood, glass, and leather, or with cotton, wool, and silk fibers, that industries founded on these materials would somehow decline because of replacement by synthetics. To some extent this has happened and will continue to occur at an increasing rate, but plastics have the following inherent limitations that need to be overcome before some longer established industries are seriously threatened.

Not structural materials

Unless reinforced with fibrous glass or other strengthening materials, plastics are not recommended for applications involving a high degree of sustained loading. The use of reinforced plastics in tank cars, boats, and certain structural housings is widespread, but to use them as replacements for automobile frames, bridge girders, or even as a base for heavy machinery is not within the realm of present-day technology or even feasible.

Nonresistant to heat

Most plastics begin to deteriorate to some extent when they are exposed to temperatures above 200°F; others are affected by dry heat at 140°F. Yet certain synthetics are unaffected by continuous exposure to temperatures to 500°F, with progress being made toward useful materials that operate at 1000°F. Deterioration becomes evident in such various ways as discoloration, warpage, melting, carbonization, reduction of strength or electrical insulation values.

Few plastics can be classified as nonburning. Some are categorized as flammable or fast burning and are excluded from any use that may endanger public safety. Certain types, classified as self-extinguishing, may in fact yield toxic by-products of combustion, which negates use in confined areas such as aircraft. Yet plastics make possible the efficient design of rocket nozzles, flame shields, and nose cones, all of which are important in space exploration when temperatures from 5000 to 25,000°F are encountered. Plastics that char and form a carbonaceous structure are used to provide the vital function of thermal insulation in rocket lift-off and reentry.

Nonresistant to sunlight

Most synthetic plastics are not recommended for continuous exposure to ultraviolet light, which can lead to deterioration just as continuous exposure to elevated temperatures degrades plastics. Actually, heat and sun hasten *oxidation*, and oxygen combining with the material causes failure. Shrinking, cracking, embrittlement, and color fade are the inevitable result of chemical change induced by weathering or environmental heat. The embrittlement of a garden hose, the cracking of patio furniture, or the roughening and haze of corrugated panels are evidence of chemical changes brought on by sun and air, aided in some cases by moisture. Outdoor signs and building fronts, however, can withstand the elements for many years while retaining their color and gloss because of the careful selection of specific materials and proper fabricating techniques.

WHAT ARE PLASTICS

Synthetic plastics are chemical compounds produced from such basic chemical elements as carbon, hydrogen, nitrogen, and oxygen. These atoms, when combined in specific ways to build long chains of molecules, result in a solid substance called a *polymer*. (Other chemicals, notably chlorine, fluorine, silicon, and sulfur, are also

1-9 Filament-wound transtage rocket engine for Apollo spacecraft. (*Reprinted by permission of Aerojet Liquid Rocket Co.*)

important for producing some polymers.) By using these few chemicals as basic building blocks, chemists are able to fashion giant molecules in an unlimited number of combinations. At least 500,000 substances based on carbon compounds are known, but only a few are capable of forming polymers of commercial significance. New polymers are announced periodically as polymer chemists synthesize new molecules or find better ways to produce raw materials.

Chemical names used to describe most synthetic plastics are often preceded by the prefix *poly* to indicate particular polymers. Some polymers are familiar to the public (e.g., polyester, polyethylene, polystyrene, and polyurethane) ; most are not. When several chemical manufacturers market the same type of polymer, trade names are applied. For example, polyethylene is known as Alathon,[1] Marlex,[2] Petrothene,[3] and so forth.

Seldom does a polymer reach the marketplace in its natural state. It may be modified to make it more flexible, self-extinguishing, or to have improved weathering. It is often colored, filled with inexpensive fillers, or strengthened with fibers to make thousands of plastic materials available to meet exacting requirements. The task of sorting out and selecting the correct plastic for an application can only be accomplished by having a thorough knowledge of basic polymers and the method (process) through which the polymer becomes a useful product.

CLASSIFICATION OF PLASTICS

Plastics are divided into three major categories :

Thermosetting types   These are plastics that undergo chemical change during processing to become *permanently insoluble and infusible*. Phenolic, amino, epoxy, and polyester resins are typical. They are also called *thermohardening* plastics.

Thermoplastic types   These are resins that soften repeatedly when heated and harden when cooled. Most *thermoplastics* are soluble in specific solvents and support flame to some degree. Softening temperatures vary with polymer type and grade, so care must be taken to avoid degrading, decomposing, or igniting the material. Nylon, acrylic, vinyl, and cellulose acetate are typical *thermosoftening* plastics.

Elastomers   Elastomers are materials that can be stretched repeatedly to at least twice their original length (while at room temperature) and which will return with force to their approximate original length when an immediate stress is released. Polymers ranked as elastomers include modified thermoplastics as well as natural and synthetic rubbers. Most elastomers attain their properties by a process of vulcanization that renders them incapable of reprocessing, as is possible with thermoplastics. Examples of elastomeric polymers are silicone, urethane, nitrile rubbers, and chlorinated polyethylene.

[1]Alathon is the registered trade name of E. I. du Pont de Nemours & Company.
[2]Marlex is the registered trade name of Phillips Petroleum Company.
[3]Petrothene is the registered trade name of U.S. Industrial Chemicals, Division of National Distillers Corporation.

SOURCES OF RAW MATERIALS

The world of plastics has been described as "the world that nature forgot." Nature has bountifully provided wood, minerals, and ores, which are made up of large, cohesive molecules and which form the basis for our structural materials. Synthetics, by comparison, must be fashioned from simpler compounds by combining them in the correct amount under proper conditions. This task is not too difficult since nature has been cooperative to some extent by joining carbon, hydrogen, and oxygen in thousands of combinations found in coal, petroleum, and natural gas. These hydrocarbons, called *intermediates*, must be separated by the chemical processes of extraction, fractionation, and distillation; phenol, calcium carbide, ethylene, propane, benzene, etc. are starting chemicals for producing synthetic resins.

AVAILABLE FORMS OF PLASTIC

In addition to the countless ways basic chemicals can be formulated to produce the desired plastic, each type of plastic can be made in several forms. However, not all types are produced in every form, thus the manner in which a particular polymer may be processed is restricted.

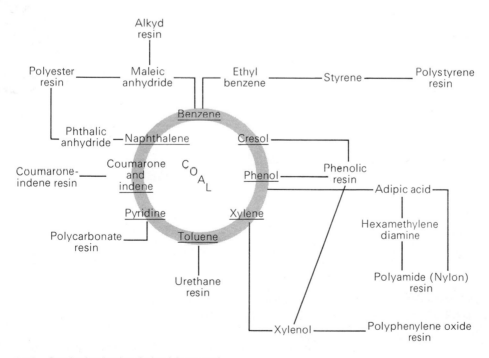

1-10   Synthetic plastics derived from coal.

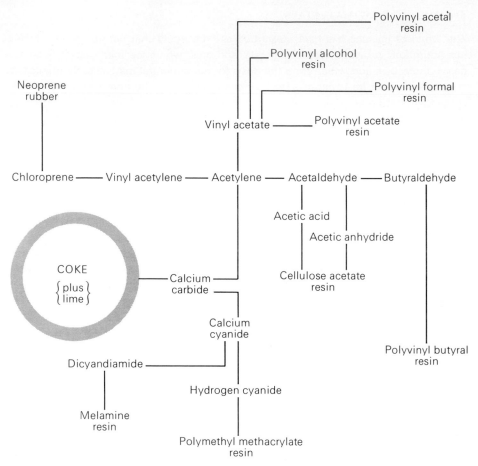

1-11   Synthetic plastics derived from coke.

Resinous types

Liquids   Resins used for adhesives, coatings, laminating, casting, and foam are available in a wide range of viscosities.

Powders and bulk   Thermosetting resins for molding are normally supplied as pow-ders. Bulkier compounds result from incorporating fibrous fillers.

Pellets   Thermoplastics used in molding processes are furnished as pellets in cubed, spherical, or cylindrical shapes.

Processed forms

Film   Thermoplastics are available in film form ranging from less than one-thousandth to ten-thousandths of an inch thick. Adhesive films are produced from thermosets for easy application.

Sheet, rod, tube, and profile shapes   Various processes are used to produce these forms from almost all type of polymers; the process is selected according to the type of material.

Fiber   Thermoplastics are produced as continuous or discontinuous (staple) fibers. They may be drawn to an exceedingly fine diameter or produced as heavier monofilament. Approximately one-third of the thermoplastic polymers are used to form fibers.

Foam   Many polymers can be foamed, but only a few thermoplastics, and even fewer thermosets, have attained commercial significance.

Laminates   This category overlaps because of the forms available in laminated shapes. Elastomers and thermosetting and thermoplastic polymers are combined with fibrous webs (paper, cloth, or felt) or used to bond wood, metallic foils, and glass sheets. Shapes available include flat and contoured sheet, rod, tube, and profiles.

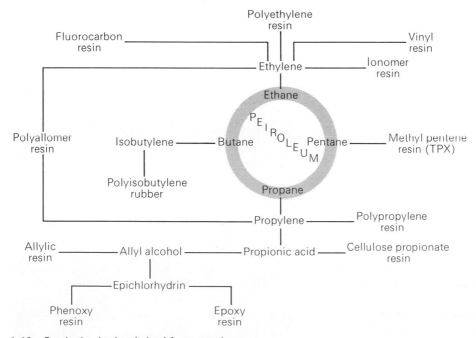

1-12   Synthetic plastics derived from petroleum.

PROCESSING METHODS

Processing refers to the method of fabricating or converting the liquid or solid form into a product. More resins are processed by molding techniques than by any other method.

Molding

Compression molding   Largely confined to processing thermosets, molding by compression is also applicable to thermoplastics. Dinnerware, ashtrays, and coffee pot handles are examples of compression molding.

Injection molding   A large segment of the plastics industry is devoted to injection-molding thermoplastics. Items ranging from tiny spacers to products weighing many pounds are molded in rapid machine cycles. More-recent developments make the production of thermoset products by injection molding possible.

Extrusion   Profile shapes, rod, tubing, film, and sheet are produced by continuous extrusion of thermoplastics. Variations of the basic extrusion process (pultrusion) utilize liquid thermosets and continuous filaments (usually glass) to produce rod, tube, and various cross sections. Fibers and monofilaments are also extruded.

Blow molding   This is the basic process for producing hollow shapes such as bottles from thermoplastics.

1-13 Compression-molded golf cart using liquid polyester with fiber glass preform. The press pictured develops 500 tons of molding pressure. (*Reprinted by permission of Dake Corp.*)

1-14   Hydraulic injection machine develops 175 tons of clamping pressure for molding thermo-plastics on rapid, automated cycles. (*Reprinted by permission of Bipel International, Inc.*)

Rotational molding   Hollow shapes are economically produced by rotating meltable, powdered polymers or fusible liquids in a heated, closed mold. Marine toilets, furniture, water tanks, and carrying cases are examples of rotational molding.

Thermoforming   Preheating a thermoplastic sheet and drawing it over or into a form provides a molding method adaptable to many plastics. Containers, signs, refrigerator door liners, and automotive parts are thermoformed

1-15   Extruder. Shape of extruded product is governed by shape of die fixed to round adapter plate. Note tip of melt screw on forward end. (*Reprinted by permission of Prodex Div., Koehring Co.*)

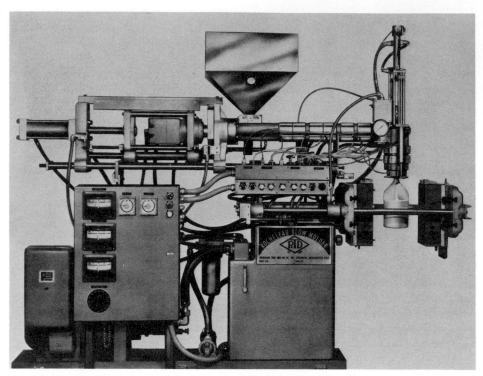

1-16   Single-station—blow-molding machine capable of producing hollow products to 7 oz in weight. (*Reprinted by permission of Rocheleau Tool & Die Co., Inc.*)

1-17   Rotocast machine produces hollow- or double-wall products from various thermoplastics. (*Reprinted by permission of McNeil-Akron.*)

1-18 Fume-exhaust hood, deep drawn, with Kydex (acrylic-polyvinyl chloride sheet). (*Reprinted by permission of Rohm and Haas Co.*)

Examination of each molding process should reveal that molds or dies are required to shape the product. Molds suitable for each process are products of fine craftsmanship created in steel or aluminum by a master craftsman, the mold maker. Other methods of producing plastic articles are no less important.

Laminating

Combining layers of fabric, paper, or mat with thermoplastic or thermosetting resins is called lamination (plywood, safety glass, and joining thermoplastic films are also included). Laminates may take the form of flat sheets, molded tubes, or wrapped rods.

1-19 Nose radome laminated with glass cloth and polyester resin, used on L-1011 Tri-Star. (*Reprinted by permission of Lockheed Aircraft Corp.*)

1-20   Four-roll inclined Z calender. (*Reprinted by permission of Stewart Bolling & Co., Inc.*)

Calendering

Essentially a rolling process to produce rigid or flexible thermoplastic sheets, calendering is also used to apply films to substrates such as paper and cloth. Wall paper, linoleum, and upholstery fabrics are made on calenders.

Casting

Either thermoplastic or thermosetting liquid resins, when poured into open molds to harden by cooling or chemical reaction, are used to cast a variety of products: statuary,

1-21   Fabrication of rigid polyvinyl chloride air duct by applying hot air to PVC welding rod. (*Reprinted by permission of Brian R. White Co., Inc. and Karl Leister Co., Switzerland.*)

novelties, tooling, and standard rod, tube, and sheet forms. Casting implies that external pressure is not required to shape the final product, although vacuum, centrifuging, and other means of densification are sometimes employed.

Variations of casting (potting or encapsulating) are used to protect electronic components with rigid or flexible resins. Foams are also cast into molds for picture frames, plaques, furniture, and packaging.

Fabrication

Standard machining operations, plus special processes such as heat sealing, ultrasonic welding, gas welding, stamping, and forging, are used to produce plastic articles when other methods are not feasible or economical because of quantity or design.

GLOSSARY

Infusible   Incapable of being melted.

Polymer   A compound formed by the reaction of simple molecules having functional groups that permit their combination to proceed to high-molecular weights under suitable conditions.

Resin   A class of solid or semisolid organic products of natural or synthetic origin, generally of high-molecular weight, and having no definite melting point.

Viscosity   The property of resistance to flow exhibited within the body of a material.

REVIEW QUESTIONS

1   Prepare a list of products that are made completely from, or contain, plastics in:
   (*a*)   the home
   (*b*)   the automobile
   (*c*)   the school
   (*d*)   recreation

2   How does the weight of plastics compare with that of aluminum? With steel? With wood?

3   How is the light weight of plastics used to advantage in:
   (*a*)   transportation
   (*b*)   packaging
   (*c*)   other ways

4   Plastics are dielectrics. What does this mean?

5   What property of plastics makes them pleasant to the touch?

6   What advantage does inherent color give plastics, when compared with metal or wood?

7   Why are plastics sometimes plated?

8   List several common products that incorporate transparent plastics.

9 What industrial or household items utilize plastics when water resistance is required?

10 List several household products that are packaged in chemical resistant plastics.

11 Why are plastics more hygienic than many other materials?

12 Why are plastics selected for use in tropical areas?

13 What factors contribute to the low cost of many plastics products?

14 Compare the styling of a boat, car body, furniture, and appliance made from plastic with one made of metal or wood.

15 Why are synthetics more versatile than wood or metal?

16 Which plastics are termed structural plastics?

17 What happens when plastics are heated above their recommended temperature limits?

18 What causes some plastics to deteriorate in outdoor use?

19 How is deterioration revealed when plastics are exposed to oxidation?

20 What are plastics?

21 Why are trade names used to differentiate plastics?

22 Explain the three major classifications of plastics.

23 What does infusible mean? What does insoluble mean?

24 To which class of materials does epoxy belong?

25 To which class of materials does nylon belong?

26 What are the three major sources for chemical intermediates used to produce synthetic resins?

27 Prepare a list describing the various forms of plastics and the applications (uses) for each.

28 What process is used to produce:
   (*a*)   pot handles
   (*b*)   monofilament
   (*c*)   bottles
   (*d*)   plywood
   (*e*)   upholstery fabric

29 Define "casting."

30 Which process best describes the machining of plastics?

PHENOLIC RESINS AND COMPOUNDS

2

Phenolic plastics are the most versatile materials in the entire family of polymers and are referred to as the workhorse of the industry. Their versatility is due partly to the low cost of raw materials and partly because of the many ways phenolic resins can be tailored to fit specific requirements. Diversified uses such as adhesives for grinding wheels and brake linings, binders for molding compounds, and laminating varnishes for a variety of products from fishing rods to rocket nozzles attest to the wide performance capability of phenolics.

SOURCES OF RAW MATERIALS

Phenol

Also known as carbolic acid, phenol is obtained as a by-product from the reduction of bituminous coal. To provide the huge quantities of coke required in the manufacture of steel, coal is heated in ovens in the absence of oxygen. Note that in Fig. 2-1 other closely related chemicals, particularly cresol and xylenol, as well as phenol, are separated from coal tar. Cresols and xylenols are sometimes used in place of, or with, phenol to develop special resin formulations. A mixture of cresol and xylenol is called *cresylic acid*. Plastics prepared from these mixtures are often referred to as *tar-acid resins* or *cresylic-acid plastics*. The term *phenolic plastics* is generally used to describe any combination of these resinous components.

Phenol, as obtained from the destructive distillation of coal, is impure because of the presence of cresols and xylenols. This can be detrimental in formulating molding compounds to meet electrical or chemical requirements or flow or cure conditions.

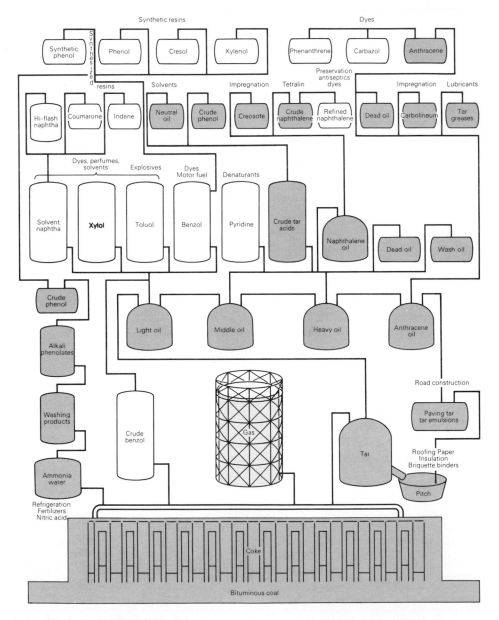

2-1   Chemicals obtained from coal.

A pure phenol is obtained synthetically from benzene, which in turn is obtained from petroleum or coal. Phenol, which is furnished as a crystalline solid that melts at 105.8°F, is a poisonous substance that causes severe burns on contact. Nonplastic uses include disinfectants and insecticides. On rare occasions, a worker may expe-rience an allergic reaction when exposed to molding compounds or laminating liquids based on phenolic resins.

In addition to its use as a reactant for phenolic resins, phenol is also an im-portant chemical constituent or intermediate in the preparation of epoxy, polyamide (nylon), polycarbonate, phenoxy, and polyphenylene oxide plastics.

Formaldehyde

Formaldehyde is normally a gas, but it is furnished to resin manufacturers as a 37 to 40% solution in water (formalin). It is produced from methyl alcohol by several methods and is important to the formation of other synthetic resins such as urea, melamine, and acetal, as we shall discuss later.

PREPARATION OF PHENOLIC RESINS

In order to produce a phenolic, phenol or cresylic acid is reacted with formaldehyde under controlled conditions. Commercially, the reaction is carried out on a large scale (20,000 lb) by "cooking" in stainless-steel kettles that have steam-heating and water-cooling devices.

2-2   Ingredients for producing phenolic resin are reacted in kettles under precise conditions of time and temperature. [*Reprinted by permission of Plastics Engineering Co. (PLENCO).*]

The reaction is started and controlled by using a catalyst and heat for a precise time, depending on the type of phenolic resin desired. Phenolic resins are thermosetting, and as such they are capable of forming insoluble, infusible substances. If the reaction is carried to completion, the process known as *polymerization* will occur. The complete polymerization of thermosets is not reversible, so it is important that the reaction be stopped at some intermediate stage to permit reheating or dissolving for processing purposes.

Consequently, thermosetting resins are classed according to the degree of polymerization they have reached:

A stage: Completely fusible and soluble (thermoplastic)

B stage: Fusible and partially soluble

C stage: Fully polymerized to an infusible and insoluble condition

An *A*-stage resin, called a *novolac*, is produced by combining less than 1 mole of formaldehyde per mole of phenol in the presence of an acid catalyst (sulfuric or oxalic). The resultant resin, being thermoplastic, can be heated for required periods to aid in the incorporation of fillers to produce molding compounds. The additional formaldehyde is obtained during the final step by adding hexa[1] in amounts up to 15 percent.

During the molding or laminating cycle, hexa breaks down to furnish the formaldehyde required for *C*-stage (cross-linked) formation, while ammonia serves as the catalyst to speed the reaction. Phenolic resins prepared in this manner are known as two-step resins.

Phenolic resins are also produced in a single step by using an excess of formaldehyde in the presence of an alkaline catalyst such as ammonia or sodium hydroxide. The resin, called a *resole*, is a *B*-stage phenolic requiring only heat to convert it to the *C* stage.

Companies engaged in coating or impregnating textiles for later use as laminate reinforcements or in formulating adhesives purchase resoles, which are necessary for applying resin in solution. Resoles are readily dissolved in alcohol to obtain required viscosity.

Whether phenolic resins are produced as one- or two-step resins, the polymerization (curing) is the same. Heat from 250 to 350°F is required to initiate the final reaction and obtain usable products.

During the initial batch reaction and continuing through the final stages of polymerization, water is evolved as a by-product. In the laminating process, reaction is slow, so the evolution of water is not as troublesome as in molding, where rapid press cycles are typical. Failure to recognize and deal with this *condensation reaction* results in formation of blistered or porous products. Detailed procedures that circumvent poor-quality products are given in Chaps. 8 and 9.

An exception to the evolution of water occurs during the production of casting grades of phenolic resin. While diminishing in usage, castable resins, which are pro-

[1]This is hexamethylene tetramine, a mixture of formaldehyde and ammonia.

UNITED STATES PATENT OFFICE.

LEO H. BAEKELAND, OF YONKERS, NEW YORK.

METHOD OF MAKING INSOLUBLE PRODUCTS OF PHENOL AND FORMALDEHYDE.

942,699. — Specification of Letters Patent. — **Patented Dec. 7, 1909.**

No Drawing. — Application filed July 13, 1907. Serial No. 383,684.

To all whom it may concern:

Be it known that I, LEO H. BAEKELAND, a citizen of the United States, residing at Snug Rock, Harmony Park, Yonkers, in
5 the county of Westchester and State of New York, have invented certain new and useful Improvements in Methods of Making Insoluble Condensation Products of Phenols and Formaldehyde, of which the following
10 is a specification.

In my prior application Ser. No. 358,156, filed February 18, 1907, I have described and claimed a method of indurating fibrous or cellular materials which consists in im-
15 pregnating or mixing them with a phenolic body and formaldehyde, and causing the same to react within the body of the material to yield an insoluble indurating condensation product, the reaction being accel-
20 erated if desired by the use of heat or condensing agents. In the course of this reaction considerable quantities of water are produced, and a drying operation is resorted to to expel it.

25 The present invention relates to the production of hard, insoluble and infusible condensation products of phenols and formaldehyde.

In practicing the invention I react upon
30 a phenolic body with formaldehyde to obtain a reaction product which is capable of transformation by heat into an insoluble and infusible body, and then convert this reaction product, either alone or compounded
35 with a suitable filling material, into such insoluble and infusible body by the combined action of heat and pressure. Preferably the water produced during the reaction or added with the reacting bodies is separated
40 before hardening the reaction product. By proceeding in this manner a more complete control of the reaction is secured and other important advantages are attained as hereinafter set forth.

45 If a mixture of phenol or its homologues and formaldehyde or its polymers be heated, alone or in presence of catalytic or condensing agents, the formaldehyde being present in about the molecular proportion
50 required for the reaction or in excess thereof, that is to say, approximately equal volumes of commercial phenol or cresylic acid and commercial formaldehyde, these bodies react upon each other and yield a product
55 consisting of two liquids which will separate or stratify on standing. The lighter or supernatant liquid is an aqueous solution, which contains the water resulting from the reaction or added with the reagents, where-
60 as the heavier liquid is oily or viscous in character and contains the first products of chemical condensation or dehydration. The liquids are readily separated, and the aqueous solution may be rejected or the water
65 may be eliminated by evaporation. The oily liquid obtained as above described is found to be soluble in or miscible with alcohol, acetone, phenol and similar solvents or mixtures of the same. This oily liquid may
70 be further submitted to heat on a water- or steam-bath so as to thicken it slightly and to drive off any water which might still be mixed with it. If the reaction be permitted to proceed further the condensation product
75 may acquire a more viscous character, becoming gelatinous, or semi-plastic in consistence. This modification of the product is insoluble or incompletely soluble in alcohol but soluble or partially soluble in ace-
80 tone or in a mixture of acetone and alcohol. The condensation product having either the oily or semi-plastic character may be subjected to further treatment as hereinafter described. By heating the said condensa-
85 tion product it is found to be transformed into a hard body, unaffected by moisture, insoluble in alcohol and acetone, infusible, and resistant to acids, alkalies and almost all ordinary reagents. This product is
90 found to be suitable for many purposes, and may be employed either alone or in admixture with other solid, semi-liquid or liquid materials, as for instance asbestos fiber, wood fiber, other fibrous or cellular mate-
95 rials, rubber, casein, lamp black, mica, mineral powders as zinc oxid, barium sulfate, etc., pigments, dyes, nitrocellulose, abrasive materials, lime, sulfate of calcium, graphite, cement, powdered horn or bone, pumice
100 stone, talcum, starch, colophonium, resins or gums, slate dust, etc., in accordance with the particular uses for which it is intended, and in much the same manner as india rubber is compounded with the above-named and
105 other materials to yield various valuable products. In compounding the condensation or dehydration product in this manner the desired materials are mixed with the same before submitting it to the final hard-
110 ening operation below described.

2-3 U.S. Patent No. 942,699. Dr. Baekeland is credited with recognizing the importance of arresting the phenol-formaldehyde reaction at an intermediate stage of chemical activity, thus developing the method for subsequent processing into useful products.

duced by reacting phenol and formaldehyde with sodium hydroxide, are cured at room temperature by the addition of 10% sulfuric acid. Converting to the *C* stage at room temperature, water is retained as microscopic bubbles within the structure. This accounts for the jewel-like transparency or translucency of pigmented materials.

Castable phenolics are available as sheets, rods, and tubes, or as special profile shapes for novelties. In liquid form castable phenolics are used for casting patterns and other special shapes.

PREPARATION OF PHENOLIC MOLDING COMPOUNDS

Phenolic resins form the basis for a wide variety of molded products that range in size from small electronic insulators to rocket nozzles weighing several hundred pounds (see Table 2-1, page 39).

A pure phenolic resin is rarely molded without modification to overcome limitations imposed by the molding process and end-use requirements. Without alteration, phenolic resins are brittle and have high shrinkage caused by polymerization and elimination of water, dimensions are difficult to control, and parts stick badly to hot-mold surfaces, making ejection difficult. In order to overcome these problems, resin manufacturers resort to the use of additives to produce a *compound*. As shown in the table below, a typical molding compound may contain:

Ingredient	Max. parts by weight
Phenolic resin (binder)	60
Filler	50
Reinforcing fiber (if used in place of filler)	80
Hardening agent	15
Plasticizer (if required)	5
Dye or pigment	5
Lubricant (release agent)	1

Lubricants are natural or synthetic waxes that are actually incompatible with the resin. During molding the wax migrates to the surface to act as a release agent. When a new mold is "broken in," molders use a special molding grade containing a higher percentage of lubricant until final traces of fine tool marks are smoothed out. Molded parts are discarded.

Phenol molding resins range in color from tan to reddish-brown because of oxidation or contact with iron in the manufacturing process. To overcome a tendency to darken further, dyes or pigments are added. As a result, compounds are limited to dark shades of blue, green, red, purple, brown, and black. Mottled effects, simulating walnut or mahogany, are obtained by blending colors. Molders rely on highly polished molds or mold textures to provide pleasing surface effects.

Inherent resin brittleness is reduced by adding plasticizers such as flake nylon or synthetic rubber (Buna-N). Special applications, e.g., pump housings or impellers, that require vibration and chemical resistance incorporate plasticizers.

2-4   Massive stretch die, cast from phenolic-liquid-casting resin and having a foamed, phenolic core. (*Reprinted by permission of Rezolin Div., Hexcel Corp.*)

The hardening agent hexa is added after all other ingredients have been blended. Final mixing is done on hot milling rolls to ensure thorough mixing. Control of flow and cure time is adjusted by varying the exposure of the compound to the milling sequence after the hardening agent is incorporated.

All thermoset resins are, by nature, adhesive. For this reason the resin is referred to as a *binder*, whose function it is to bind all other ingredients of a compound together. An analogy would be cement, which holds sand, gravel, and perhaps reinforcing rod together as a composite.

FILLERS

Phenolics are popular mainly because of the wide variety of fillers used to reduce cost or to provide specific properties such as heat resistance or impact strength. There is a practical limit governing the amount of a particular filler that can be incorporated in a resin matrix. Too little resin results in exposure of filler particles, thus leading to

2-5   Fan, baffle, and frame insulator of this grinder are molded from phenolic to provide protection from moisture, impact, and electrical shock. (*Reprinted by permission of Durez Div., Hooker Chemical Corp.*)

water absorption, low strength, and poor gloss. Conversely, too much resin results in resin-rich areas that, without benefit of filler, will craze, crack, or warp.

Porous fillers such as wood flour and cotton are thoroughly impregnated with resin and are undetectable; impermeable fillers such as fibrous glass are merely coated.

Origin

Fillers can be classified according to origin:

Organic	Inorganic
Cellulosic:	Mineral:
Wood flour (mainly spruce)	Asbestos
Cotton linters	Mica flake or powder
Alpha cellulose	Talc
Paper pulp board	Pumice
Macerated fabric	Glass fibers or textiles
Agricultural waste products:	Quartz fibers
Walnut shell flour	Silica fibers or textiles
Sisal fibers	Diatomaceous earth
Bagasse (sugarcane fiber)	Barytes
Corn husk fiber	Aluminum oxide
Coconut fiber	Boron fibers
Seed hulls	Refractory clay
Cork	
Synthetic:	Metals, powdered:
Nylon filament	Lead
Rayon yarn	Copper
Orlon* yarn	Aluminum
Dacron† fiber	Iron
Teflon‡ fiber	Molybdenum disulfide
Carbonaceous:	
Carbon fibers or textiles	
Graphite fibers or textiles	
Carbon black	
Graphite flake	
Bitumen	

*Trade name for Du Pont acrylic.
†Trade name for Du Pont polyester.
‡Trade name for Du Pont fluorocarbon.

Experimentally, many unlisted fillers have been combined with phenolic resins in an attempt to utilize agricultural wastes. Peanut shells, wheat chaff, straw, bamboo, palm fronds, coffee, and dehydrated vegetable fiber (carrot and rhubarb) have been evaluated.

Types

Molding compounds can also be classified according to the particular properties exhibited by the molded product. Each filler, or combination, results in a different compound. The wide variety of combinations, plus the amount of hardener and degree of *B* staging to control cure time, makes the task of selecting the proper phenolic compound difficult at the outset of a new molding application. To assist the molder in selecting the proper grade, a system of classification has been standardized by the industry.[1]

General-purpose compounds   To fulfill the majority of molding jobs, from production of pot handles, knobs, and toaster bases to meeting demands for electrical components such as terminal strips and relay bases, general-purpose (GP) phenolic is specified. The filler selected for general-purpose use is purified and finely divided wood flour. Cotton linters are also used.

The term wood flour may lead one to visualize huge stockpiles of sawdust or wood chips at a sawmill ready to be mixed with a liquid or powdered phenolic. Actually, because raw wood contains too many impurities and thus can't meet exacting needs, it can't qualify as a filler. Natural impurities such as lignin and sap would interfere with resin polymerization. To prepare wood flour, wood chips must be cooked in strong acids. This cooking causes a digestion process, which removes the undesirable ingredients. The wood should then be pulverized to make the cellulose as fibrous as possible so that intimate blending of resin and filler is assured.

Coarse-particle, untreated wood chips are used in the manufacturing of low-cost particle board, which is widely used for shelving and partitions because appearance is secondary.

General-purpose compounds are selected to meet the following conditions:

High surface gloss

Good mold filling, especially when fine detail such as engraving is molded

Reasonable mechanical strength

Low cost (18 to 22 cents/lb)

Heat resistance to 250°F continuous, or to 400°F for short periods

Water and chemical resistant

Good electrical insulation at common voltages

Medium- and high-impact compounds   Fibrous fillers tend to interlock and thus transmit shock when dropped or struck. Cotton fibers, sisal, and paper pulp improve impact strength up to 5 times that of general-purpose grades; rayon yarn, nylon monofilament, and glass fibers increase breakage resistance up to 20 times.

When molding impact materials it is not always possible to prevent coarse fibers from surfacing and thus causing a mottled appearance. Organic fibers, particularly sisal, are poor in water resistance; glass fibers tend to permit water penetration between the glass and resin interface. Most industrial products such as tool handles, gears, and structural components are not concerned with these limitations.

[1] The Society of the Plastics Industry, Inc. (SPI) classification is listed in Table 2-3 at the end of this chapter.

A disadvantage to consider seriously is the exposure of raw fiber ends that develop when molded parts are trimmed or machined. The compression molding process requires a slight excess of compound beyond that required to fill out the mold and ensure density. This excess, called *flash*, is removed by tumbling parts in a barrel, then hand filing or sanding. This is a relatively simple matter with general-purpose compounds since flash is brittle. The flash on impact grades is tough, requiring removal by die cutting, lathe turning, or other machine processes. Removal of flash from impact grade products is not only expensive, but it removes the pure resin protective film and exposes coarse filler particles. This presents a dull, rough surface that cannot be buffed or polished to the original luster imparted by the mold. Also, this exposure of unprotected fibers lessens water and chemical resistance and is detrimental in electrical applications.

Molding impact grades entails greater expense because the required higher molding pressures necessitate more rugged steel molds and presses of higher capacity. These compounds have greater bulk than powdered general-purpose grades and require molds with adequate loading space to contain the uncompressed material. Molds designed for general-purpose grades can be used for impact grades if the bulk is reduced by cold pressing into tablets (preforms).

It is possible to combine general-purpose powders with high-impact grades when difficult mold filling is a problem. This is accomplished by strategically locating preforms within a mold. A special design problem, such as reinforcing a critical section, is solved by using uncured impregnated paper prelocated in the mold and surrounded by molding powder.

Mineral   Asbestos is used as filler for applications where higher temperature is encountered. Asbestos is recommended for *intermittent* use to 450°F; resin deterioration caused by accelerated oxidation occurs at a *continuous* exposure of 400°F.

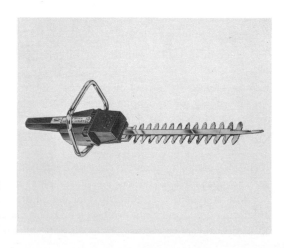

2-6   A hedge-trimmer housing molded from medium-impact phenolic. [*Reprinted by permission of Plastics Engineering Co. (PLENCO).*]

Long-fibered asbestos provides intermediate impact strength as well as heat resistance, and short fibers of asbestos, called *floats*, are used for chemical applications. Asbestos may not always be satisfactory for critical electrical insulators because of the presence of conductive minerals and water of crystallization.

Mica in sheet form is commonly used as insulation because high temperature (1000°F) would char an organic material. Heating elements in toasters and electric irons are typical examples of products that use mica insulation. Pulverized and compounded with a phenolic resin, mica provides designers of electronic equipment with dependable insulation in surroundings to 400°F. Referred to as *low-loss phenolic*, mica-filled resins are used to mold insulators in radio and television circuits. Mica imparts a natural lubricity to a compound, which makes release from hot molds simpler. This property is also advantageous in items such as small gears and slide pins for clocks.

Fibrous glass is a popular choice when extremely high-impact strength at relatively low cost is required. Used in chopped fiber or shredded textile form, glass reinforcements expand the use of phenolic resins into applications such as helmet liners, rocket nozzle backup sections, tool handles, and gun stocks. Hundreds of products that require impact resistance, dimensional stability, improved heat resistance, vibration dampening or sound deadening incorporate glass fibers to meet these requirements.

Compounds containing glass pose special problems in molding when bare glass is exposed to hot-mold surfaces. This is caused by special finishes applied to glass filaments during manufacture. Excessive molding pressure forces resin from the fiber, causing sticking and difficult removal.

Silica and quartz fibers combined with phenolic are used in important areas of space vehicles and ballistic missiles. These products have short-term requirements

2-7 Various types of molding powders and bulk compounds for compression and transfer molding. (*Reprinted by permission of Los Angeles Trade-Technical College.*)

that generate a need for molded parts to withstand temperatures to 30,000°F. The longevity of rocket exhaust nozzles and reentry surfaces is increased by combinations of silica or quartz fibers and phenolic resin molded to shape or machined from molded blocks.

In the aerospace industry, laminates are preshaped from phenolic-impregnated silica or quartz-woven tapes, and then compression molded to polymerize the resin. Large rocket nozzles and flame shields are produced in this manner.

Carbonaceous   If rayon fibers or textiles are pyrolyzed at a high temperature in an inert atmosphere (nitrogen), the resulting product is carbon or graphite in the original fibrous or textile form. The graphite form results when heating is above 4500°F, producing a more crystalline structure and slightly higher carbon content (99 percent). Carbon fibers are less crystalline and contain approximately 97 percent carbon.

Because of their high cost ($18 to $35/lb), carbon or graphite reinforcements have been limited largely to use in critical sections of rocket throats, exit cones, and blast tubes for solid propellant motors. Both types of carbonaceous fibers are used in bulk form or as preformed molded laminates.

Ordinary flake graphite combined with a phenolic binder provides lubricity to moving parts difficult to lubricate or operate in water, such as bearings or pulleys in marine equipment.

Metallic   Powdered metals find specialized, though not extensive, use in unusual applications. For example, lead is used as an x-ray barrier when molded into components for medical equipment. Iron particles, surrounded by phenolic resin and molded to shape, are used for wireless coils, motor brushes, and resistors. In these cases the phenolic binder is only a temporary vehicle that is burned away after molding to expose the conductive iron particles. Molybdenum serves a similar purpose as graphite in furnishing lubrication value to a molded part.

2-8   Carbon fibers converted into yarn, felt, mat, and tow, from which molded and laminated parts are produced. (*Reprinted by permission of Kreha Corp. of America.*)

TABLE 2-1   Applications for molded phenolics

Purpose	Use
General	Pot handles, knobs, camera parts, iron handles, closures, appliances, switches, relays, radio tube bases
Impact resistant	Gears, gunstocks, portable tools, pulleys, welding rod holders
Electrical grade	Condensers, spacers, insulators, coil forms, supports
Heat resistant	Ashtrays, rocket blast tubes, igniter parts, pump parts
Chemical resistant	Photo-developing equipment, labware, washing machine parts
Special use	Oil-less bearings, slide cams, x-ray parts, cores, bushings

Synthetic   Nylon filaments or chopped nylon fabrics are selected to provide impact strength or vibration dampening in applications where light weight is important. Aircraft assemblies involving attachment blocks, cable spacers, and pulleys are representative uses.

Powdered Teflon is combined with phenolic for molded bushings for guide rods of hydraulic presses.

Odorless grades   Phenolic compounds are not recommended for general use where an odor of phenol would be objectionable, as in lids for cooking pots. Special odorless grades are available for covers of steam vaporizers, percolator parts, and closures for medicine and perfume bottles.

2-9   Large-bearing—bearing-retaining ring and other parts illustrate industrial uses for phenolic resin combined with paper, cotton, and other fabrics. (Filament-wound coupling and terminal board are epoxy-glass.) (*Reprinted by permission of Spaulding Fibre Co., Inc.*)

TABLE 2-2   Applications for laminated phenolic

Purpose	Use
Paper base	Terminal strips, engraved instrument panels, washers, insulating sleeves
Fabric base	Gears, bearings, helmets, radio, television mounting boards
Asbestos base	Higher temperature applications such as valve parts, bearings, cams, washers
Decorative grades	Special pressed serving trays and other items, often with metal inlay

Laminating

Phenolic resins dissolved in alcohol or water are used as varnishes to impregnate webs of paper, cotton, nylon, glass, or asbestos in felt, mat, or woven form (see Table 2-2). As with molding compounds, a wide choice of phenolic resin types combined with these reinforcements provides a wide variety of flat panels or molded shapes. Industrial grade laminates predominate since color limitation precludes use as decorative surfaces. Laminated phenolic products are available as sheets, rods, bars, and tubes.

MISCELLANEOUS USES FOR PHENOLIC RESINS

Adhesives

Special formulations of phenolic resins account for large-volume usage in many diversified fields. Lamp bulbs are cemented to their bases with phenolic glue. Asbestos brake linings are bonded to brake shoes with heat-resistant phenolic, and plywood and structural wood laminates utilize phenolic where dark color is not detrimental.

2-10   Resin-bonded diamond wheel (valued at $10,000) produced with heat-resistant phenolic resin. [*Reprinted by permission of Plastics Engineering Co. (PLENCO).*]

Both hot-set and room-temperature curing systems are used. Resinous adhesives, based on *resorcinol*, harden at room temperature when activated by para-formaldehyde (a concentrated formaldehyde in powder form) to furnish waterproof, heat-resistant glue lines. The use of high-frequency welding equipment has greatly accelerated production of furniture, laminated beams, and building trusses.

Coatings

Synthetic resin varnishes formulated with phenolic have long been used in place of shellac for interior and exterior surfaces and floors. Combined with drying oils, these *oleo-resinous* coatings polymerize through oxidation of the oil to form a tough, durable surface.

Phenol-furfural resins

A class of phenolics, called *phenol-furfural* resins, is used as a binder for molding compounds or as a coating or impregnant for porous products. Furfuraldehyde is reacted with phenol in place of formaldehyde and blended with fillers for molding purposes. Phenol-furfural resembles phenol-formaldehyde in appearance and is always black in color. These compounds are selected when molding requires a long duration of flow, as, for example, a large television cabinet or furniture drawer. Improved chemical resistance is also a factor, which is why it is selected for acid tanks or containers used in textile or paper processing.

Liquid furfural resins are used to strengthen plaster statuary and lamp bases; they are also applied to wood or metal tanks for corrosion resistance. Polymerization

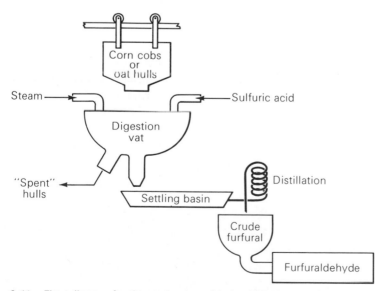

2-11   Flow diagram for the production of furfuraldehyde.

2-12 This plaster novelty impregnated with furfural resin has improved gloss, chip resistance, and impact strength. (*Reprinted by permission of Furane Plastics, Inc.*)

occurs at room temperature after catalyzing with hydrochloric acid. Objects to be coated or impregnated may be brushed, dipped, or sprayed.

Modifiers

Even though phenolic resin was the earliest thermosetting synthetic plastic to attain commercial prominence, it continues to challenge the polymer chemist seeking ways to improve its known characteristics. One important modification involves the copolymerization of phenolic and silicone resins to produce phenyl-silane. Although generally produced for laminated structures for aerospace use, the resin can be applied to glass, silica, quartz, or carbonaceous fibers to provide bulk molding compounds.

Unmodified phenolic resins rated as high-temperature plastics can be used in structural applications and retain 90 percent of their initial strength after exposure for 100 h at 500°F when tested at that temperature. Phenyl-silanes, by comparison, retain up to 50 percent of room-temperature values after 200-h exposure at 600°F when tested at that temperature.

Phenol   Formaldehyde   Phenol formaldehyde

2-13   Reaction of phenol and formaldehyde.

TABLE 2-3 Classification of rigid molding materials—Society of the Plastics Industry, Inc.

SPI grade* number	Basic Requirements						Mechanical Properties M			Chem. Prop. C
	Heat distortion temp., °F 264 psi fiber stress	Impact strength izod ft-lb per-in. notch	Tensile strength psi	Recommended service temp. °F no load	Flexural strength psi upper limit	Compressive strength psi upper limit	Modulus of elasticity in tension psi × 10⁶	Hardness Rockwell	Specific gravity	H₂O absorption % 24 h wt. gain + solubles lost %
	Min.	Min.	Min.	Max.	Min.	Min.	Min.	Min.	Max.	Max.
PF† 22033	220	0.3	3000	212	7000x	15000x	0.40	M-50	1.35	1.50
PF 22044	220	0.4	4000	212	7000x	15000x	0.40	M-55	1.35	1.50
DAP† 22054	220	0.5	4000	300	8500x	20000x		M-108	1.45	0.20
PF 22084	220	0.8	4000	212	7000x	15000x		M-55	1.35	1.50
PF 22204	220	2.0	4000	212	7000x	15000x		M-57	1.35	1.50
PF 23034	230	0.3	4000	250	8000x	15000x	3.0	M-100	2.00	0.07
PF 25034	250	0.3	4000	250	6000x	15000x	0.50	M-50	1.70	0.3
PF 25406	250	4.0	6000	250	9000x	15000x	0.90	M-100	1.45	1.75
MF† 26025	260	0.2	5000	250	6500x	30000x	1.1	M-116	1.48	0.60
MF 26035	260	0.3	5000	300	7400x	25000x	1.95	E-90	1.80	0.15
MF 26035-1	260	0.3	5000	250	5800x	9000x			1.46	0.60
DAP 26174	260	1.7	4000	350	11000x	25000x			1.40	0.20

TABLE 2-3 Classification of rigid molding materials—Society of the Plastics Industry, Inc. (*Continued*)

SPI grade* number	Basic Requirements					Mechanical Properties M				Chem. Prop. C
	Heat distortion temp., °F 264 psi fiber stress	Impact strength izod ft-lb per-in. notch	Tensile strength psi	Recommended service temp. °F no load	Flexural strength psi upper limit	Compressive strength psi upper limit	Modulus of elasticity in tension psi × 10^6	Hardness Rockwell	Specific gravity	H$_2$O absorption % 24 h wt. gain + solubles lost %
	Min.	Min.	Min.	Max.	Min.	Min.	Min.	Min.	Max.	Max.
UF† 27026	270	0.2	6000	170	8000x	30000x	1.3	E-94	1.55	0.80
UF 27026-1	270	0.2	6000	170	7500x				1.55	
PF 27034	270	0.3	4000	300	8000x	15000x	3.0	M-100	2.00	0.10
MF 29025	290	0.2	5000	210	9500x	40000x	1.30		1.50	0.35
PF 29026	290	0.2	6000	300	9000x	22000x	0.80	M-110	1.45	0.80
PF 29034	290	0.3	4000	350	9000x	15000x	1.50	M-100	2.0	0.20
PF 29036	290	0.3	6000	300	9000x	22000x	0.80	M-108	1.45	0.80
PF 29046	290	0.4	6000	300	9000x	22000x	0.80	M-100	1.45	1.20
PF 29075	290	0.7	5000	250	9000x	20000x	0.90	M-93	1.45	1.50
PF 29206	290	2.0	6000	250	9000x	18000x	0.90	M-104	1.45	1.75
DAP 30024	300	0.2	4000	400	7000x	20000x		M-99	1.65	0.2
PF 30025	300	0.2	5000	300	7000x	18000x	0.07	M-105	1.90	0.6
MF 30035	300	0.3	5000	275	7500x				1.50	0.55
MF 30058	300	0.5	8000	250	12000x	30000x	1.6	M-119	1.53	0.60
DAP 30065	300	0.6	5000	400	9000x	30000x			1.60	0.50

Material	Grade No.									
MF	30105	300	1.0	5000	250	11000x	25000x	1.6	M-114	1.53	0.50
PF	30206	300	2.0	6000	350	10000x	15000x		M-110	1.74	0.50
PF	33036	330	0.3	6000	400	10000x	20000x			2.00	0.20
EA†	35023	350	0.2	3000	250	7500x	18000x	1.4	60b	1.9	0.50
EA	35023-1	350	0.2	3000	250	7500x	18000x	1.8	60b	2.1	0.15
PF	35024	350	0.2	4000	400	7000x	15000x	1.0	M-115	2.0	0.20
PF	35024-1	350	0.2	4000	400	7000x	15000x	1.0	M-95	1.68	0.50
EA	35033	350	0.3	3000	300	7500x	16000x	2.2	60b	2.28	0.15
EA	35034	350	0.3	4000	250	9500x	21000x	1.8	58b	1.90	1.00
DAP	35057	350	0.5	7000	400	12000x	25000x			1.60	0.20
MF	40027	400	0.2	7000	210	10000x	40000x	1.3	M-118	1.53	0.40
MF	40406	400	4.0	6000	300	10000x				2.00	0.21
EA	40806	400	8.0	6000	300	14000x	24000x	2.2	70b	2.08	0.10
S†	50023	500	0.2	3000	550	6500x	16000x		M-85	2.0	0.40
S	50033	500	0.3	3000	550	6000x	12000x			2.0	0.50
S	50304	500	3.0	4000	550	8000x	13000x			2.0	0.50

*Explanation of SPI grade numbers: Example: PF 30206 (30-20-6)
30 = Heat distortion (deflection) temperature of 300°F
20 = 2.0 ft-lb. impact strength
6 = 6,000 pounds per square inch (psi), minimum tensile strength

†Code for materials
DAP = Diallyl phthalate
EA = Ester-alkyd
MF = Melamine-formaldehyde
PF = Phenol-formaldehyde and/or rubber modified
UF = Urea-formaldehyde
S = Silicone

45

GLOSSARY

Catalyst   A substance that markedly speeds up the polymerization (cure) of a resin when added in minor quantity as compared with amounts of primary reactants. Also referred to as an initiator.

Condensation reaction   Chemical reaction in which two or more molecules combine, with the resulting separation of water or other simple substance. If a polymer forms, the process is called polycondensation.

Mole   The gram-molecular weight of an element or compound that is frequently used in chemical calculations. Also called a gram molecule, a mole of any element or compound has a weight in grams equal to the molecular weight of the substance.

Novolac   A phenolic-aldehydic resin that, unless a source of methylene groups is added, remains permanently thermoplastic.

Plasticizer   A chemical agent added to resins to make them softer and more flexible.

Polymerization   A chemical reaction in which similar small molecules are linked together to form large molecules.

Resole   Single-stage resin produced when equal or greater amounts of formaldehyde are used per mole of phenol in the presence of an alkaline catalyst.

Resorcinol   A polyhydric phenol obtained from coal.

REVIEW QUESTIONS

1   Why are phenolic resins called the workhorse of the plastics industry?
2   By what other name is phenol known?
3   Name two other related chemicals, often combined with phenol to produce tar-acid resins.
4   What are the two major sources for phenol?
5   What is formalin?
6   Describe the three stages of polymerization through which a phenolic resin passes to become thermoset.
7   What is the purpose of using hexa?
8   What is a resole?
9   What is meant by "condensation reaction," and what is its significance in processing?
10   What is the difference between a casting resin and a molding compound?
11   What are the color limitations of phenolic molding compounds?
12   List several products made from general-purpose molding compounds.
13   Which type of fillers improves impact strength?
14   Which filler is referred to as "low loss"?
15   In what areas of industrial use are quartz, silica, and carbon fibers used?
16   Name three mat or textile reinforcements used to produce phenolic laminates.

17 Besides molding, casting, and laminating, what other uses for phenolic resins can be mentioned?

18 What is the purpose of using phenol furfural as a binder for molding compounds?

19 Compare phenyl-silane resin with phenol formaldehyde.

20 The trade name Resinox SC 1013 applies to which type of material?

AMINO PLASTICS

3

Thermosetting resins derived from ammonia compounds are referred to generically as *aminoplasts*. Their development has led to further expansion of the plastics industry into new areas of molding and laminating and has brought about great changes in the textile, furniture, and paint industries.

UREA-FORMALDEHYDE RESINS

Almost 20 years passed before the second thermosetting resin, capable of competing with phenolics, was perfected. In the United States, urea resin was developed by Drs. H. John and F. Pollack, under Patent Nos. 1,355,834 and 1,458,543 issued in 1928.

One factor limiting more widespread use of phenolic resins has been the narrow choice of colors. Urea resin, in contrast, is a clear, water-white substance comparable in clarity with glass. It was first suggested as a replacement for lenses and dials, but poor scratch resistance plus a tendency to develop cracks upon aging ruled against these applications. Later it was found that, by adding 30 to 40 percent pure cellulose filler, urea resin stabilizes to a point where predictable mechanical strengths can be obtained. Thus we find that pure urea resin cannot be used as a casting material. Instead, it is marketed as an adhesive for wood, a baked-on enamel for appliances, an impregnant to make fabrics crease- and crush-resistant, and a binder for molding and laminating purposes.

3-1 Urea-formaldehyde switch plates molded on automatic compression presses. (*Reprinted by permission of Dake Corp.*)

Preparation of urea crystals

Urea is a white crystalline solid that is water-soluble and melts at 269.6°F. It was first synthesized by the German chemist Friedrich Wohler in 1828. Urea is presently used as a major ingredient In commercial fertilizers because of its slow release of nitrogen. To become a useful plastic material, urea crystals are first transformed into urea resin in a method similar to that used for producing phenolic resin. Urea polymers provide an excellent example of how modern synthesis can develop a solid substance from four common gases.

Preparation of urea resin

The formation of urea resin results from the reaction of urea crystals with formaldehyde. Depending on the proportions of ingredients, type of catalyst, and reaction temperature, the resultant partial polymers are further treated to satisfy various end uses: (1) dissolved in water or alcohol for impregnation of paper or textile fibers; (2) combined with other resins for use as surface coating; (3) blended with water-activated catalysts for cold-setting adhesives; or (4) blended with additives for molding purposes.

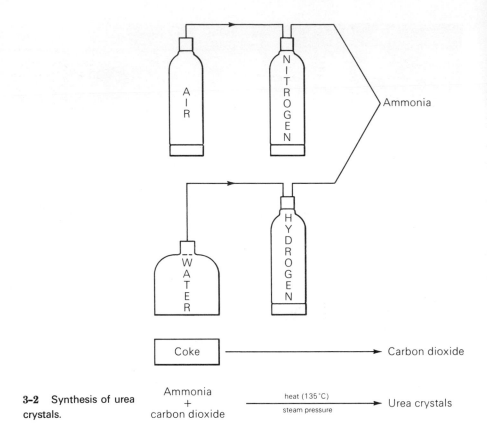

3-2 Synthesis of urea crystals.

$$\text{Ammonia} + \text{carbon dioxide} \xrightarrow[\text{steam pressure}]{\text{heat (135°C)}} \text{Urea crystals}$$

Preparation of molding compounds

Unlike phenolic compounds that combine with an infinite variety of fillers, urea resin is restricted to only two: alpha cellulose and wood flour. There are several reasons for this limitation, although theories about its chemical explanation differ:

1 The urea molecule, somewhat smaller in size than phenolic, is capable of penetrating into the cellulose structure through pore spaces.

2 The water-soluble urea may be carried through the cell wall by capillary action.

3 A direct chemical attraction occurs between substituent atoms.

Irrespective of which phenomenon takes place, the result is a substantially stronger molded article that retains a high degree of translucency.

Urea-formaldehyde molded items are produced by compression-molding methods, generally at temperatures substantially lower than those used for phenolic. In fact, the same mold may be used to mold the same item (e.g., electrical plugs) from either phenolic or urea. Since urea compounds are slightly more costly, dark-colored articles are molded from phenolic; pastel-colored items are made of urea.

COMPARISON OF UREA AND PHENOLIC RESINS

Heat resistance

Urea plastics are not recommended for use in environments where temperatures continuously exceed 170°F because the resin cannot retain its original strength due to embrittlement. Since it is thermosetting, and therefore infusible, it will not soften; it decomposes (carbonizes) at 400°F. Phenolics, it will be remembered, can contain heat-resistant fillers that sustain integrity for long periods at 400°F and to 500°F for short periods.

Water resistance

Some synthetics are adversely affected by water absorption caused by direct immersion or humidity. Ureas have a tendency to absorb moisture, which excludes their use for some products. Slight amounts of moisture result in swelling, and when the water evaporates the product shrinks because it returns to its original dimensions. If this cycle is repeated often enough, alternate expansion and contraction create minute surface cracks called *crazing*. In time, these fine, hairline cracks propagate through the entire wall section of the piece. For this reason urea is not recommended for items such as dinnerware, pot handles, or cutlery, where the combination of hot water and drying would accelerate failure.

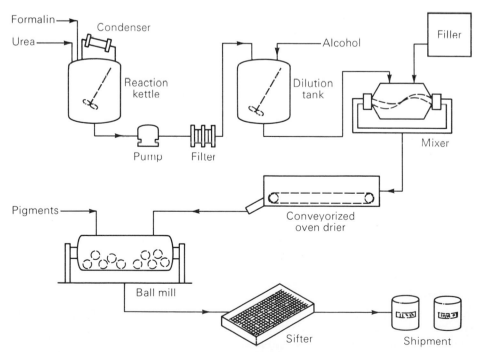

3-3   Simplified flow diagram for preparation of urea molding compounds.

Taste and odor

Chemical resistance of all thermoset resins is responsible for their wide use in industrial and household applications. What warrants comparison between different types are the relative merits of urea and phenolic for products where taste and odor are involved; the danger of imparting a phenol or urea odor or taste must be avoided.

The process of polymerizing thermoset resins is not an exact one in the sense that each urea or phenolic molecule can combine perfectly with the proper proportion of aldehyde. Consequently, there may be unreacted (free) phenol or urea to an extent of 2 or 3 percent, although for practical purposes the molded or laminated product is considered cured. Thus a taste or odor of the plastic can be transmitted to food with which it comes in contact. This problem is more likely to occur with phenolic unless special odorless grades are used for items such as refrigerator parts or bottle caps.

The excellent oil and grease resistance of urea, and its availability in many colors, makes urea the outstanding candidate for cosmetic closures and containers and for household appliances.

Electrical insulation

Phenolic and urea plastics are used for hundreds of electrical units, but each resin has major differences that must be recognized. For such general needs of the electrical industry as switch bases, terminal strips, instrument cases, and knobs, phenolic is chosen for its lower cost and superior impact strength. If color increases sales appeal, urea is used, despite its higher cost.

In situations where there is a danger of electrical flashover (arcing), urea is far superior to phenolic. An overloaded electrical circuit or high-voltage surge proves more damaging to phenolic insulation than to its urea counterpart. Heat generated by the sustained spark (arc) results in a carbon formation as the phenolic resin surface decomposes; urea chars but does not carbonize. Once a carbon path is established from one terminal to another, the scarred phenolic surface provides a path for leakage or current at much lower voltage, usually necessitating replacement. Urea, by comparison, retains its insulation value much the same as before the initial flashover.

UREA RESINS AS ADHESIVES

The plywood industry and furniture makers rely on urea-formaldehyde resins for fast bonding of wood veneers. Resin producers furnish several types of glue to suit the method employed in wood-working industries:

Hot-press adhesives

A glue-spreading unit transfers the adhesive to wood veneers that are stacked into a hot (260°F) press for 10 to 15 min at moderate pressure (150 to 300 psi). Heat activates the catalyst, which results in polymerization.

High-frequency adhesives

A portable unit is used to send electronic impulses through the structure. In this process (known as wood welding), the dielectric resin hardens in an instant as the pulsating current generates frictional heat within the glue line. Electrodes of various contours enable furniture assemblers to glue complex shapes.

Cold-set adhesives

When mixed with water, powdered urea resin provides a slower setting glue, which is often necessary in fabricating large structures such as roof trusses. With this method a latent catalyst such as ammonium chloride premixed into the resin breaks down to form an active catalyst (hydrochloric acid) when water is added.

UREA RESINS FOR TEXTILE TREATMENT

A revolution in the garment and textile trades occurred in the early 1930s when crush-resistant fabrics were offered to the public. Cotton, linen, and synthetic blends of cloth, after being printed or dyed, are passed through a resin bath so that they become impregnated with a urea solution. Similar to the way in which the small urea molecule penetrates alpha cellulose pulp and wood flour, the resin becomes an indistinguishable part of the fabric. As the material is drawn through heated drying towers and squeezed between heated cylinders, the excess water is evaporated and the resin cured. Softness of the fabric is not lost because of the early stage of poly-merization at which the resin is introduced (*A* stage). When a more advanced urea is used (*B* stage), the change in the fabric (hand or drape) is quite noticeable. The fibers become stiffer and "polished" as a result of a superficial coating rather than impregnation.

THIO-UREA

Conventional urea formaldehyde, modified with sulfur (*thio* is the Greek word for sulfur) is available as a resin for molding or adhesives. The presence of sulfur imparts better water resistance, though at some sacrifice in molding quality because of the corrosion of mold parts.

MELAMINE-FORMALDEHYDE RESINS

In 1940 a new series of thermosetting resins was introduced by American Cyanamid Company. The chemistry involved in producing basic melamine is complex, although molding and laminating processes closely parallel those used in working with urea and phenolic materials.

Depending on the end use, the partial polymer is extracted from the reaction kettle and further modified for use as adhesives, coatings, paper treatment, molding,

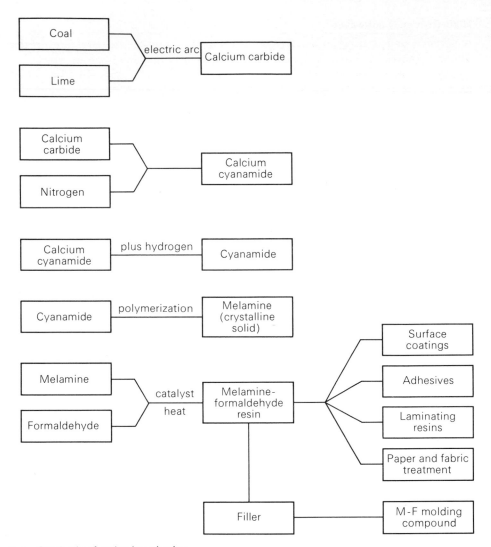

3-4   Synthesis of melamine plastics.

or laminating resins. Like urea, melamine is a clear, water-white solid capable of accepting any degree of pigmentation. Unlike urea, melamine is capable of combining with a wide variety of fillers to extend the range of molded and laminated products. Besides alpha cellulose, which is used for general-purpose filler, molding compounds are made with shredded cotton fabric or fibrous glass for impact strength, asbestos for heat resistance, and mineral fillers for electrical properties.

Perhaps the most successful application for alpha-cellulose-filled melamine is dinnerware. Compression molded in solid colors or with a melamine-impregnated decorative paper overlay, melamine dinnerware has long been a household standard. Light weight and breakage resistance have provided outlets for dinnerware used by institutions and aboard ship and aircraft. Military and naval requirements specify impact grades made with macerated cotton cloth.

Melamine, although not as scratch resistant as chinaware, exhibits the hardest surface of any synthetic resin. Heat resistance of organic-filled compounds is 300°F, with excellent water resistance, so that cycling through dishwashers or steam cleaners is not harmful. Staining by tea, coffee, and lipstick presented problems with earlier formulations, but this difficulty has been largely overcome.

Other uses for alpha-filled melamine include housings for electric shavers, utensil handles, business machine housings, and components for kitchen appliances. Filled with wood flour, melamine is used in ignition systems where arc resistance at a moderate temperature is required. Aircraft connectors, molded in millions annually, are made from mineral-filled compositions. Glass-filled—melamine-molded items include heavy-duty circuit breakers and switch components.

Combined with cellulosic fillers, melamine-phenolic resins meet the need for light color, heat stability, and moisture resistance at an intermediate cost.

MELAMINE LAMINATING RESINS

Melamine is an almost universal choice for impregnated paper used as the top sheet for decorative panels and engraved signs. Applied as a clear-liquid penetrant to previously printed sheets or as a pigmented resin to plain sheets, the resin-treated overlay is hot pressed over less-expensive, phenolic-impregnated core sheets. A special overlay sheet of clear-treated stock can be applied to either of the two major styles of paneling to increase mar resistance.

Panels are used for tables, counter and sink tops, wall covering, desk tops, furniture sections, and similar uses requiring highly decorative, durable finishes. Plasticized grades are furnished for applications where coving is necessary, as in

3–5   Melamine-phenolic-molded products. [*Reprinted by permission of Plastics Engineering Co. (PLENCO).*]

coved sink installations. Moderate heating makes the cured resin sufficiently thermo-elastic to permit the forming of large curves. Other uses of melamine laminates include engraving stock laminated in sandwich form, with contrasting outer plies cut to expose the core sheet.

Besides presenting a hard, mar-resistant surface, melamine laminates are superior to any other applicable resin in offering a composition that at the same time is impervious to alcohol, oils, grease, mild acids, detergents, and also is fire resistant. Industrial laminates using cotton, glass, or asbestos textiles for reinforcement are widely used for circuit boards, mounting panels for electrical gear, and chemical equipment.

MELAMINE ADHESIVES

Several million pounds of melamine powder and liquid adhesives are produced annually. Although more expensive than urea glues, melamine adhesives have better water resistance, which is why they are used for exterior grades of plywood. When a more economical adhesive is required, melamine can be used to modify urea in up to equal proportions while still meeting less-critical plywood specifications.

Melamine adhesives are applied and cured by the same methods as those used for urea.

3-6   Kitchen area with extensive use of melamine decorative paneling. (*Reprinted by permission of Fabricon Products Div., Eagle Picher Corp.*)

AMINO RESINS FOR PAPER TREATMENT

The importance of impregnating paper with amino resins rivals the development of treating textiles for shrinkage control and crease resistance. Small amounts of urea or melamine resins added to the paper pulp during manufacture greatly increase the wet strength of the final product. When wet strength is most important (e.g., paper towels, napkins, tissue, and lightweight paper bags), urea is used as the impregnant. Grease and stain resistance is improved with melamine, so melamine is added to the pulp to produce butcher paper, maps, chip board, wall paper, and multiwall bags for industrial use.

In many cases, urea or melamine could be used for the same application. With the list of new products increasing yearly, it would not be surprising to see a greater use of disposable clothing that could withstand at least several wash cycles.

AMINO RESINS FOR SURFACE COATINGS

Urea and melamine, specially formulated with butyl alcohol and formaldehyde, are used in combination with alkyd resins for industrial and commercial bake enamels. The urea alkyds are generally restricted to indoor products such as refrigerators, kitchen cabinets, and washers. Automotive finishes and machinery enamels utilize

3-7 Bar top made with printed overlay sheet. (*Reprinted by permission of Fabricon Products Div., Eagle Picher Corp.*)

the melamine formulations. These amino-modified alkyds impart hardness and durability to the coated surface and enable the finish to withstand abrasion or discoloration
from chemicals. Coatings, which are applied by spray or dip methods, are cured at
temperatures between 200 and 300°F to 1 h.

GLOSSARY

Alpha cellulose   A purified form of cellulose derived from wood pulp.

Arcing   Sustained spark (flashover) caused by excessive current or short circuit between adjacent conductors.

Latent catalyst   A catalyst that requires heat, moisture, or other means of activation.

REVIEW QUESTIONS

1   Why can't urea resin be used for casting?
2   What four gases are used to produce urea resin?
3   Which material, urea or phenolic, offers the wider color choice?
4   Which material, urea or phenolic, has the higher heat resistance?
5   Why is urea not recommended for use in wet environments?
6   What advantage does urea plastic exhibit, when compared with phenolic, for arc-resistant
applications?
7   How does urea resin impart crush resistance to fabrics?
8   List four fillers used with melamine for molding compounds.
9   Which plastic, urea or melamine, exhibits better heat resistance? Better water resistance?
10   List four applications for molded melamine.
11   Sketch or describe the composition of a decorative panel.
12   List several applications for urea- and melamine-impregnated paper.

POLYESTER AND ALKYD RESINS

4

Polyesters, a large and important family of synthetic resins, were initially developed from 1938 to 1942. This development has led to entirely new processing techniques in molding and laminating large and often complex structural products. (Polyesters were so important to our defense efforts in World War II that they were classified "secret" materials.) The introduction of polyesters that could polymerize quickly at room temperature without external pressure brought about wholly new industries and opened new markets for plastic products.

In 1968, 600 million lb of polyesters were used by industry. The reasons for this steadily increasing volume are numerous:

1   Low cost (25 to 55 cents/lb; see Table 4-2 at end of chapter)

2   Ease of producing from abundant raw materials

3   Wide range of formulations to meet a variety of requirements

4   Crystal clarity to light straw color; can be pigmented to any desired hue

5   Available as liquids, pastes, or flake

6   Adaptability to many molding and laminating methods

7   Ability to be polymerized with a variety of catalysts

4-1   Radome on Lockheed
L-1011 Tri-Star produced from
glass-cloth–polyester-resin
laminate. (*Reprinted by permission of Lockheed Aircraft
Corp.*)

  8   Permits passage of microwaves for radar applications

  9   Resists deterioration by sunlight, oxidation, saltwater, fungus, and many chemicals

10   Excellent mechanical, thermal, and electrical properties

11   Ability to be cast in a variety of simple molds

12   Ability to be made self-extinguishing

     Polyesters are produced in thermoplastic varieties as transparent film (Mylar[1] or Scotchpak[2]) or as textile fiber (Dacron[3] or Fortrel[4]). The importance of thermosetting types is derived from the nature of the curing process—it does not produce gaseous by-products commonly associated with condensation-type polymerization. It should be recalled that as phenolic and amino resins harden, chemical by-products are split off from the main reaction. These condensates (water, ammonia, and excess formaldehyde) require that high molding pressures be available to counteract internal vapor pressures which build up within the product. Heavy-duty molds and presses are necessary to mold these resins and overcome high pressures developed by expelled volatiles.

     By comparison, polyesters undergo a chain formation process called *addition polymerization*, in which no condensate is formed during the cure. For this reason they are termed *low-pressure resins* since no external pressure is necessary for casting or laminating, and only low pressures (50 to 200 psi) are needed for many compression-molded parts. When little or no pressure is required, tooling can be made from

[1]Mylar is the registered trade name of E. I. du Pont de Nemours & Company.
[2]Scotchpak is the registered trade name of 3M Company.
[3]Dacron is the registered trade name of E. I. du Pont de Nemours & Company.
[4]Fortrel is the registered trade name of Celanese Plastics Company.

4-2   Tub-shower combination produced from polyester-glass laminate. Note smooth, glossy surfaces and absence of joints, seams, or crevices. (*Reprinted by permission of Owens-Corning Fiberglas Corp.*)

lighter materials (wood, plaster, foam, aluminum, rubber, and plastics). Size and shape are unlimited.

This is not to imply that first-class, matched metal tooling and hydraulic presses are not widely used. The several parts that make up the Corvette auto body are mass produced by compression molding, although early versions were made on cast plastic and aluminum tools by laminating methods.

4-3   Scotchpak polyester film eliminates need for a cooking utensil, reduces shrinkage, eliminates cleanup, and withstands oven temperatures to 500°F. (*Reprinted by permission of 3M Co.*)

PREPARATION OF POLYESTER RESINS

An *ester* results from a reaction between an alcohol and an acid (*esterification*). Ingredients (500 to 6,000 gal) are reacted in stainless-steel kettles, heated gradually from 350 to 450°F. A batch is cooked until the desired viscosity, color, and acid number are reached. The acid number is a measure of unreacted acid groups determined by laboratory methods.[1] At this point, reaction is slowed by cooling from 200 to 350°F so the polyester syrup can be thinned by dissolving in a reactive monomer.

Many combinations of alcohols and acids provide hundreds of polyester resin types. For a resin to be capable of polymerizing to form a three-dimensional thermosetting structure, the alcohol selected must be *polyhydric* [contain at least two hydroxyl (OH) groups] ; the acid must be *dibasic* (contain molecules capable of furnishing two hydrogen ions). Numerous difunctional hydroxyl compounds, referred to as *diols*, and organic acids are available to the chemist. As a result, a polyester can be tailored for use in a wide variety of fabricating conditions and satisfy a wide

[1]Acid number is determined by the number of milligrams of potassium hydroxide required to neutralize an acid in 1 g of reacted resin. It provides a clue to the degree of esterification (chain formation).

4-4 Echo II "Satelloon" 135 ft in diameter, fabricated from aluminized polyester film. Shown being inflated for testing. (*Reprinted by permission of G. T. Schjeldahl Co.*)

range of product requirements. The following table shows the advantages and sources of various polyesters:

Type	Source	Features
Acids or anhydrides		
Maleic anhydride	Coal	Low cost
Fumaric acid	Maleic acid	Low cost; toughness
Adipic acid	Coal (phenol)	Flexibility
Sebacic acid	Castor beans	Resilience
Phthalic anhydride	Coal (naphthalene)	Low cost; good, average properties
Isophthalic acid	Coal (xylene)	Toughness; heat and chemical resistance
Chlorendic acid	Petroleum	Self-extinguishing
Tetrahydrophthalic anhydride	Petroleum	Tack-free surface
Diols (from petroleum)		
Propylene glycol		Low cost; hardness
Ethylene glycol		Low cost; usually with other diols
Dipropylene glycol		Lower water absorption
Diethylene glycol		Higher impact strength
Tripropylene glycol		Increased flexibility
Triethylene glycol		Increased toughness
Butylene glycol		Chemical resistance
Cross-linking monomers		
Styrene	Coal; petroleum	Low cost; compatibility
Vinyl toluene	Coal	Heat-deflection resistance
Acrylic	Coal	Light stability
Diallyl phthalate	Coal; petroleum	Heat and chemical resistance
Diallyl isophthalate	Coal; petroleum	Higher heat resistance (400°F)

Probably no other area of polymer chemistry affords the researcher so many molecular "building blocks" with which to construct a synthetic resin. Figure 4-6 illustrates a typical reaction that would occur in a batch process. Note that a condensation reaction is involved in the initial stages of polymerization with the evolution of water, and compare this with the formation of a phenolic resin as discussed earlier.

An examination of the chemical reaction reveals several sites of unsaturation ($C=C$ and $C=O$). To produce the three-dimensional thermoset resin structure, a compatible monomeric substance such as styrene is added. Monomer addition, called *thinning*, controls workable viscosity and resin flow during molding, laminating, or coating operations. Styrene monomer is a low-viscosity liquid capable of polymerizing to the thermoplastic, polystyrene. This monomer, also known as vinyl benzene, has the structural formula $C_6H_5CH=CH_2$ and is also unsaturated. Under the proper conditions of either catalysis, catalyst plus heat, or ultraviolet light, the

4-5   Corvette underbody molded from polyester and glass being removed from the press. (*Reprinted by permission of Automotive Products Div., North American Rockwell.*)

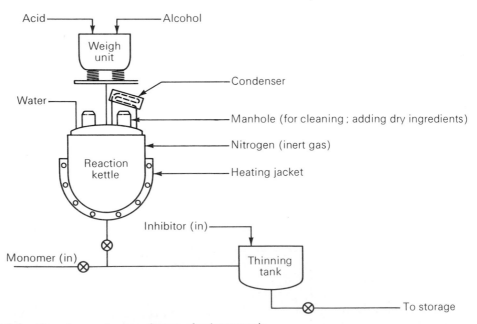

4-6   Flow diagram for manufacture of polyester resin.

4-7 Structural formula. Propylene glycol + maleic anhydride = polyester resin + water.

$C=C$ linkage of the monomer forms a network with the $C=C$ bonds of the polyester, resulting in a hard, unmeltable, cured resin. Let $P =$ an unsaturated polyester, $S =$ styrene monomer. (See Fig. 4-8.)

Monomer, added by the resin manufacturer, is usually present in amounts to 40 percent. High-viscosity resins, usually with much lower monomer proportions, are often thinned by the user to: (1) reduce viscosity for spraying; (2) improve wetting of close-weave fabric or mat; (3) impregnate paper; (4) facilitate addition of catalysts; and (5) thin the resin for filler addition.

Normally styrene additions above 10 percent are detrimental because of the possibility of forming polystyrene molecules within the polyester matrix. Liquid styrene should be handled carefully to avoid hazards such as fire, which could be created from static that arises when the styrene is poured from a metal drum. Static electricity can be drawn away by grounding the container.

ALLYLIC RESINS

Allylic resins, developed originally by Food Machinery and Chemical Corporation, offer these specific advantages over conventional polyesters: (1) greater stability in storage; (2) lower shrinkage during cure with less tendency to warp; (3) better hot strength, which is important in removing parts from molds and in service; (4) good chemical and solvent resistance; and (5) better performance in microwave (radar) applications over a wide temperature range.

4-8 Schematic of styrene and polyester reaction.

The most widely used of the number of allyl resins produced are those based on a reaction between allyl alcohol (from propylene) and phthalic anhydride (from naphthalene) resulting in a *diallyl phthalate resin* (DAP). Shortages of phthalic acid led to development of isophthalic acid (from xylene) in 1957. This in turn brought about improved allylics, with resins showing better impact, heat, and chemical resistance. These DAIP resins (*diallyl isophthalate*) are widely used in the manufacture of truck cabs, trailer and camper components, and chemical equipment.

DAP and DAIP resins, in contrast to unsaturated polyesters, are available as a monomeric liquid or prepolymerized as a finely pulverized powder soluble in the monomer. Resins can be prepared in any viscosity by mixing the two types to form adhesives, body putty, molding compounds, and laminating resins. A further benefit results from the fact that allylic resins, unlike the unsaturated polyesters, do not require a reactive monomer such as styrene to effect a cure. Allylic monomers shrink only 12 percent, compared with shrinkage of 18 percent for styrene and between 8 and 10 percent for polyester resins. Combining prepolymer with monomer reduces shrinkage to as low as 4 percent, which is advantageous in laminating and molding. Because of low shrink and absence of vapor pressure, allyls are used to vacuum-impregnate porous castings of aluminum and magnesium.

Molding compounds represent a substantial market for DAP and DAIP resins and are formulated to meet a number of military, electronic, and commercial specifications. Compounded with Orlon, Dacron, asbestos, or glass fibers, many grades meet demands of compression and transfer molders.

4-9 Hoods for agricultural trucks typify applications of isophthalic resins. (*Reprinted by permission of Organic Materials Div., Koppers Co., Inc.*)

Molding compounds exhibit excellent flow and mold filling, usually at lower pressures. Thin sections cure as quickly as 30 sec. Colors are unlimited although the number of standard colors is kept to a minimum. Fillers impart the following features :

Type	Characteristics
Orlon	Electrical properties
Dacron	Impact resistance ; stiffness in thin sections
Asbestos	Heat resistance ; lower shrinkage
Glass fiber	Heat resistance ; highest impact strength ; lowest shrinkage

Molded products are serviceable from 300 to 450°F, depending on fillers and whether the base resin is DAP or DAIP.

Decorative laminates produced by continuous laminating compete favorably with melamine. Even though resins are more expensive, paper or fabric impregnated with allylic can be laminated and cut into panels on a continuous basis. Melamine, by comparison, requires 1,000 psi during cure, plus a long cure time.

DAP monomer upgrades conventional polyesters in proportion to quantity added, improves pot life, and contributes to heat resistance and other properties.

4-10   Transformer case compression molded from DAP resin containing glass fibers. (*Reprinted by permission of Dake Corp.*)

FIRE-RETARDANT POLYESTERS

Polyesters in liquid form and when polymerized support combustion unless they are compounded with flame-retardant chemicals or are originally synthesized from chlorine-containing reactants. Molding compounds and laminating resins made with general-purpose polyesters are rendered self-extinguishing by incorporating small amounts of chlorinated paraffin with antimony trioxide. The combination of wax and antimony exerts a *synergistic* effect that becomes less effective because of migration accelerated by exposure to heat or ultraviolet. Resins with these additives are opaque white. Allyl resins synthesized from diallyl chlorendate are self-extinguishing.

Clear, styrenated polyesters with chemically combined chlorine are available;[1] they meet stringent military specifications and commercial applications where *minimum* flame spread is important.

CURING THE POLYESTERS

The popularity of polyesters can be attributed to the many ways curing can be effected. The curing mechanism is so simple that the home craftsman can use polyesters to make laminated products or castings, as well as repair laminated structures.

In order to be polymerized, polyesters undergo distinct phases before conversion to the hard, dense, fully cured stage. Resins are furnished as low- or medium-viscosity liquids or as flake that is soluble in a suitable monomer. To prevent premature gellation, the manufacturer adds inhibitors (hydroquinone; catechol). To overcome the effect of the inhibitor and to speed the reaction, *initiators* (hereafter called *catalysts*) are added in specific amounts.

The wide variety of catalysts, in conjunction with *promoters* (accelerators), enables the processor (or do-it-yourselfer) to select the system that fits the job.

Room-temperature catalysts

Methyl ethyl ketone peroxide (MEKP):   This is a colorless liquid with three types of reactivity: standard, fast, and extra fast. MEK peroxide diluted with 60% dimethyl phthalate (DMP) is the most widely used catalyst.

Benzoyl peroxide: This is available in granular or powder form or in equal parts with tricresyl phosphate (TCP) as a paste. Solid types are extremely shock- and heat-sensitive, thus requiring extreme caution in handling and storing. Paste types are safe.

[1] Hetron is the registered trade name of Durez Division, Hooker Chemical Corporation.

Liquid resin → (catalyst + promoter) → Gel (semisolid) → (exotherm) → Partially polymerized structure → (room temp. aging or heat) (several hours to several days) → Fully polymerized structure

4-11   Stages in polyester polymerization.

Cyclohexanone peroxide:　When diluted with 50% dibutyl phthalate (DBP), this catalyst reduces exotherm and provides low color; it is a paste type.

Cumene hydroperoxide:　This is a colorless liquid or a 50 percent paste with DBP.

Room-temperature catalysts, unless used in conjunction with a promoter (see Table 4-1), polymerize a resin too slowly. Small amounts of promoter make major differences in cure rate.

Promoters

Cobalt naphthenate:　This is a metallic soap, liquid, of purple color, 6 percent cobalt metal, and it results in pink-tan color in clear castings.

Cobalt octoate:　This is a metallic salt containing 12 percent metal. Its use results in faster gel times with less discoloration, and it is less expensive than cobalt naphthenate on a use basis.

Manganese naphthenate:　This is a liquid, 6 percent metal, effective with cumene hydroperoxide for gradual cures with lower exotherm.

Dimethyl aniline:　This liquid is excellent for room-temperature cure when used with benzoyl peroxide.

Choice of type of room-temperature curing of a polyester is based on eight factors: (1) time available to use the activated resin before gellation; (2) thickness of product; (3) permissible cure time; (4) resin content as a percentage of total product weight; (5) inclusion of fillers (amount, plus degree of heat conductivity); (6) heat conductivity of mold; (7) surrounding temperature; and (8) effect of colorants (some accelerate, others inhibit cure rate).

The system providing a moderate or long gel time, plus rapid cure, is generally desirable. However, the fastest cure does not necessarily yield the best product. Overcatalyzed resins are discolored, develop craze marks or voids, or are low in mechanical properties. Comparisons of catalyst-promoter systems are based on gel time and peak exotherm.

Elevated-temperature catalysts

Laminated, molded, and cast products are cured in ovens, autoclaves, and by various other methods described in Chap. 12. The advantages of using an elevated-tempera-

TABLE 4-1　Typical combinations for room-temperature cure

Catalyst	Promoter
MEK peroxide	Cobalt naphthenate or cobalt octoate
Cyclohexanone peroxide	Cobalt naphthenate or cobalt octate
Cumene hydroperoxide	Manganese naphthenate
Benzoyl peroxide	Dimethyl aniline

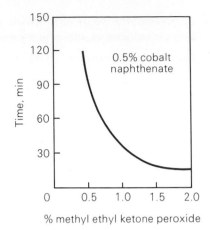

4-12 Time-to-peak-exotherm temperature vs. catalyst concentration. (*Data by permission of Durez Div., Hooker Chemical Corp.*)

ture curing system are : extended pot life (often by many months) ; tack-free surfaces ; and denser, void-free products due to pressure applied.

Moderate-temperature catalysts (180 to 250° F)   Moderate-temperature catalysts are benzoyl peroxide, cumene hydroperoxide, MEK peroxide (standard reactivity), and lauroyl peroxide. Resins are catalyzed with or without the promoters mentioned above. Addition of small amounts of promoter reduces the temperature required for cure.

High-temperature catalysts

Dicumyl peroxide :   A free-flowing powder in combination with calcium carbonate.

Tertiary butyl perbenzoate (TBP) :   A colorless, relatively stable liquid widely used in molding compounds (premix), pre-impregnated fabrics, and roving (prepreg), particularly with DAP and DAIP resins. Contributes to long storage life.

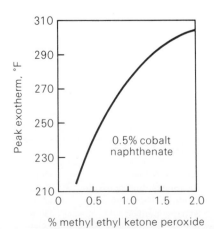

4-13 Peak-exotherm temperature vs. catalyst concentration. (*Data by permission of Durez Div., Hooker Chemical Corp.*)

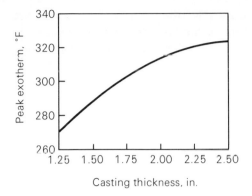

4-14   Peak-exotherm temperature vs. casting thickness. Peak exotherm from room temperature of 73°F. 1% MEK peroxide and 0.25% cobalt naphthenate. (*Data by permission of Durez Div., Hooker Chemical Corp.*)

Casting thickness, in.

2,5-dimethyl hexane, 2,5-diperbenzoate: A colorless liquid that provides greater stability than TBP, with greater economy. High-temperature catalysts are used for cure to 350°F, depending on the application.

The dangers in working with flammable polyesters, catalysts, and promoters cannot be overemphasized and will be discussed further in subsequent chapters. For example, the shock sensitivity of some pure catalysts in dry form is so high that dropping a packet of them can cause an explosion. Spilled, dry catalyst run over by a shop cart can ignite from friction, and prolonged inhalation of vapors from liquid catalysts, increased by evaporation of monomers, leads to severe throat irritation. Peroxides or resins containing peroxides cause severe irritation of the eyes. If this occurs, wash eyes with generous amounts of cool water and see a doctor immediately. Never mix any catalyst with any promoter because some combinations are highly explosive. Always add promoter to resin, *mix well*, and then add catalyst just before using. Use separate dispensers for each component.

Never dispose of a batch of unused catalyzed resin in a trash container unless it has gelled, exothermed, and cooled. A delayed reaction can cause a fire when combustibles are present.

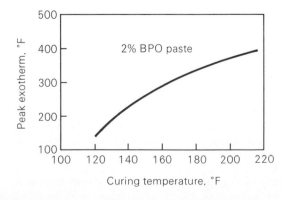

2% BPO paste

4-15   Peak-exotherm temperature vs. curing temperature. (*Data by permission of Durez Div., Hooker Chemical Corp.*)

Curing temperature, °F

Store peroxides in a cool place, preferably below 60°F, safe from people ignorant of their danger and safe from heat or fire hazards. As temperatures rise, catalysts decrease in shock resistance. In case of fire, know and use recommended fire-fighting procedures; peroxides explode violently. Store catalyst so that if packages leak their contents will not come in contact with promoters.

Thixotropic agents

Polyester resins are produced in a wide range of viscosity. For most uses, other than press molding, resins have low viscosity for casting and afford good wet-out of reinforcements. This creates problems of drainage from vertical surfaces. Adding particulate fillers does not provide an answer since the result is a paste that does not flow or allow air removal.

A true thixotropic additive[1] causes a liquid to be mobile when stirred, but immediately reverts to a gel when agitation ceases. Original (apparent) viscosity is dependent on the amount of thixo used.

ALKYD RESINS

The term *alkyd* is coined from two words, alcohol and acid, and is used to describe polyesters other than those previously discussed. Based on reactions between alcohols and acids, alkyd resins are widely used as enamels, hot-melt adhesives, and printing inks.

Alkyds for surface coatings are produced in thousands of formulations that harden by oxidation of drying oils (linseed, tung) or by baking. Straight alkyds are used for building interior and exterior paint or as bake finishes for toys, metal furniture, and venetian blind slats. Alkyds are modified with many synthetic resins to fit numerous uses such as:

Modifier	Application
Phenolic	Rail cars, trucks, farm equipment
Amino	Steel cabinets, stoves, refrigerators
Epoxy	Primers, drum coatings

ALKYD MOLDING COMPOUNDS

Polyester resins modified with DAP or DAIP prepolymers are available as a dry granular powder or in sheet, rope, or slug form.

Fillers include clay, asbestos, glass fibers, or cellulose when light weight is required. Alkyds cure by addition polymerization without evolution of volatiles.

[1]Cab-O-Sil (fumed silica) is the registered trade name of Cabot Corporation, Santocel is the registered trade name of Monsanto Company, and Asbestos #244 is the registered trade name of Union Carbide Corporation.

4-16 Rotor, distributor cap, and coil top, all molded from granular alkyd, offer exceptional arc resistance and dimensional and heat stability under severe conditions. [*Reprinted by permission of Plastics Engineering Co. (PLENCO).*]

Granular types cure as quickly as 20 sec, requiring fast acting (350-in./min closing speed) presses. Mineral-filled resins are used for switch components and housings, auto ignition coils, and similar applications where arc resistance, low-moisture absorption, and dimensional stability at elevated temperature are required. Glass-reinforced grades add impact strength and improved heat resistance.

Alkyds are available in opaque, light colors and can be formulated for fire retardance.

Rope and sheet forms have limited shelf life unless stored below 50°F to prevent catalyst decomposition. When TBP is used as a catalyst, the decomposition product benzaldehyde can be detected by an almond odor.

TABLE 4-2 Cost comparison of various materials

Material	Cost/lb
Alkyd:	
Granular	0.40–0.60
Rope	0.60–0.75
Polyester resin:	
General purpose	0.22–0.28
Heat resistant	0.35–0.40
Diallyl phthalate resin:	
Monomer	0.57
Prepolymer	0.85
DAP molding compounds	0.75–3.50
Diallyl isophthalate resin:	
Monomer	0.90
Prepolymer	1.75

GLOSSARY

Anhydride   An oxide that, when combined with water, gives an acid or a base.

Autoclave   Pressure vessel utilizing steam for both heat and pressure.

Difunctional   A carbon compound having two sites of reactivity.

Exotherm   Heat liberation.

Gellation (gel)   Transition from liquid phase to a semisolid (rubbery) condition.

Inhibitor   A substance that slows or prevents chemical activity and is used to prolong storage life or to reduce rate of exotherm.

Matrix   That which encloses anything.

Monomer   A simple molecule capable of reacting with similar molecules to form a polymer.

Particulate   Made of small particles such as pulverized clay or short fibers.

Pot life   Working time of a catalyzed resin.

Prepolymer   Solid phase of a resin advanced to an intermediate stage of polymerization but still soluble in its monomer.

Synergistic effect   The effect obtained by combining two materials, thus increasing the efficiency of both.

Unsaturated compound   A compound having more than one bond between two adjacent atoms and capable of accepting other atoms at that point to reduce it to a single bond, thus satisfying it (saturation).

REVIEW QUESTIONS

1   Compare the addition reaction with the condensation reaction.
2   List six reasons for the popularity of polyester resins.
3   What benefits are derived by utilizing methods that require little or no pressure to mold or laminate a product?
4   What is an ester?
5   How is the acid number of a resin determined? What is the significance of the acid number?
6   For what purposes is styrene monomer added to polyester resin?
7   What advantages do allyl resins exhibit when compared with conventional polyesters?
8   What precaution should be observed when handling styrene monomer?
9   List several advantages of using a room-temperature catalyst.
10   Which catalyst is heat- and shock-sensitive?
11   What is the difference between a catalyst and a promoter?
12   How can a conventional polyester be made fire retardant?
13   What is meant by synergistic effect?
14   What factors must be considered in choosing the correct catalyst and promoter combination?

15   What does peak exotherm mean?

16   How can exotherm be reduced?

17   What processes employ TBP as a catalyst?

18   List the dangers that must be recognized in working with catalysts and promoters.

19   Define thixotropic. For what purpose is a thixotropic agent used?

20   List several applications for alkyd resin coatings. List applications for alkyd molding compounds.

EPOXY PLASTICS

5

Almost anyone who has tried to repair a broken lamp or other household item is familiar with "goof-proof" epoxy adhesives, available in two tubes at grocery and hardware stores. By mixing equal amounts from tubes *A* and *B* and applying them to the joint, good bonds can be obtained, often between dissimilar materials. The adhesive properties of epoxies, coupled with low curing shrinkage, heat and chemical resistance, plus high mechanical strength account for a rate of growth unparalleled by any thermoset resin, except perhaps polyesters. Epoxies have invaded more areas of application in a shorter period than any synthetic resin.

Reasons for the popularity of epoxies are many and varied:

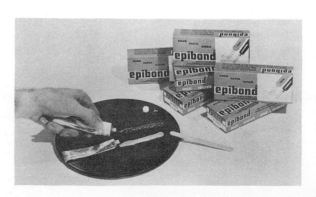

5-1 Epoxy adhesive kit containing tubes of resin and hardener fulfills many industrial and household applications. (*Reprinted by permission of Furane Plastics, Inc.*)

Durability: Cast epoxy tools are used to shape metal in automotive, aircraft, and other metal-working trades. Molds for blow molding, thermoforming, and laminating withstand combinations of heat and pressure better than other castable resins.

Low shrinkage: Large tools and shrink fixtures can often be made "net" without compensating for cure shrinkage when making original patterns. Encapsulating delicate glass components without crushing or distorting ultrafine lead-in wires is a good reason for selecting epoxies for electronic applications.

Excellent adhesion: Although among the highest-priced adhesives ($0.60 to $8.00/lb), they form bonds with high peel strength and low stress and are easily applied. Bonds of dissimilar materials such as stainless steel to phenolic honeycomb, polyester to glass, and aluminum to wood are examples of epoxies' versatility.

Heat resistance: Formulations of adhesives, laminating resins, and potting compounds retain more than 50 percent of original mechanical values at 400°F and 30 percent at 500°F. Adhesives used for bonding aluminum skins in aircraft exterior sections provide smoother contours and reduce stress around rivets. Laminated glass fabrics are used for pipe, tanks, and other structures that operate at elevated temperatures.

Chemical resistance: Because they are thermosetting, with a dense, closely knit chemical structure, epoxies are inert to solvents, acids, and alkalies. Saltwater does

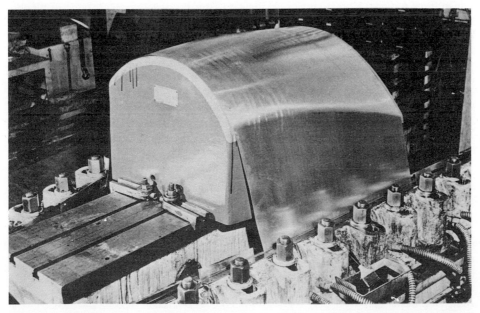

5-2   Epoxy stretch die in operation. (*Reprinted by permission of Rezolin Div., Hexcel Corp.*)

not attack epoxy readily, so that it may be used in many marine products, as well as in coatings for such underwater structures as pilings.

Ease of use: Resins can be formulated to cure rapidly at room temperature. Maximum heat resistance and mechanical strengths are realized with systems that harden with moderate heat. Adhesives, casting resins, and laminating resins can be formulated to cure with or without externally applied heat. Patching compounds, sealants, and coatings are examples of rapid curing resins that harden at room temperature. One interesting application is in repair of battle-damaged or ruptured steam lines aboard ships. Wrapping with glass cloth saturated with catalyzed resin makes the line operational within one hour.

There have been many interesting and unusual applications of epoxies, since the original concept by Dr. P. Castan in Switzerland during the 1930s. Castan's goal was to develop a synthetic resin for artificial dentures or to develop an adhesive to cement teeth directly to the jawbone. Simultaneously in the United States, S. O. Greenlee of the Devoe and Raynolds Company developed epoxy for use as a coating. Coatings are a major outlet for epoxies, but epoxy is just beginning to replace acrylic for use in dentures. Original patents issued to Ciba Products Company of Switzerland and to Devoe and Raynolds formed the basis for cross-licensing other manufacturers. (Some of the major resin producers are listed in Appendix B.)

Despite a variety of resins available as basic types, the demand for specialty resins is so great that a substantial number of companies have emerged as *formulators*. These firms provide a great service to the end-user by developing proprietary blends of resins, hardeners, and fillers. Through constant research and disseminating knowledge through seminars, in-plant training programs, and publications, the formulator is responsible for epoxy's acceptance by many industries. A partial list of formulators is found in Appendix B.

5-3   Coil encapsulated by transfer molding. Right-hand example illustrates flow and adhesion. (*Reprinted by permission of Hysol Div., Dexter Corp.*)

PREPARATION

The word *epoxy* is coined from the Greek *ep* meaning "between" or "over," plus *oxy*
for "oxygen," to describe the epoxide or basic unit. The epoxide unit $HC \overset{O}{\underset{}{\triangle}} CH_2$
is a three-member ring compound that can be opened by hydrogen-bearing molecules
to form three-dimensional chains in an addition-type reaction.

The major source for epoxide is epichlorhydrin, which is derived from the
propylene fraction of petroleum. It is interesting to note that when epichlorhydrin
was produced from glycerin a century ago, investigators reacted it to form resinous
substances, but without commercial success.

Epoxy technology began in 1945 as a culmination of efforts to form a polymer
based on the reaction of epichlorohydrin with bisphenol *A* (from acetone and phenol).
A second group, termed *epoxy novolacs*, is produced when bisphenol *A* is replaced
with phenol or resorcinol; a third type is derived from peracetic acid. Epoxies serve
many industrial uses in their own right but are also used to modify and upgrade
thousands of adhesive, coating, and sealant formulations when blended with phenolic,
nylon, polyvinyl acetals, and polysulfide rubber.

CURING

Many industrial users prefer to purchase specific multicomponent systems (resin and
curing agents) from reliable formulators rather than employ chemically oriented
personnel or maintain extensive research and testing facilities. It is then only necessary
to select a system that meets the need and follow mixing and curing instructions.
Even highly reactive systems that cure rapidly at room temperature can be shipped as
a single-component system (frozen epoxies) in refrigerated containers and trucks.
Manufacturers of electronic products utilize frozen epoxies to eliminate mixing in the
preparation of encapsulated or potted modules, coils, and resistors.

Epoxy resins are cured at room temperature or with applied heat. Improved

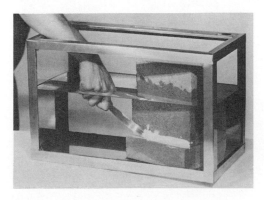

5-4   This epoxy formulation bonds to wet
concrete and cures under water. (*Reprinted
by permission of Furane Plastics, Inc.*)

properties and retention of strength at high temperatures are generally obtained with curing agents that require external heat both during the initial cure and a postcure cycle. Comparisons of various resin-hardening systems are conducted by a standardized heat-deflection test discussed in detail in Chap. 33.

Basic epoxy resins are furnished either as a liquid or as an easily melted solid. Solid resins have higher molecular weight, are less expensive, and are used primarily for coating formulations, prepreg, and molding compounds. All epoxies have extensive storage capability until activated.

Epoxy, like polyester, cures by addition, without formation of gaseous by-products, and with evolution of exothermic heat. For this reason it is used in many similar processes such as contact or vacuum-bag laminating, filament winding, and low-pressure molding.

The selection of a curing mechanism depends upon the working time available to perform the task, the properties desired in the product, the amount of exotherm that can be tolerated, and the degree to which heat can be dissipated during cure. An example is a massive casting or thick laminate that could warp the tool or deteriorate its surface by repeated usage.

The epoxy is cured as a homopolymer by adding a catalyst or by copolymerizing with a hardener that becomes part of the final structure. We have shown previously that precise ratios of catalysts, accelerators, and polyesters are important to quality and speed of reaction. By comparison, the ratio of catalyst to epoxy is not normally as critical as for polyester, but manufacturers' recommendations should be followed closely for best results. It will be recalled that polyester catalysts are used in amounts from less than 1 to 2 percent. Catalyst ratios for epoxies vary from 5 to 20 percent, depending on type and on the particular epoxy resin, but slightly less or greater amounts are not, as a rule, detrimental. Exotherm varies, affecting cure rate, but expected properties are realized when the resin fully polymerizes.

Hardeners, on the other hand, must provide interconnecting links in the epoxide

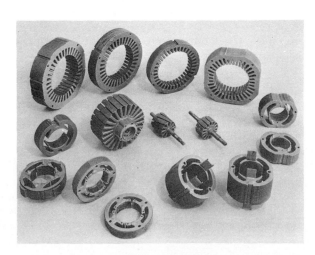

5-5 Stators and rotors coated with finely powdered epoxy sprayed onto heated part. (*Reprinted by permission of Hysol Div., Dexter Corp.*)

chain. The amount of hardener used is based on stoichiometric ratio calculated to provide the desired properties. Slight variations may be tolerated without affecting final results. Proportions of some hardeners may be as high as 125 parts per hundred parts of resin. Less amounts than those called for can cause uncured or poor-quality resin structures.

The substantial number of curing systems precludes all but a discussion of basic types. Readers who are interested in more detailed discussions should consult the Suggested Readings at the end of the book, as well as many informative brochures available from basic resin manufacturers and formulators.

Primary aliphatic amine hardeners

The important characteristics of this group of chemicals are that working time (pot life) is extremely short and that high exotherm prevents their use for large castings or thick laminates. Most are irritating to the skin, some cause long-lasting brown stains, and some cause burns similar to an acid burn (see Table 5-1).

Anhydrides

Anhydride hardeners are characterized by low exotherm requiring heat cure. They are often used with amines to speed gel and cure time (see Table 5-2).

Secondary cyclic amines

Secondary cyclic amines are used in conjunction with primary amines to modify pot life, exotherm, and cure rate (see Table 5-3).

MODIFYING THE EPOXY RESIN

Many epoxies are somewhat brittle, thus nullifying their excellent adhesion. Bonded materials may peel or split when flexed. Some encapsulating resins may not have the impact strength required. By incorporating a modifying agent that can "stretch out"

5-6   Wound-coil bobbins before and after encapsulation by low-pressure transfer method. (*Reprinted by permission of Hysol Div., Dexter Corp.*)

TABLE 5-1 Primary amine hardeners

Name	Form	Hazards	Amount required*	Pot or shelf life†	Typical cure cycle	Heat-deflection temp., °F‡	Uses
Aliphatic types:							
Diethylene triamine (DETA)	Liquid	Highly toxic; causes burns	8–12	30 min	7 days at room temperature or gel at room temperature plus 2 h at 212°F	176	Fast cure; thin castings, small laminates, or repair
Triethylene tetramine (TETA)	Liquid	Same as for DETA	10–13	30 min	Same as for DETA	245	Fast cure; good properties at room temperature
Triethylamine	Liquid	Toxic; can cause dermatitis or allergy	10	Extensive; to 8 h	8 days at room temperature or gel at room temperature plus 2 h at 212°F	220	Fast cure
Aromatic amines:							
Metaphenylene diamine (MPDA)	Solid	Stains; less toxic	14	2 h	3 h at 310°F	330	Excellent heat and chemical resistance
Diamino diphenyl sulfone (DDS) or (DADS)	Solid		30	2 h	24 h at 250°F plus 4 h at 350°F	290–375	High-temperature casting and laminating
Boron trifluoride complex (BF₃—400)	Liquid		1–5	Extensive	2 h at 221°F plus 4 h at 392°F	250–350	Prepreg
Piperidine	Liquid	Skin sensitizer	5–7	6–10 h	2–15 h at 302°F	212–248	General purpose

*Parts per hundred parts of resin.
†Based on 1 lb mixed at room temperature.
‡After full cure; often in conjunction with postcure.

TABLE 5-2 Anhydride hardeners

Name	Form	Hazards	Amount Required*	Pot life†	Typical cure cycle	Heat-deflection temp., °F	Uses
Phthalic anhydride (PA)	Solid	Causes burns	30–45	Extensive	4–24 h at 302°F	300	Bulk casting; low exotherm seldom used alone
Pyromellitic dianhydride (PMDA)	Solid	Toxic; causes burns	Varies; to 60	Extensive	5–20 h at 430°F	540	High heat resistance
Methyl nadic anhydride (MNA)	Liquid	Causes burns	Varies; to 80, plus DMP-30, 2%	2–3 h	2 h at 400°F plus 4 h at 300°F	275	Tensile strength; filament winding
Hexahydrophthalic anhydride (HHPA)	Solid	Causes burns	Varies; to 80, plus DMP-30, 1%	Extensive	2 h 212°F plus 2 h at 300°F	260	Clear castings; good impact strength; encapsulating

*Parts per hundred parts of resin.
†Based on 1 lb mixed at room temperature.
‡After full cure; often in conjunction with postcure.

TABLE 5-3 Secondary cyclic amines

Name	Form	Hazard	Amount required*	Pot life†	Typical cure cycle	Heat-deflection temp., °F‡	Uses
Tridimethyl aminoethyl phenol (DMP-30)	Liquid	Skin sensitizer	6–10	Less than 1 h	Several hours at room temperature	. . .	Adhesives and coatings
Dimethyl amino phenol (DMP-10)	Liquid	Skin sensitizer	14–16	Moderately short	Several hours at room temperature	. . .	Small castings

*Parts per hundred parts of resin.
†Based on 1 lb mixed at room temperature.
‡After full cure; often in conjunction with postcure.

the epoxide portion of the chain, the rigid resin becomes tough and more flexible. Several groups of chemicals are used to improve a base resin when required.

Versamids

These polyamide liquids, manufactured by General Mills, are used in ratios of 50 to 100 phr,* depending on the type. Their polymers are not to be confused with linear polyamides (nylons) discussed in Chap. 19. Versamid types 125, 115, and 100 are derived from vegetable oils and are epoxy hardeners. They are nontoxic, develop low exotherm, and reduce shrinkage. As a result, massive castings or thick laminates requiring long pot life are possible.

Polysulfides

Polysulfide rubbers were developed in 1926 during research intended to find a suitable antifreeze liquid. Instead, the result was a series of synthetic rubbers called *thiokols*,[1] which are used as polymers in their own right. Polysulfide polymers cured with peroxide catalysts are used as sealants, flexible membranes in building construction, and as a patch compound where water and chemical resistance at wide temperature fluctuations are required. These rubbers are formulated to meet rigid aircraft, military, and building specifications.

Polysulfides used to make epoxies flexible are furnished as liquids that combine with the epoxy under the influence of an amine hardener usually present at 10 percent phr. The more polysulfide used, the more flexible the cured resin becomes. Epoxy flooring, providing skidproof and chemical-resistant qualities, is an accepted product in high-traffic commercial and industrial buildings

Reactive diluents

Many liquid epoxies are highly viscous, making them difficult to pour, let alone wet fabrics for laminates or penetrate intricate areas of an encapsulated item. The use of common solvents for resin thinning is not favored unless a solvent can be flashed off prior to cure, as in coating fabric for prepreg or in deposition of thin films or coatings.

Incorporating a *reactive diluent* improves fluidity of a resin, which not only makes application easier, but permits the addition of fillers when required. Reactive diluents react with other epoxide units in the resin to become a part of the final structure.

Reactive diluents include monofunctional types such as allyl, butyl, cresyl glycidyl ethers,[2] polyfunctional vinylcyclohexene dioxide, and 1,4-butane diol diglycidyl ether.

Epoxy formulation for specific applications obviously requires experimentation based on sound knowledge of chemical reactions, plus an appreciation of the toxicogenic aspects of many curing agents.

*This is parts per hundred parts of resin.
[1] Thiokol is the trade name of Thiokol Chemical Corporation.
[2] RD-2 is the registered trade name of Ciba Products Company.

Safety

The ratio of employee accidents, sickness, or discomfort is almost directly related to the respect generated within a company toward hazards, particularly chemical ones. Hazards can be reduced almost to zero if supervisors evolve a comprehensive educational and safety program that is constantly implemented, reviewed, updated, and enforced. Employees should be given a set of safety instructions in printed form and be followed up periodically for compliance.

Solid epoxies are not considered toxic, but when heated or dissolved they create vapors. Amine hardeners are more corrosive than anhydrides, causing itchy skin, rash, sensitivity, edema, and, in severe cases, deterioration of internal organs. Anhydride hardeners, being acid, can cause burns. When possible, special types of hardeners, called *safety hardeners*, should be used to avoid the toxic hazards or dermatitis effects that can occur with specific chemical types. Safety hardeners have been developed by the epoxy formulators to meet the demands of industries that use quantities of epoxies for tooling, laminating, adhesives, and coatings.

These circumstances are not intended to convey the idea that epoxies are dangerous to handle. Good personal hygiene, shop cleanliness, and adequate ventilation can eliminate hazards. Many of my acquaintances and I have handled epoxy systems for more than 20 years. In that time, only very few cases of allergic reactions have occurred. One burn case reportedly occurred when an employee sat on some spilled hardener.

On rare occasions, a person hypersensitive to a particular epoxy ingredient suffers skin rash or puffiness around the eyes at the first exposure, but more often allergies are acquired by repeated exposures. If the affected person ignores the first signs of irritation and persists in whatever practices led to it, he may incur more serious damage. The general idea persists that fair-complexioned people are more susceptible to allergic reaction from epoxies, but the writer has known Black and Chicano technicians to be allergic on short exposure, albeit less commonly.

Most epoxy formulations are nonhazardous. However, a few are extremely hazardous when handled carelessly, and some compounds fall between these two extremes. The Epoxy Resin Formulators Division of The Society of the Plastics Industry (SPI) has published information on the classifying and labeling of epoxy products for industrial use and for use by those in the health professions (see Tables 5-4 to 5-9).

General safety

Good general safety practices in aircraft, electronic, and laminating plants can be summarized in four points:

1   Keep work areas clean.
    (*a*)   Cover benches, tables (and floor when feasible) with vinyl, polyfilm, or waxed paper.
    (*b*)   Place film on weighing equipment.

(*c*)   Provide swing doors, if possible, to avoid contact with door knobs.

(*d*)   Wipe up spills *immediately*.

2   Provide remote smoking area.

3   Work only in well ventilated areas with positive exhaust of vapors.

4   Keep unauthorized personnel out of the area unless properly safeguarded.

TABLE 5-4   Guide for classifying epoxy products according to their skin-irritating, sensitizing, and carcinogenic (in animals) potentialities

Hazard category (class)	Description	Definition
1. Practically nonirritating	The product is nonirritating or produces only transient mild skin irritation following prolonged or repeated contact.	The undiluted product produces *no* irritation when applied to the skin of rabbits* for 24 h *and* it produces no irritation (or very slight irritation) when applied to rabbits for 8 h/day for 5 consecutive days.
2. Mildly irritating	The product produces only irritation of the skin following prolonged or frequent contact.	The undiluted product fails to produce severe edema† or severe erythema‡ when applied to the skin of rabbits* for 24 h.
3. Moderately irritating	The product may produce injury with some persons when in contact with the skin for a prolonged period of time or following a relatively few short-term contacts.	The undiluted product fails to produce severe edema† or severe erythema‡ in 4 h but does so when applied to skin of rabbits* for 24 h.
4. Strong sensitizer	The product is a strong sensitizer.	A strong sensitizer is a substance that produces an allergic sensitization in a substantial number of persons who come in contact with it.
5. Extremely irritating	The product is capable of producing substantial injury when in contact with the skin for a relatively short period of time.	The undiluted product produces severe edema† or severe erythema‡ when applied to the skin of rabbits* for 4 h *or* the undiluted product produces injury in depth (destruction or irreversible change in structure of tissue) when applied to the skin of rabbits* for 24 h.
6. Suspected carcinogen in animals	The product can cause a carcinogenic response when applied topically to or inhaled by experimental animals.	The product is a material that causes tumors or cancers in a statistically significant number of mice* in a lifetime of skin painting by standard test procedures, *or,* by animal studies, causes depression of blood-forming organs.

*Or other suitable laboratory animals selected at the discretion of the testing or research team.
†Severe edema—is swelling 1 mm or more (Draize score of 4).
‡Severe erythema—is bright or beet redness (Draize score of 4).

TABLE 5-5 Guide for labeling epoxy products according to their skin-irritating, sensitizing, and carcinogenic (in animals) potentialities

Hazard category (class)	Composition of warning labels*
1. Practically nonirritating	**For Industrial Use Only!** Avoid gross contamination of skin. **SPI Classification 1**
2. Mildly irritating	**For Industrial Use Only!** **Caution!** **May Cause Skin Irritation** Avoid skin contact. If contact occurs, wash with soap and water at the first opportunity. **SPI Classification 2**
3. Moderately irritating	**For Industrial Use Only!** **Warning!** **May Cause Injury to Skin Following Prolonged or Repeated Contact** Prevent prolonged or frequent skin contact. If contact occurs, wash at the first opportunity with soap and water. **SPI Classification 3**
4. Strong sensitizer	**For Industrial Use Only!** **Warning!** **May Cause Skin Sensitization or Other Allergic Responses** Avoid inhalation of vapor. Use good ventilation particularly if heated or sprayed. Prevent contact with skin. If contact occurs, wash immediately with soap and water. **SPI Classification 4**
5. Extremely irritating	**For Industrial Use Only!** **Danger!** **Causes Serious Skin Burns** Do not get on skin or clothing. In case of contact, immediately flush skin with plenty of water for at least 15 min. Remove and wash contaminated clothing before reuse. **SPI Classification 5**
6. Suspected carcinogen in animals	**For Industrial Use Only!** **Danger!** **Prolonged or Repeated Contact of Liquid or Breathing of Vapors or Mists May Cause Delayed and Serious Injury** Do not handle or use until the manufacturer's safety precautions have been read and understood. **SPI Classification 6**

*In addition to designation of the SPI classification and the statement "For Industrial Use Only," the name
and address of the producer or distributor is required on the labels of *all* epoxy products.

Important note: Epoxy products that are both irritants and sensitizers (as defined in Table 5-4) should bear
the signal words Caution!, Warning!, or Danger!, reflecting the higher degree of hazard and should otherwise be labeled in accordance with both hazards. Epoxy products that present any of the additional hazards
defined in Table 5-6 and 5-8 should also be labeled to reflect these hazards and the illustrative labels shown
above must be modified to do so. The complete label should then be used in addition to or in combination
with any label required by law.

TABLE 5-6 Guide for classifying epoxy products according to their toxic potentialities and their local effects on the eyes

Hazard category (class)	Definition
Highly toxic by ingestion	A product that has a median lethal dose (LD_{50} value) of 50 mg or less per kilogram of body weight when administered orally to albino rats weighing between 200 and 300 g each.
Toxic by ingestion	A product that has a median lethal dose (LD_{50} value) of more than 50 mg but not more than 1,000 mg/kg of body weight when administered orally to albino rats weighing between 200 and 300 g each.
Highly toxic by inhalation	A product that has a median lethal concentration (LC_{50}) of 200 parts per million or less of vapor* or 2 mg/liter or less of mist† or dust‡ when administered by continuous inhalation for 1 h or less to albino rats weighing between 200 and 300 g each.
Toxic by inhalation	A product that has a median lethal concentration (LC_{50}) of more than 200 parts per million but not more than 2,000 parts per million by volume of vapor or more than 2 mg/liter but not more than 200 mg/liter of mist or dust when administered by continuous inhalation for 1 h or less to albino rats weighing between 200 and 300 g each.
Highly toxic by percutaneous absorption	A product that has a median lethal dose (LD_{50}) of 200 mg or less per kilogram of body weight when administered by continuous contact for 24 h or less with the bare skin of albino rabbits weighing between 2 and 3 kg each.
Toxic by percutaneous absorption	A product that has a median lethal dose (LD_{50}) or more than 200 mg/kg but not more than 2,000 mg/kg of body weight when administered by continuous contact for 24 h or less with the bare skin of albino rabbits weighing between 2 and 3 kg each.
Corrosive—eyes	A product that causes destruction or irreversible alteration of living tissue of the eye at the site of contact. A product would be considered corrosive if in undiluted form it produces blindness, destruction, or irreversible change in the structure of the cornea or conjunctiva when applied to the eyes of rabbits for 24 h.
Irritant—eyes	A product is an irritant to the eyes if available data on human experience indicate that it is an irritant to the eyes, or if a positive test result is obtained when the product is tested by the method described in Appendix A (Sec. 191.12 of the Code of Federal Regulations).

*Vapor: A gaseous form of a substance that is normally in the solid or liquid state. A vapor diffuses.
†Mist: Suspended liquid droplets generated by condensation from the gaseous to the liquid state or by breaking up a liquid into a disbursed state such as by splashing, foaming, or atomizing.
‡Dust: Solid particles generated by handling, crushing, grinding, rapid impact, detonation, or decrepitation. Dusts do not diffuse in air but settle under the influence of gravity.

TABLE 5-7 Guide for labeling epoxy products according to their toxic potentialities and their local effects on the eyes

Hazard category (class)	Signal word	Statements of hazards	Precautionary statements
Highly toxic by ingestion*	Danger!	Poisonous if swallowed	Do not take internally. Wash thoroughly after handling.
Toxic by ingestion	Warning!	Harmful if swallowed	None.
Highly toxic by inhalation*	Danger!	Poisonous if inhaled	Do not breath (vapor–dust–mist). Use with adequate ventilation.
Toxic by inhalation	Warning!	Harmful if inhaled	Avoid breathing (vapor–dust–mist). Use with adequate ventilation.
Highly toxic by percutaneous absorption*	Danger!	Poisonous if absorbed through skin	Do not get in eyes, on skin, or on clothing. Remove contaminated clothing immediately. Wash thoroughly after handling.
Toxic by percutaneous absorption	Warning!	Harmful if absorbed through skin	Avoid contact with eyes, skin, or clothing; wash thoroughly after handling.
Corrosive—eyes	Danger!	Causes severe eye burns	Do not get in eyes. **First Aid Instructions:** In case of contact, immediately flush eyes with plenty of water for at least 15 min; get medical attention.
Irritant—eyes	Warning!	Causes eye irritation	Avoid contact with eyes. **First Aid Instructions:** In case of contact, immediately flush eyes in plenty of water.

*Labels also require the word "Poison," the skull and crossbones symbol, and the statement "First Aid Instruction." Labels must also contain one or more of the following first aid instructions: (a) "If swallowed, induce vomiting by inserting finger down throat, or by giving a teaspoon of salt in a glass of warm water; repeat until vomit fluid is clear; call a physician; or, Do Not induce vomiting—in the case of petroleum distillates." (b) "If inhaled, remove to fresh air. If not breathing, give artificial respiration. Call a physician." (c) "In case of contact, immediately flush skin with plenty of water. Wash contaminated clothing before reuse."

Personal safety

1   Avoid direct contact with resins, hardeners, and catalysts.

   (a)   Use vinyl, rubber, or disposable polyfilm gloves and apron.

   (b)   Wipe affected areas of skin with dry towel *first*, followed by soap and water. Use denatured alcohol to remove any last traces of resin from the skin.

   (c)   Avoid the use of solvents such as acetone to remove catalyzed resin from the hands. These solvents have a twofold harmful effect: As they evaporate, they carry away essential skin oils, making the skin more susceptible to lesions and infections, and they dilute the hardener, causing deeper pore penetration (even to the blood stream). Spread of a harmful chemical is increased over a greater area.

TABLE 5-8 Guide for classifying epoxy products according to their flammable and pressure-generating properties

Hazard category (class)	Definition
Extremely flammable liquid	A product that has a flash point at or below 20°F as determined by the method described in Appendix B (Sec. 191.13 of the Code of Federal Regulations).
Extremely flammable solid	A product that ignites and burns at an ambient temperature of 80°F or less when subjected to friction, percussion, or an electric shock.
Flammable solid	A product that ignites and burns with a self-sustained flame at a rate greater than one-tenth of an inch per second along its major axis when tested by a method described in Appendix C (Sec. 191.14 of the Code of Federal Regulations).
Flammable liquid	A product that has a flash point of above 20°F to and including 80°F as determined by the method described in Appendix B (Sec. 191.13 of the Code of Federal Regulations).
Pressure-generating substances	A product falling within any of the following categories: (1) It explodes when subjected to an electric spark or to the flame of a burning paraffin candle for 5 sec or less; or (2) it expels the closure of its container when held at or below 130°F for 2 days or less; or (3) it erupts from its open container at a temperature of 130°F or less after having been held in the closed container at 130°F for 2 days; or (4) it decomposes to release gas in its containers.

TABLE 5-9 Guide for labeling epoxy products according to their flammable and pressure-generating properties

Hazard category (class)	Signal word	Statements of hazards	Precautionary statements
Extremely flammable liquid	Danger!	Extremely flammable	Keep away from heat, sparks, and open flame. Use with adequate ventilation.
Extremely flammable solid	Danger!	Extremely flammable	Keep away from heat, sparks, and open flame. Do not grind or subject to friction.
Flammable liquid	Warning!	Flammable	(Same as for extremely flammable liquid).
Flammable solids	Warning!	Flammable	Keep away from heat, sparks, and open flame.
Pressure-generating substances	Warning!	May contain gas under pressure	Keep away from heat. Store in a cool place. Loosen closure cautiously.

(*d*)   Change to clean work clothing often.

(*e*)   Avoid long sleeves and tight collars while working.

(*f*)   Use barrier (emolient) creams.

(*g*)   Use approved safety glasses or mask when required. Avoid contact lenses; volatiles can condense and concentrate between lens and eye.

(*h*)   Wash eyes with quantities of cool water if accidentally splashed. Do not rub. See a doctor at once.

(*i*)   Consult a doctor if coughing, excessively dry respiratory passages, or congestion persists.

(*j*)   Avoid inhaling vapors of solvents or heated resins.

2   Become familiar with the particular types of resin systems used so that preventive or first aid measures can be taken.

3   Check first before using any container whose label is missing or illegible. Obtain positive identification by chemical analysis or other means to avoid hazards caused by using an incorrect substance.

4   Avoid dusts of clay, asbestos, silica, glass, or other fillers by using a respirator when mixing large quantities or frequent small quantities.

Fire hazards

Like polyesters, epoxies can generate high exotherms, particularly with room-temperature catalysts. In many instances, an industry works with both the epoxy and the polyester resins in the same shop. Under no circumstances should you *ever* permit a polyester catalyst to come into contact with an epoxy hardener. When mixed, peroxides and amines present an explosion hazard because of the violent decomposition of the peroxide by the amine. These two reactants should be stored separately. Take the following precautions when working with epoxies:

1   Do not throw unused, catalyzed epoxy resins into trash containers with combustibles.

2   Conduct work away from areas considered flammable.

3   Do not leave work unattended while curing takes place.

4   Keep ingredients on hand in small batches commensurate with normal production.

5   Use safety cans for solvent liquids.

Storage

Storage should be arranged so that reactive materials are separated in case of leakage. Labels should be legible on all containers. Broken packages should be removed and the area cleaned.

APPLICATIONS OF EPOXY RESINS

The major markets for epoxy resins are coatings, adhesives, and laminating or tooling resins, although substantial quantities are also used for encapsulating, potting, and molding. Encapsulation is an old art long used with pitch, wax, and shellac for earlier transformers and coils. Synthetic resins, namely epoxy, silicone, and special grades of vinyls, now dominate the field.

Encapsulating processes are further broken down into specific categories:

Casting liquid polymers into molds utilizes removable molds having the necessary cost and performance factors. Silicone and fluorocarbon elastomers, metal, glass, and injection molded or thermoformed molds are used.

Potting components involve a permanent container of metal, molded or formed plastic, glass, ceramic, or rubber.

Imbedment techniques are variations of casting, potting, or molding and may involve liquid, powder, or foam encapsulants.

The epoxy resins used may be transparent, opaque, rigid, flexible, filled, or foamed. Depending on viscosity and production factors, encapsulants are poured, injected manually or from dispensing units, or used in connection with transfer-molding techniques. Entrained air is removed by vibrating, vacuum, or centrifuging prior to cure. In the transfer-molding method, production is based on soft flow powders or liquid epoxies.

GLOSSARY

Aliphatic   Saturated molecule having an open-chain structure

Amine   Indicates presence of —NH$_2$ or —NH group

5-7   Typical transformer before and after encapsulation with liquid epoxy, formulated for superior thermal conductivity. (*Reprinted by permission of Emerson and Cuming, Inc.*)

Denatured alcohol   Industrial ethyl alcohol denatured by adding a substance that renders it unfit for human consumption

Homopolymer   A polymer consisting of a single type of repeating monomeric units, exclusive
of minor irregularities

Polysulfide   Polymer containing sulfur and carbon linkages

Prepreg   Resin-coated or preimpregnated fabric, mat, or roving produced in a ready-to-use
condition for laminating

Stoichiometric   Pertaining to weight relationships in chemical reactions

REVIEW QUESTIONS

1   List several applications for epoxy casting resins.
2   What properties of epoxies make them valuable as adhesives?
3   What is the difference between a basic epoxy resin manufacturer and an epoxy formulator?
4   Sketch and describe the epoxide unit.
5   What chemical is reacted with bisphenol *A* to produce an epoxy?
6   What are the benefits derived from using a frozen epoxy?
7   Which processes utilize solid forms of epoxy resins?
8   Epoxies cure by addition polymerization.
    (*a*)   What does this mean?
    (*b*)   Which other thermoset resins also cure by this mechanism?
9   What factors must be considered in selecting the correct epoxy system for a particular
application?
10   What are the differences between a catalyst for an epoxy and an epoxy hardener?
11   What is the result of using less than the recommended proportion of hardener?
12   What is meant by pot life?
13   What safety precautions should be taken when working with chemicals with which you are
not familiar?
14   What is the purpose of using Versamid?
15   What benefits are derived from adequate safety practices by the employee? By the employer?
16   Prepare a list of safety practices to be enforced in a plant.
17   What would you do if a hazardous chemical was accidentally splashed in your eye?
18   What would be the effect of combining a peroxide catalyst with an amine?
19   What are some of the major markets for epoxies?
20   What is the difference between encapsulating and potting?

SILICONES

6

Polymers based on the element silicon were the subject of concentrated research that began in 1935. Several companies, notably General Electric, Union Carbide, and Dow-Corning (a company formed jointly by Dow Chemical and Corning Glass Works), devoted efforts to finding ways of combining the inorganic silicon atom with the organic carbon atom.

Silicone dioxide (SiO_2), which occurs as sand or quartz, is the most abundant element in the earth's crust. It has been used for centuries to make glass; in this form its chemistry is basically unaltered. Silicon is separated from its oxide and the pure metal transformed into a fine powder by heating in the presence of carbon in an electric furnace.

Like carbon, silicon has a *valence* of four, indicating a great ability to combine with a number of other elements, as it does in nature to form silicates. But unlike carbon, it is incapable of forming *double bonds* to create long silicon-to-silicon polymer skeletal structures.

Fortunately, nature originally combined silicon with oxygen to form quartz, which forms the basic building block, or backbone, of silicone polymers. The recurring unit found in all silicones, whether fluids, oils, greases, or resins, is called the *siloxane link*, designated

$$
\begin{array}{ccc}
| & & | \\
-\text{Si}-\text{O}-\text{Si}- \\
| & & |
\end{array}
$$

to indicate points at which other compounds may be introduced.

PREPARATION

Silicon is first reacted with methyl chloride to form a mixture of varying methyl-chlorosilanes that are subsequently rectified to obtain dimethyl dichlorosilane.

$$Si + CH_3Cl \xrightarrow[\text{copper catalyst}]{\text{heat at 500°F}} \begin{cases} Si\ (CH_3)\ Cl_3 \text{ monomethyl trichlorosilane} \\ Si\ (CH_3)_2\ Cl_2 \text{ dimethyl dichlorosilane} \\ Si\ (CH_3)_3\ Cl \text{ trimethyl chlorosilane} \end{cases}$$

Silicon   Methyl
powder   chloride
          (gas)

Methyl chlorosilane mixture

The mixture is separated by fractional distillation to obtain starting monomers for different silicone end products. Reacting the methyl chlorosilane mixture or any of its constituents with water results in a direct substitution of hydroxyl (OH) groups for the chlorine. Depending on the number of hydroxyls, a wide range of molecular weights is obtainable.

A typical reaction using dimethyl dichlorosilane and water forms a linear polymer useful as a silicone oil. The number of siloxane units is controlled by adding trimethyl chlorosilane as a *chain terminator* as shown by the end group trimethyl silane. By careful selection of monomers and by controlling the reaction, fluids are produced over a wide viscosity range to serve as base liquids with a wide span of boiling and freezing points. Aircraft hydraulic fluids, heating fluids for high-temperature processes, antifoam agents, household and industrial lubricants, and mold release agents in foundry and plastics molding operations illustrate the divergent uses for silicone liquids.

Based on linear polymers, fluids and oils can mix with clay, colloidal silica, and other fillers to produce high-temperature greases, wipe-on mold release agents, shoe polish, commercial floor wax, and similar products. The lubrication of the speedometer cable in the modern automobile is a good example of the use of silicone grease, which provides lubrication at subzero-degree temperatures without hardening.

Silicone resins prepared for molding or laminating can be furnished as flexible or rigid thermosetting types.

Dimethyl   Water   Hydroxyl   Silicone oil
dichloro   dimethyl
silane   silane

6-1   Structural formula for silicone oil.

MOLDING COMPOUNDS

For conventional transfer and compression-molding processes, several types are furnished as granular or bulk compounds. Curing is done by addition polymerization without volatiles evolving, hence lower pressures are used. In addition to molding and curing at temperatures to 360°F for as long as 20 min, postcuring is recommended to obtain maximum properties. Bulk compounds contain short glass fibers; powder types have a silica filler. Storage life is limited to several months.

The major outlets for molded products are applications in electronic, aircraft, or missile uses where temperatures to 750°F can be sustained. Connectors, terminal strips, coil forms, and housings are serviceable in areas of high humidity, as well as in subzero-degree temperatures.

SILICONE LAMINATING RESINS

Thermosetting silicones are transparent liquids that may be diluted with toluene or xylene to impregnate reinforcements. Silicone laminates for use at temperatures above 450°F are reinforced by inorganic materials such as glass, quartz, high silica, or as-

6-2 High-performance ignition system is protected by silicone-rubber-insulated spark-plug wire and molded angle boots. (*Reprinted by permission of Dow-Corning Corp.*)

6-3 Interior of Lunar Module Spacecraft showing perforated ceiling covers fabricated of rigid silicone laminate. (*Reprinted by permission of Dow-Corning Corp.*)

bestos in filament or fabric form. Laminates require a minimum pressure of 30 psi to develop acceptable strength. Wet lay-up and vacuum-bag processes are not practical because of the difficulty of solvent removal.

Most laminated structures are made from preimpregnated materials supplied by a coating specialist, rather than hand-impregnated, as is practical with polyester and epoxy resins. This is because close control of resin content, residual solvent, and catalyst, as well as the proper degree of drape and tack are necessary, and such controls are impossible to maintain on a production basis by manual methods.

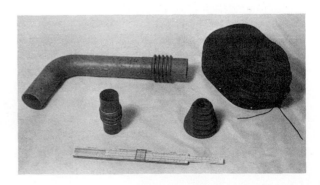

6-4 Flexible ducts and boots laminated with silicone rubber. (*Reprinted by permission of Arrowhead Products Div., Federal-Mogul Corp.*)

Curing is accomplished at 350°F at pressures of 30 to 2,000 psi, depending on whether an autoclave, hydroclave, or press is used. An afterbake based on a time schedule and heat graduation to 450°F is necessary to realize the best properties of the laminate.

Silicone laminates are nonburning and retain much of their initial strength after long-term exposure to 500°F. Products are serviceable to 700°F, which is several hundred degrees higher than heat-resistant phenolic laminates.

Interlaminar shear strength is lower than phenolic or epoxy laminates. A tendency to delaminate makes it necessary to clamp parts firmly and to use sharp tools when machining.

SILICONE ELASTOMERS

Elastomeric silicones are used to preimpregnate glass fabrics for production of flexible laminates by vacuum-bag, lagging, or matched-die molding. Vulcanizing is conducted at 350 to 375°F to produce hot air ducts, boots, and covers.

Silicone casting rubbers are available as viscous liquids. Room-temperature cures are effected as quickly as 10 min with 1 percent stannous octoate catalyst. Other formulations provide longer working times (to 4 h) through the use of dibutyl tin dilaurate activators. Silicone rubber called RTV (*room-temperature vulcanizing*) provides a simple means of making molds for reproduction of cast objects in plastic, ceramic, low-melting metals, or as molds for encapsulating electronics components. Prototypes and limited production are conveniently made in single- or multiple-piece molds.

Normally release agents are not required in making or using an RTV mold. Casting silicone elastomers in RTV, particularly when a mold is slightly undercured, may result in sticking. Paraffin wax dissolved in xylene is an effective parting agent if release presents any problem.

6-5 Flexibility, heat resistance, and reproduction of fine detail are shown in production of candles using RTV molds. (*Reprinted by permission of Dow-Corning Corp.*)

6-6   Circuit-board components
are easily viewed through transparent, flexible silicone. (*Reprinted by permission of Emerson
and Cuming, Inc.*)

ENCAPSULATING GEL

In addition to elastomers used to protect delicate components from heat, shock, and
humidity, soft gels are used to provide a transparent covering for complex electronic
parts.

SILICONE FOAMS

Foamable silicones are available in several forms to suit the processing method. One
powder type requires only heat (320°F) to start foaming action. Foams may be used

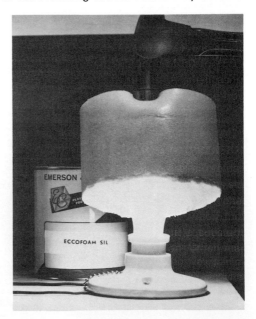

6-7   Resilience of silicone foam is apparent
in this demonstration. (*Reprinted by permission of Emerson and Cuming, Inc.*)

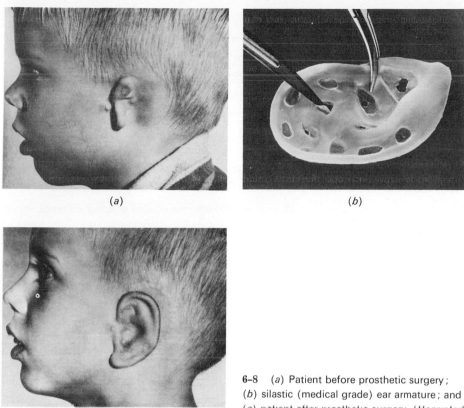

(a)

(b)

(c)

6-8   (a) Patient before prosthetic surgery;
(b) silastic (medical grade) ear armature; and
(c) patient after prosthetic surgery. (Reprinted
by permission of Dow-Corning Corp.)

as foam-in-place encapsulants, molded shapes, or made into slabs. High-density foams to 16 pcf (pounds per cubic foot) are used in aircraft electronic systems where operation at 400°F would be detrimental to other foams. Low-density foams are produced with two-component systems that cure in a few hours after mixing and expansion. Silicone foams are used in applications where heat resistance and light weight are paramount, such as missile and aircraft components.

Special grades of silicone elastomers, both as a solid and as a fine-celled foam, have been highly successful as body implants and for external prosthesis. Human tissue does not reject such devices probably because of the inertness of the silicone to body fluids.

GLOSSARY

Double bond   Unsaturated compounds containing a double linkage between carbon atoms (for
example, ethylene).

Drape   The ability of a fabric to form compound curves.

Elastomeric   The capability of a material to stretch a minimum of twice its original length at room temperature and recover original dimensions.

Fractional distillation   The process whereby a complex material is separated into its component parts by heating (or catalysis) and subsequent condensation.

Interlaminar shear strength   The ability of a laminate to resist peeling or separation at the resin-fabric interface.

Linear Polymer   A long-chain molecular structure, as contrasted to one having side chains or branches.

Tack   The degree of stickiness.

Valence   The property of an atom which is measured by the number of hydrogen atoms (or its equivalent) that atom can combine with, if negative, or displace in a reaction, if positive.

REVIEW QUESTIONS

1   What common substance is the major source for silicon dioxide?
2   Sketch the siloxane link.
3   Where are silicone liquids used?
4   Which other resins, previously studied, cure by addition polymerization?
5   What chemicals are used to dilute a silicone resin to reduce its viscosity?
6   Why is it impractical to impregnate fabrics "by hand" when using silicone resins?
7   Why is it necessary to use sharp cutting tools and clamps when machining silicone laminates?
8   What catalysts are used to cure silicone rubbers?
9   What is meant by the term elastomer?
10   What does RTV mean?
11   What release agent is recommended for silicone elastomer molds?
12   Where are silicone foams used?

PREPARATION OF MATERIALS FOR MOLDING

7

In the earliest days of compression-molding plastic products, molds were of simpler design. They were built to be pried apart manually with pry bars in order to gain access to the part. This type of hand molding will always be needed for complicated parts, but the trend is constantly moving toward fully automated press cycles. For example, a mold containing 100 cavities is capable of producing 125,000 phenolic bottle caps per day with no attention paid to the process except to ensure an adequate supply of molding powder.

This chapter reviews methods by which a constant flow of materials, in proper condition for molding, is supplied to the press room.

HANDLING AND STORAGE OF MATERIALS

Even a small molding operation with only a few presses may require storing a hundred varieties of molding resins. We have already seen how a wide range of compounds can be formulated by changing type of fillers, color, and flow characteristics.

Molding compounds are shipped in cartons, bags, and fiber drums. Liquid resins are delivered in drums or pumped from tank trucks or rail cars to storage silos above or below ground.

Thermoset compounds, though dry, are advanced only to the *B* stage and thus are susceptible to chemical changes. Phenolic, urea, melamine, alkyd, and epoxy are stable for long periods at 70°F, and some polyesters require storage between 0 and 40°F to prevent premature gelling or catalyst decomposition. All molding compounds must be stored in dry surroundings because small traces of moisture adversely affect cure of molded parts.

Thermoplastics are fully polymerized; thus they require no special storage except to be kept dry and safe from fire hazards. Because polymers may be kept many years without chemical change, we shall concern ourselves only with thermosets.

Moisture absorption causes premature hardening or loss of flow in urea, delayed curing of phenolic, and, in all materials, porosity, blisters, or surface blemish. Refrigerated compounds should be allowed to reach room temperature in closed containers to prevent condensation.

Fire hazards are minimal with all compounds, except with polyester premix, which may contain solvents or flammable resins. Inhalation of dust is to be avoided, as in any situation.

PREPARATION OF MATERIALS

Compounds used by molders arrive in many physical forms: Powder, putty, rope, premix, bulk, molding board, nodules, diced squares, cut strands, and preforms. These can be used singly or in specific combinations.

When using powder, the press operator weighs out the required amount into a container or scoops material with a measured container. Multicavity molds are charged from a loading tray constructed of wood or masonite that has cavities holding correct volumes of loose powder or the proper preform.

Unpelletized materials are used when (1) material must be distributed to intricate mold sections by hand; (2) material will not preform automatically because of

7-1   Eighty-cavity compression mold. (*Reprinted by permission of Hull Corp.*)

bulk or tack; (3) short runs do not justify the necessary preform die, and (4) preforming presses are not available.

PREFORMING

Preforming (also called pelletizing or tabletting) reduces the bulk powder to a compressed tablet (pill) of exact shape and weight for the job. Just as it is easier to take an aspirin tablet than to measure an exact dosage of powder, so too is it easier for the press operator to use preforms. Thus resin manufacturers and a few private enterprises offer a preforming service as an aid to molders.

Preforms reduce the bulk factor of a material (see Table 7-1), so that less loading space in a mold is required. Bulk factor is defined as:

$$\frac{\text{Molded density of the part}}{\text{Apparent density of loose powder}}$$

Stated another way:

$$\text{Bulk factor} = \frac{\text{Volume of loose powder}}{\text{Volume of the molded part}}$$

Molding compounds contain a high percentage of voids that disappear as the mold closes under pressure to compress the powder. With a shallow mold such as an ashtray, it may not be practical to use a compound of higher than 2.50 bulk factor without preforming. Loose material may spill over the parting line causing a shortage of the charge, interference with proper mold closing, and excessive flash, possibly with out-of-tolerance vertical dimensions.

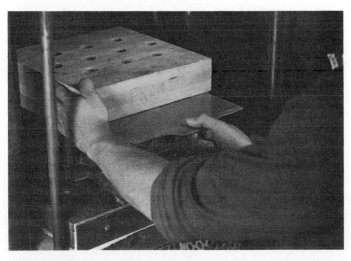

7-2 Loading tray for 10-cavity mold. (*Reprinted by permission of Los Angeles Trade-Technical College.*)

TABLE 7-1   Bulk factor* of thermosetting molding compounds

Material	Bulk factor
Alkyd:	
Granular	1.8–2.4
Putty	1.1–2.0
Asbestos	1.8–2.5
Diallyl phthalate (DAP):	
Asbestos	2.2–2.3
Orlon	4.2
Dacron	4.2–7.0
Glass fiber	2.2–7.0
Epoxy:	
Unfilled	2.0
Mineral	1.8–2.8
Glass fiber	2.5–3.0
Melamine:	
Cellulose	4.0–7.0
Asbestos	2.1–2.5
Macerated fabric	5.0–10.0
Glass fiber	5.0–10.0
Phenolic:	
General purpose	2.1–4.4
Asbestos	2.0–14.0
Mica	2.1–2.7
Macerated fabric	3.5–18.0
Glass fiber	2.0–10.0
Powdered metal	2.0–4.0
Polyester (premix):	
Mineral-filled	2.0–2.8
Glass fiber	4.0–18.0
Silicone:	
Asbestos	6.0–8.0
Mineral	1.7–2.0
Glass fiber	6.0–9.0
Urea:	
Cellulose	2.2–3.0

*Bulk factor is also known as compression ratio.

Advantages

1   Bulk factor is reduced.

2   Consistent, accurate weight ensures less waste, plus a more uniform product.

3   Loading mold cavities is more rapid.

4   Heat is conducted faster; press can be brought to full pressure sooner, thereby reducing cycle.

5   Contamination is minimized.

6   Mold construction is less expensive.

Classes of materials

Uniform   Coarse, granular, free flowing.

Nonuniform   Contains "fines" in addition to coarser granules and requires agitation to preform automatically.

Fluffy   Flock or short fiber asbestos and nodules are difficult to preform except manually or with a horizontal-type preformer.

Bulky   Rayon cord, diced fabric, macerated cloth, or long-fiber reinforcements must be individually weighed and preformed manually (similar to cold molding).

Equipment

Single punch types   Designed for shorter production runs and specially shaped tablets, presses are available to 200 tons.

Horizontal type   Used for both granular and bulky compounds, the horizontal preformer produces as much as 20 tablets a minute.

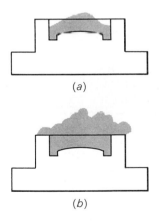

(a)

(b)

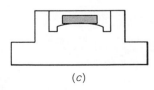

(c)

7-3   Comparison of effect of bulk factor on mold charge. (a) Bulk factor of 2; (b) bulk factor of 10; and (c) preform.

7-4   Single-punch preform machine. (*Reprinted by permission of Stokes Div., Pennwalt Corp.*)

7-5   Horizontal-type preform machine. (*Reprinted by permission of Bipel International, Inc.*)

Rotary type   With multiple dies, rotary machines produce as much as 2,700 pre-forms a minute, depending on size.

Manual performing   A fast-acting, manually controlled hydraulic press can produce preforms at a rate of 60 to 100 per hour, depending on material. Soft, puttylike resins can be packed, much as a snowball, to reduce bulk.

Preforming operations raise dust that covers the area. Thus equipment is housed away from the molding, storage, and finishing areas. Respirators are necessary in most instances, and care is taken against explosion hazards.

Preheating

Molding cycles are reduced considerably when material can be preheated prior to charging a mold. However, not all molding jobs benefit from preheating; for example, when sections are extremely thin or when location of mold charge is time consuming. The fact that plastics are poor heat conductors must also be considered in production.

Any method by which molding compound can be brought above room temperature without causing premature curing can be used.

Electronic (high-frequency) preheat   The most rapid and efficient method of preheating is electronic. In this process, molding compound, usually preformed, is placed between platens (electrodes). Alternating current between 27 and 40 MHz* is directed through the material. Since plastics are excited in the directions of current flow, electrical energy is converted into heat. This is brought about by molecular "rubbing" as the polarity of molecules is reversed many times per second. Heat is built up within the center of the preform more than at the outer surfaces, creating an ideal situation for molding.

*MHz (megahertz) = 1 million cycles per second.

7-6   Rotary preform machine. (*Reprinted by permission of Stokes Div., Pennwalt Corp.*)

7-7 High-frequency electronic preheater. (*Reprinted by permission of Mytron Corp.*)

When the heated preform is placed in the hot mold or in the transfer chamber of a transfer press, the outer edges are brought to molding temperature quickly, so that an even flow and cure result.

In addition to faster molding cycles, items produced from dielectrically heated material are more completely cured, have less entrapped volatile matter, and are more dense. Aftershrinkage is reduced, thus creating a more stable condition. This improved cure brings out the highest mechanical and electrical properties the material can exhibit.

It is possible to place loose compound in a shallow wood or cardboard tray for electronic preheat.

Oven heating   Methods for drying both thermoset and thermoplastic materials and raising the temperature to improve the molding process involve various approaches. In both instances, care must be exercised to avoid precuring thermosets or melting thermoplastic pellets into a solid mass.

Preheating thermosets by an air-circulating oven or in a cabinet-drawer type is slower and more likely to produce a nonuniform condition unless agitation is provided. Materials are placed in depths not exceeding an inch, while close control of oven temperature and timing are maintained.

Infrared method   Penetration of infrared rays below the powder surface permits use of stationary or rotating trays or the use of a conveyor belt placed directly below lamps. Stationary types can heat a layer to 2 in. deep; conveyors limit depth to 0.25 in.

Hot plate   Preforms are heated on an electric or steam-heated plate or simply placed on the press platen. The press operator turns the preform regularly and rotates several preforms in sequence with the cycle.

Hopper heaters Commonly used with extruders and injection machines for thermoplastics, hopper drier-heater units force warm air through the bottom by means of a perforated fin or cone heater. Compression and plunger presses operated on semi- or fully automatic cycles may utilize hopper heaters.

GLOSSARY

Apparent density Ratio of the volume of loose powder (uncompressed) to its weight (expressed in grams per cubic centimeter).

Cold molding A process by which molding compounds (usually bituminous resins and inorganic filler) are compressed to shape and subsequently are cured in ovens after removal from the mold.

Hopper Conical reservoir that contains the molding material and from which it falls by gravity into the molding machine. Metering devices are sometimes used to dispense correct volume or weight.

Molded density Ratio of the volume of a molded part to its weight (expressed in grams per cubic centimeter).

Polarity Residual electrical forces surrounding a molecule.

Premix A premix molding compound is a bulk mixture of resin, fillers, and/or reinforcements usually prepared by the molder immediately prior to use.

REVIEW QUESTIONS

1 Which resins exhibit long storage life?
2 What effect does moisture have on phenolic? On urea?
3 Why should refrigerated compounds be allowed to reach room temperature before using?
4 When is a loading tray used?
5 When are molding powders preferred to preforms?
6 When is preforming necessary?
7 What are the advantages of using preforms?
8 Explain the four classes of materials as they apply to preforming.
9 What are the three types of preforming machines?
10 What are the advantages of electronically heated preforms?
11 Explain how loose powder can be preheated electronically.
12 What other methods of preheating materials are used?

COMPRESSION-MOLDING PROCEDURES

8

Often in the course of industrial development materials are discovered before machinery is available to process them. This situation existed when the first thermosetting resin, phenol formaldehyde, was patented in 1909. Aside from its use as an insulating varnish for paper, the resin, when combined with wood particle filler, was evaluated as a molding compound. Companies then engaged in compression-molding natural rubber for steering wheels, electrical plugs, and appliance parts were instrumental in finding ways to handle phenolic plastic. Today many firms mold plastics, in addition to rubber, because the molding methods for both are similar.

Molding by compression is the oldest method of producing thermoset plastic articles. The process is primarily intended for thermoset resins, but it is adaptable to molding thermoplastics. Normally compression molding of thermoplastics is not economically competitive with the more rapid injection-molding process because a mold requires heating to soften the resin, followed by cooling to solidify the molded shape. One molding cycle could take as much as 30 min, but the injection cycle takes only seconds. Large, thick slabs are compression molded from specific thermoplastics, which are then machined into shapes not otherwise obtainable with molds. Typical applications for compression-molded parts include appliance bases, knobs, ashtrays, missile components, battery cases, pot handles, bottle caps, rocket nozzles, dinnerware, wall switch plates, gears, cams, pulleys, washing machine agitators, and electrical and electronic components.

The basic process consists of placing a predetermined weight of molding compound into a heated mold cavity, closing the mold, and exerting pressure by means of a compression press. The press and mold remain closed until the compound becomes rigid enough to maintain its molded shape. Hardening (cure) is caused by a chemical change induced by the heat of the mold. When the cure, which is first determined by trial, is completed, press and mold are opened, and the part is removed while hot. Part removal is manual or automatic, depending on mold design.

This description is oversimplified. Many details must be considered in order to mold parts that have no defects because, unlike thermoplastics, thermosets cannot be reclaimed for reuse.

In this chapter we shall discuss details of press operation, installation of molds, establishing a molding cycle, and correcting faults.

THE COMPRESSION PRESS

A compression press is a rugged piece of equipment (see the table of hydraulic press parts and their functions below). A press will provide many years of dependable service if properly maintained, which is why it is not uncommon to see presses built over 40 years ago still in operation daily.

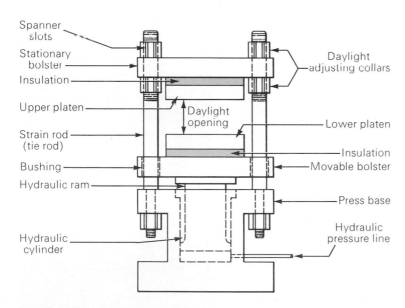

8-1   Typical hydraulic compression press.

Press part	Function
Stationary bolster	Heavy cast steel plate providing "backstop" to hold compressive force
Upper and lower platens	Thick steel plates with heating channels and for fastening molds
Insulation	Asbestos board to reduce heat loss
Strain rods (tie rods)	Provide guides for platen movement; tie press together
Adjusting collars	Hold upper bolster in position; adjust for daylight opening between platens
Daylight opening	Maximum distance between upper and lower platens
Movable bolster	Floating plate attached to ram; moves upward to close the mold
Bushings	Bronze wear surfaces for accurate guidance of bolster
Hydraulic ram	Cylindrical steel piston to transmit pressure from a hydraulic system
Press base	Heavy steel support; contains hydraulic cylinder and piston

The press shown in Fig. 8-1 is an *upstroke* type. The majority of presses are of this type. Some molding is conducted on a *downstroke* press in which the upper bolster moves downward toward a stationary bolster that is at or near floor level. Downstroke machines are useful for extremely large parts, such as boat hulls, rocket nozzles, and auto and aircraft sections, which require a long stroke and daylight

8-2 Downstroke press. Note convenience of plug-in panel for electrically heated molds. Press has 150-ton rating. (*Reprinted by permission of Dake Corp.*)

opening. Long-stroke upstroke presses require a deep pit to contain a long press base, hence press installation and maintenance are more costly.

Rating

A compression press is rated according to the number of tons of clamp pressure it can exert by means of hydraulic or hydromechanical pressure. A hydraulic automobile jack is rated at 1 to 5 tons, but a compression press (operating on the same principle) is built with pressure capacities of ten to several thousand tons. A modern molding plant operates a range of presses to accommodate a wide variety of molds. Attention must also be given to platen size to accommodate the mold to be used, as well as to the heating capacity necessary to obtain the most economical cycle.

Calculating press tonnage

A press develops tonnage by fluid pressure furnished by one of several means:

1   Hand pump (lever-operated)
2   Motor-driven pump
3   Accumulator system

    The accumulator system is explained in Chap. 13, which deals with the development of heat and pressure for molding. When it is an integral part of the press

8-3   Four-thousand-ton compression press for large moldings requiring high molding pressure.(*Reprinted by permission of Haveg Industries, Inc., Reinhold Aerospace Div.*)

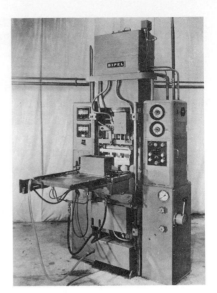

8-4   Twenty-five-ton compression press, fully automatic. Feeder box and parts stripper project in front.
A typical cycle would be: (1) mold opens with top
ejectors actuated to keep molded parts in cavity half;
(2) bottom ejectors raise parts out of cavities; (3)
stripper comb advances and stops beneath pieces
being held up on ejector pins; (4) bottom ejectors
retract, leaving parts on comb tray; (5) table advances to dump parts out rear of press as tray tips
there. At same time, feeder board dumps powder into
mold cavities; (6) table retracts out of press; and
(7) mold closes and cure cycle commences. (*Reprinted by permission of Bipel International, Inc.*)

system, the motor-driven pump is a common means of developing pressure. Regardless of the method used, the power that furnishes fluid pressure to the ram is called
line pressure. Line pressure is delivered to the ram at a fixed value between 1,000 and
3,000 psi and is a function of pump design.

It is necessary to know the diameter of the ram and the line pressure to determine
press tonnage. A gauge at the pump or at the press indicates maximum line pressure
that can be developed.

Formula :   Press capacity (tons) $= \dfrac{\text{ram area (sq in.)} \times \text{line pressure (psi)}}{2,000 \text{ lb/ton}}$

8-5   Manually operated bench press, 20-ton capacity. Self-
contained pump, electrically heated platens, with provision for
water cooling. Valuable machine for production of small parts and
evaluation work. (*Reprinted by permission of Pasadena Hydraulics,
Inc., Div. of Rucker Co.*)

8-6   Motor, pump, and hydraulic system of self-contained compression-transfer press. (*Reprinted by permission of Los Angeles Trade-Technical College.*)

Sample problem: A press has a ram 8 in. in diameter. The maximum line pressure delivered to the ram is 3,000 psi. What is the press capacity?

Solution: $8 \times 8 \times 0.7854$ $= 50.26$ sq in. (area)
$50.26 \times 3,000$ $= 150,786$ lb
$150,786 \div 2,000 = 75$ tons (capacity)

Presses are constructed with standard capacities of 25, 50, 75, 100, 150, 300, and to several thousand tons. The majority of compression molds are used in the range of 50 to 150 tons because of the preponderance of small products over large items. A small ashtray, for example, can be made on a 25-ton press, whereas an adding machine housing may require 300 tons.

THE COMPRESSION MOLD

Compression molds are machined from special grades of tool steel to withstand high pressure and resist wear from abrasive plastics. Other metals such as aluminum, kirksite,[1] and meehanite[2] are used for low-production requirements, for experimental molds, or when molding pressures are low, as in molding with liquid resin. Molds represent a substantial investment and must be protected from careless damage or corrosion. When major accidents occur, molds cannot be repaired, and many months'

[1] This is an alloy of zinc, aluminum, copper, and magnesium.
[2] Meehanite is a variety of cast iron.

production is lost while a new mold is produced. The following table explains the functions of various parts of the compression mold shown in Fig. 8-7:

Mold part	Function
	Upper half
1. Adapter plate	Provides means of clamping mold to upper press platen
2. Force retaining plate	Provides support for force
3. Force (also called plug, male, or core)	Shapes the inside of a hollow part or one face of a solid part
4. Guide pin (leader pin)	Ensures true alignment of mold halves
5. Heat source	Steam channels or electric cartridges
6. Socket head cap screw	Fastens mold parts together (many required)
	Lower half
7. Parting line	Top surface of cavity retainer block
8. Cavity	Shapes outside of a hollow part or opposite face of a solid part
9. Cavity retainer block	Supports cavity
10. Guide-pin bushing	Sleeve to align guide pins
11. Adapter plate	Provides means of clamping mold to lower platen
12. Heat source	Steam channels or electric cartridge
13. Socket head cap screw	Fastens mold parts together (many required)

One mold differs from another in size, shape, or number of cavities. Each mold, however, has many features common to all molds, as shown in Fig. 8-7. Mold designs vary according to the shape of product, the critical aspects of part dimension, and how the part is to be cleaned of flash (excess material).

Basic mold types

Compression-mold designs fit into three basic categories:

Flash   This type of mold is least expensive. Flash molds are used for shallow parts in which excess material is allowed to flow over a cutoff edge without restraint. Accuracy of the mold charge is not as important as with other types, although part density may be nonuniform.

Positive   A positive mold is one in which the force telescopes into the cavity without restriction. This type of mold is used for high-bulk compounds and where a vertical flash is desired for finishing purposes. Part dimensions parallel to the direction of molding pressure (horizontal planes) are controlled only by the weight of the molding charge and rate of press closing, correlated with pressure and temperature.

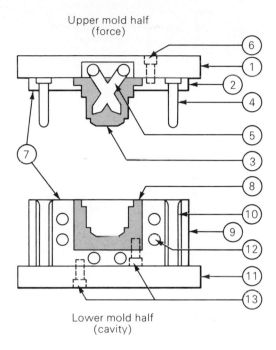

Upper mold half
(force)

Lower mold half
(cavity)

8-7   The compression mold.

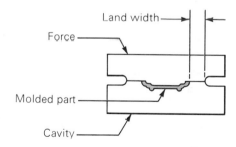

Land width

Force

Molded part

Cavity

8-8   Flash mold.

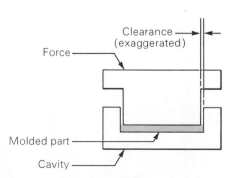

Clearance
(exaggerated)

Force

Molded part

Cavity

8-9   Fully positive mold.

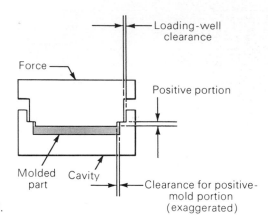

8-10 Semipositive mold.

Semipositive (landed positive)   This type combines the features of both positive and flash molds. Excess material is flashed out over the internal land until the force nears the end of its travel. At this point the flash hardens rapidly to seal the remaining material in the cavity. Closer control of dimensions is realized with a semipositive mold. Flash is horizontal to the sidewall of the part, making deflashing operations easier in many cases.

Semipositive molds are further divided into two types based upon the location of the lands. Figure 8-11 illustrates the internal land that is machined into the cavity at the top surface line of the molded part.

External lands, or pads, are located on the parting line outside the cavity proper and serve the same purpose as internal lands, while providing a vertical flash. During the final closing of the mold, lands absorb pressure at the same rate as the material in the cavity, so their areas must be included in calculating molding pressure.

Calculating molding pressure

Molding pressure means the pressure per square inch exerted on the material in the mold cavity (see Table 8-1). Each type of compound, as well as the design and size

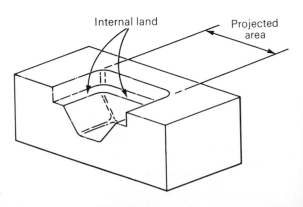

8-11   Internally landed mold.

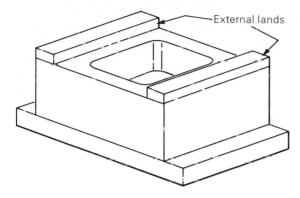

External lands

8-12   Externally landed mold.

of the part to be made, must be carefully considered when estimating the pressure needed to mold the product successfully. Inadequate pressure results in porous, blistered, or incomplete pieces; excessive pressure places undue strain on mold and

TABLE 8-1   Recommended molding pressures for compression-molding thermosets

Type of filler	Molding pressure (psi)
Phenolic	
Woodflour	1,500–3,500
Cotton flock	1,500–4,000
Macerated cotton	2,000–5,000
Tire cord	2,000 5,000
Sisal	2,000–5,000
Rubber	1,500–3,000
Mica	1,500–3,000
Asbestos	2,000–4,000
Mineral	2,000 3,000
Glass:	
Diced squares	2,000–10,000
Chopped roving	2,000–6,000
Bulk	1,000–6,000
Urea	
Alpha cellulose	4,000–8,000
Alpha cellulose plus plasticizer	2,000–4,000
Melamine	
Alpha cellulose	2,000–8,000
Asbestos	2,000–8,000
Mineral	2,000–6,000
Glass:	
Chopped roving	2,000–8,000
Bulk	2,000–6,000

TABLE 8-1 Recommended molding pressures for compression-molding thermosets (*continued*)

Type of filler	Molding pressure (psi)
Alkyds	
Mineral (granular)	1,000–1,500
Glass:	
Extruded rope	600–1,000
Putty	400–800
Bulk	1,000–2,500
Diallyl phthalate (DAP)	
Synthetic (nylon, orlon, dacron)	500–2,000
Glass fiber	500–2,000
Asbestos	500–2,000
Polyester	
Glass fiber (premix)	50–500
Sisal	50–500
Mineral (clay)	50–300
Epoxy	
Mineral	100–1,000
Glass fiber	100–2,000
Silicone	
Asbestos	1,000–8,000
Glass fiber	1,000–5,000
Mineral	1,000–5,000

press. The materials supplier can only suggest a range of pressures to be used. It is for the molder to determine the correct pressure, as well as all other conditions of the cycle. In order to determine whether the mold can be used in a particular press, the required molding pressure should be closely estimated. First it is necessary to find the projected area of the mold cavity or the total projected area of all cavities in a multicavity mold. Projected area is the area of the cavity (or cavities) measured at at the parting line.

Formula: Mold pressure (tons) $= \dfrac{\text{molding pressure (psi)} \times \text{projected area (sq in.)}}{2{,}000 \text{ lb/ton}}$

Sample problem: A part measuring 6 in. long and 4 in. wide is to be molded at a pressure of 2,500 psi. What pressure must be exerted by the press?

Solution: 6 × 4   = 24 sq in. (projected area)
         24 × 2,500 psi = 60,000 lb (total pressure)
         60,000 ÷ 2,000 = 30 tons

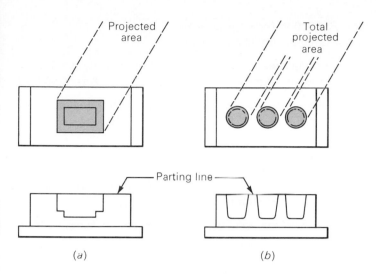

8-13   Projected areas. (*a*) single-cavity mold ; (*b*) multicavity mold.

Sample problem :   A multicavity mold contains three identical cavities each measuring 1.50 in. in diameter. What mold pressure is required if the material is to be compressed at 5,000 psi?

Solution :   1.50 × 1.50 × 0.7854 = 1.77 sq in. (area of one cavity)
1.77 × 3   = 5.3 sq in. (total projected area)
5.3 × 5,000   = 26,500 lb (total pressure)
26,500 ÷ 2,000   = 13 tons

Practice problem :   What press pressure (tons) is required to develop 2,000-psi molding pressure on five cavities when each cavity is 3 in. in diameter?   *Ans* : 35.3 tons

The formula illustrated is basic to the initial estimates of mold-pressure requirements. In actual practice, press capacity is considered at 60 to 75 percent of rated maximum so that pressure could be increased if trial runs prove this to be necessary. When pressure is miscalculated, the mold may have to be placed in a larger, more expensive machine or a softer flow grade of compound may have to be used. The latter may reduce the originally desired mechanical properties, or make final part dimensions incorrect.

By inversion of the basic pressure formula, molding pressure can be determined.

Formula :   Molding pressure (psi) = $\dfrac{\text{press pressure (lb)}}{\text{projected area (sq in.)}}$

Sample problem :   A pressure gauge indicates a press pressure of 15 tons exerted on a mold with a cavity measuring 10 × 2 in. What molding pressure is applied to the plastic compound?

Solution: 15 × 2,000 = 30,000 lb (press pressure)
              30,000 ÷ 20 = 1,500 psi (molding pressure)

Practice problem: What molding pressure (psi) is exerted on a cavity 8 × 3 in. if
                      the press exerts 48 tons?   *Ans*: 4,000 psi

Until now we have been calculating pressures applied to molds used for shallow parts to 1 in. in depth. Greater pressure is required to move material up the steep walls of the cavity. For shallow-draw parts, pressures are figured on the basis of projected area alone. For deep-draw products, 500 to 750 psi is added for each additional inch of cavity depth beyond the first inch. This rule of thumb is used by the molder to estimate the size of press needed for producing items such as television housings, missile cones, and washing machine agitators.

Sample problem: A washing machine agitator is to be molded of asbestos-filled
                    phenolic (see Fig. 8-15). What press tonnage must be developed
                    if the part measures 10 in. in diameter and is 14 in. high?

Solution: 10 × 10 × 0.7854 = 78.54 sq in. (projected area)
              78.54 × 9,500 psi = 746,130 lb (mold pressure)
              746,130 ÷ 2,000 = 373 tons

A mold this size would be used on a 500-ton press since a press is normally used at 75 percent capacity.

Practice problem: A blender base molded in melamine measures 6 in. in diameter
                      at the parting line and is 8 in. deep. Assuming an initial pressure
                      for the first inch of depth at 4,000 psi and 750 psi for each
                      additional inch of depth, how many tons of pressure are re-
                      quired?   *Ans*: 130.7 tons

Installation

Most compression molds are fastened to the press for the duration of the production run. At completion of the manufacturing order, the mold is removed to the storage

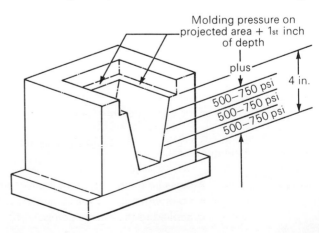

Molding pressure on
projected area + 1st inch
of depth

plus

500–750 psi
500–750 psi
500–750 psi

4 in.

8-14   Method of calculating
pressure for deep cavity.

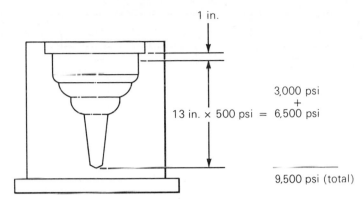

1 in.

3,000 psi
+
13 in. × 500 psi = 6,500 psi

——————
9,500 psi (total)

8-15   Calculating pressure for washing-machine agitator.

area until needed. The following is a step-by-step outline on how to install an average two-part (male—female) mold.

Check press platen alignment   If press platens do not close in parallel alignment, the total press tonnage applied would be concentrated on a relatively small portion of the mold, resulting in unequal flow of material in the cavity or in mold damage.

Procedure: If the press platens can be brought in contact without a mold installed, raise the movable platen to within a few thousandths of the stationary platen. This is accomplished by moving a valve lever at the press or activating a "close press" electrical switch. Use a feeler gauge to check each corner of the platens. If press daylight exceeds ram travel (sometimes referred to as *ram stroke*), close press to maximum. Use a machinists' square plus feeler gauge or an inside micrometer of appropriate length. Inside calipers may also be used. When platens are misaligned, check stationary bolster to see if adjusting collars are properly set. Each should be equidistant from tops of strain rods.

Adjust press daylight   Molds of various heights can be accommodated on a particular press by adjusting daylight opening. In situations where both mold halves are fastened to the press, the minimum daylight necessary is equal to the height of the closed mold plus twice the depth of the part to be molded. In certain operations, only the upper half of the mold is clamped in the press, and the cavity is guided and held in position by channels. Daylight is then figured at mold height (closed) plus cavity depth or guide-pin length, whichever is greater.

Removing the cavity section at the end of each molding cycle makes loading of metal inserts or unscrewing mold side cores easier.

Relocate the upper bolster   *Procedure*: Place the mold on the lower platen and close the press on minimum pressure. If the press cannot fully close, build up remaining daylight with flat steel plates. Increase pressure to 10,000 lb. With pressure applied,

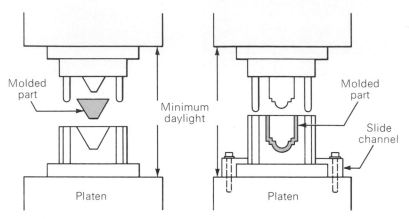

8-16   Setting a mold for proper daylight.

loosen the lower adjusting collars on each tie rod by spinning them one full turn. If collars do not spin easily by hand, use a spanner wrench or a drift pin and hammer. *Avoid damaging spanner slots or holes. If collars are still immovable, increase pressure another 10,000 lb.* When collars are free, lower press slowly until the stationary bolster drops slightly away from the upper retaining collars. With all adjusting collars now free, the press daylight can be increased or reduced by spinning the collars to the required spacings. Whether daylight is made smaller or larger, the final lockup is the same. When increasing the daylight, raise the stationary bolster to

8-17   Twenty-five ton compression-transfer press. Note length of threaded portion of tie rods for maximum daylight opening and T slots for mounting molds. (*Reprinted by permission of Pasadena Hydraulics, Inc., Div. of Rucker Co.*)

within $\frac{1}{16}$ in. of the nearest upper collar. (It is assumed that all collars will not be in exactly the same position.) Spin all collars down against the bolster, *hand tight only.* Apply press pressure of 10,000 lb, and with pressure on, spin the lower collars up against the bottom side of the bolster. It is never necessary to use wrenches or any force to tighten collars when press pressure is used as indicated.

To decrease daylight, drop the upper bolster to within $\frac{1}{16}$ in. of the nearest lower adjusting collar, and spin the upper collars down, hand tight, against the bolster. Apply 10,000 lb pressure and spin lower collars up, hand tight only. There is justifiable reason for adjusting the collars hand tight. At 10,000-lb pressure, the strain rods are stretched. When pressure is removed, their relaxation pulls the collars extremely tight. Using wrenches places more strain on one rod than another, and it may crack when molding pressure is applied.

Install   Press platens have a series of tapped holes that provide the means for attaching a mold. An alternate method is to use T slots milled into the platen when a mold adapter plate is designed for their use. Small molds that can safely be lifted by hand are merely placed on the center of the bottom platen. *Never pick up a mold by lifting with the top adapter plate unless both halves are strapped or bolted together.* The bottom half may slide away and cause injury. Larger molds are hoisted with a lift table or forklift.

Locate the closed mold as near to platen center as possible, and close the press. Various types of clamps can be used to hold the mold on the platens. Hold-down bolts should be long enough to thread into the platen a minimum of $1\frac{1}{2}$ times their diameter. Bolts can be allen, square-, or hex-head type. Before clamping, lubricate bolt threads with antiseize compound[1] to make later removal easier.

Select platen holes that provide for a minimum of two clamps each on opposite sides of both the upper and the lower mold sections or one clamp per side if more convenient. Small molds may require only two clamps per mold half on opposite sides. Position clamp bolts as close to the mold as possible. Spacer blocks are required for horseshoe and slotted clamps and occasionally with the knee type to keep clamps parallel to the platen. Position spacers at the extreme end of the clamp for maximum leverage and support. Press opening pressure is high. If a mold is clamped incorrectly, clamps cannot hold the mold, especially if the molded part tends to stick.

With the press closed and pump off for safety, clamp upper half first. This is

[1]A high-temperature compound such as Fel-Pro should be used.

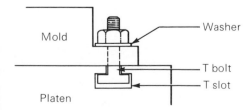

8-18   Method of clamping mold with T bolts.

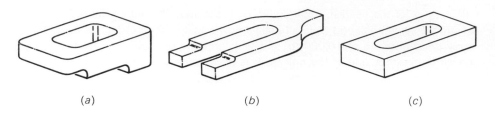

(a)   (b)   (c)

8-19   Mold clamps. (a) Knee; (b) horseshoe or U; and (c) slotted plate.

best for shallow molds so that the mold can be opened a short distance if bolt-head clearance is needed to clamp lower half. *Never disengage a mold beyond guide-pin engagement before clamping both halves.* When both halves are fastened securely, open the press slowly and observe the operation. With press power off, inspect all cavity and force surfaces for any cured plastic that may have been left in the mold from previous use. *Never allow a piece of cured material (or an entire piece) to remain in a cavity.* When molding compound is placed on top of the hardened resin, the total molding pressure is concentrated on that area, which in terms of pressure per square inch could amount to several hundred thousand pounds. The mold can be damaged.

Cleaning the mold

Use sharpened brass, copper, or aluminum to remove any foreign material from a molding surface. Brass is preferred because of its relative softness as compared with the molding surfaces. *Never use steel or iron to scrape a mold.* Even though molds are heat-treated to improve their surface resistance against wear and distortion, they can be easily gouged or scratched. If a hammer is required to help chisel an obstinate case, use a brass block or brass hammer. A steel hammer may accidentally strike a molding surface.

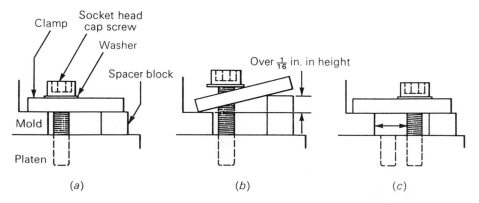

(a)   (b)   (c)

8-20   (a) Correct clamping method. (b) and (c) Incorrect clamping methods. Besides being incorrect, they are unsafe.

Air nozzles are used to assist cleanup and to blow molded parts out of cavities. Safety glasses are mandatory to prevent eye injury from flying chips or granules. An air nozzle should have a copper tube extension flattened almost closed to increase velocity, while providing a scratch-free probe. For final inspection, use a mirror to examine all portions of cavities.

Mold-heating devices

Molds are heated by one of several methods, depending on heating costs, size of plant operations, and other considerations (perhaps geographical in regard to availability of boiler fuel). Chapter 13 discusses various plant layouts and development of heat and pressure for molding.

The most popular methods of heating molds are with high-pressure steam or electricity. With either method, a mold may be heated directly or indirectly.

Indirect Molds for shallow articles such as cafeteria trays are effectively heated by conducting heat from press platens. The expense of boring long holes in the mold is eliminated and mold installation is faster.

Direct Molds for deep-draw parts such as a rocket nozzle must be heated from within to obtain uniform heat distribution. The mold designer uses one or more methods to accomplish this.

Electric These devices operate on the same principle (resistance heating) as a toaster or electric iron. Resistance wires encased in a mica or ceramic insulating sheath and protected by an outer brass shell attain a temperature of 1000 °F. Heaters operate on an on-off principle controlled by thermostats or electronic controllers mounted on the press or wall.

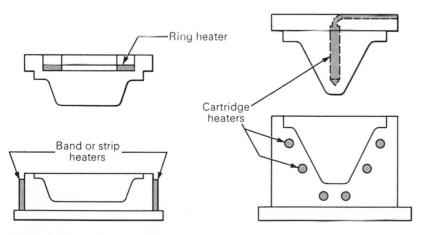

8-21 Methods of heating shallow- and deep-draw molds.

8-22 Indicating-pyrometer-controller for plastics-molding presses. (*Reprinted by permission of West Instrument Div., Gulton Industries, Inc.*)

Electrical heaters are plugged into the supply outlet after checking the insulation for worn or broken areas. Steam connections are made in order to have the inlet to the top port unless channels enter and exit on the same plane. Mold halves are separately controlled and may be varied as much as 20 deg when required.

The important temperature is the heat of the cavity walls and the molding surfaces of the force. Do not rely on thermostat readings entirely. Electronic controls furnish or cut off current on demand only when a thermocouple sensing unit indicates that more or less heat is needed. This is doubly important when molds are heated indirectly, since much of the heat input is lost by radiation from mold surfaces and from absorption by the molding compound. With steam heat, the temperature remains constant at constant boiler pressure and is self-compensating.

Setting mold temperature

Regardless of the type of material to be molded, a range of temperature must be considered (see Table 8-2). Thin-wall sections are generally molded at the upper end of the recommended range (except for deep-draw items that require more flow time), and thick-sectioned parts are molded at the lower end. A product designed with both thick and thin sections is molded at the low or middle range to prevent overcuring the thin section.

The wide range of compounds available to the molder, coupled with an infinite

TABLE 8-2 Recommended temperature range for thermosetting molding compounds

Material	Temperature, °F
Phenol-formaldehyde	290–350
Melamine-formaldehyde	280–350
Urea-formaldehyde	275–310
Polyester	180–300
Diallyl phthalate	250–325
Epoxy	290–390
Silicone	300–370

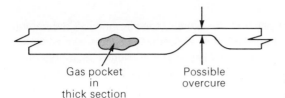

Gas pocket   Possible
in   overcure   **8-23**   Curing problems with uneven
thick section   sections.

variety of sizes and shapes of products to be molded, obviously makes any attempt at charting an exact set of temperature-pressure relationships quite presumptuous. The most efficient cycle is established through trial runs in which temperature and pressure may be changed. Press closing rate and cure time are adjusted to compensate for the variation.

During the time the molding material is in contact with the hot mold, a complex series of changes take place:

1   The compound absorbs heat, which softens the resin.

2   When sufficiently soft, the compound flows (much like hot asphalt) under the influence of molding pressure.

3   Gases develop as a result of chemical changes in the resin system. In the condensation type of polymerization, the gases are water in the form of steam, plus ammonia and formaldehyde. Resins that cure by addition may yield monomeric substances, solvents, or other vapors connected with their compounding. Air trapped during mold closing is always a factor, even if other gases are absent.

4   The molding becomes smooth (glossy if the mold is highly polished), dense, permanently hard, and insoluble.

5   The product becomes free of blisters and porosity.

6   The molded article is mechanically strong and has the thermal, chemical, or electrical properties for which it was designed.

7   Shrinkage occurs because of polymerization of the resin.

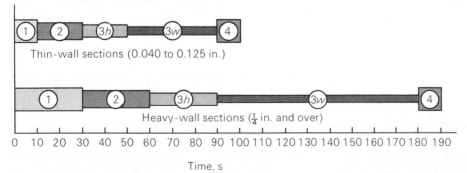

8-24   Typical molding cycles. (1) Press closing; (2) flow period (including degassing when used); (3h) cure time with preheat; (3w) cure time without preheat; and (4) press opening.

Estimating the mold charge

Molding compounds vary in weight per unit of volume (specific gravity) depending on the type of resin, type and percentage of fillers, and other additives. An asbestos-filled resin will be heavier than a wood-flour-filled material, hence requiring more material by weight to occupy the same volume. It is possible to calculate the amount of material to fill out a mold if the net volume can be ascertained from an engineering drawing or model of the part. For simpler shapes, the volume of the force can be deducted from the volume of the cavity. The net figure in cubic inches is converted to pounds, ounces, or grams[1] :

$$\frac{\text{Specific gravity} \times 62.4 \text{ (lb/cu ft of water)}}{1{,}728 \text{ (cu in./cu ft)}} = \text{lb/cu in.}$$

lb/cu in. × volume (cu/in.) = net weight of molded part (lb)

Manufacturers' bulletins should be consulted for the correct specific gravity of a particular compound.

For complex parts too difficult to calculate, it is safer to undercharge the cavity and mold short parts, gradually increasing the charge until the correct charge is determined. Overcharging a mold can result in damage to delicate mold sections, especially on a new mold tryout in which correct closing speed, pressure, and temperatures have not yet been established.

In actual molding practice, molds are charged with a slight excess of compound to ensure fill-out and part density. The excess is squeezed out of the cavity as the mold closes. The squeeze-out, called *flash*, amounts to 1 to 5 percent above the net weight of the molded product. Flash is removed by hand or automatically. Because it is waste, it is kept to a minimum.

Loading the mold

The press operator locates the charge of material (powder, preforms, or in combination) directly into the cavity. For parts that have a large projected area, loose material is spread evenly to obtain uniform heating and flow. Preforms are stacked or spread out. For difficult shapes that have deep mold recesses, preforms can be pushed into them in either the cavity or the force.

Many products contain "molded-in" inserts of metal, vulcanized fiber, laminated sheets, or other design features. These are located on holding pins in the cavity, force, or both, prior to loading the cavity.

Molding the part   As soon as the charge is placed, the press is closed to exert pressure. Press closing requires from a few seconds to more than a minute. Complete mold closing takes place as the material flows in response to heat and pressure. Initial travel of a mold can be rapid (60 to 120 in./min), but it must be slowed as

[1] 1 lb = 454 g.

guide pins enter bushings and compression begins. Final mold closing is reduced to a rate of 10 to 20 in./min to avoid mold damage. When molding material is preheated outside of the press, it is desirable to close the mold without reducing ram speed, since the compound is already softened.

Degassing the mold By the time the mold is closed, several changes have taken place: the temperature of the material has been raised to almost that of the mold; chemical reaction has started; and volatile gases have developed.

If volatiles cannot escape through mold vents or between clearances of cavity and force, the mold is momentarily opened ($\frac{1}{16}$ to $\frac{1}{2}$ in.), which drops molding pressure to zero. Press re-closing can be immediate or delayed for several seconds. After this "breathing" part of the cycle, the mold is again pressurized for the duration of the cure. On difficult jobs, such as with slower reacting materials or thick sections, degassing can be accomplished more than once, but only in the early phase of the cycle. Usually visual inspection of surface or interior sections indicates the correctness of the degassing sequence.

Condensation resins are more troublesome in formation of blisters or porosity. However, any compound is likely to contain moisture and all materials and mold cavities contain air, which may make degassing necessary. Since breathing a mold is included in the cure cycle, production time is not increased.

8-25 Large internal blisters typical of very inadequate cure. An occurrence of numerous external blisters indicates excessive mold temperature or moisture. (*Reprinted by permission of Los Angeles Trade-Technical College.*)

Curing the part   Each thermosetting material requires curing (resin polymerization) after the initial flow and mold filling occur. Cure time depends on the type of material, part thickness, mold temperature, and preheating (if used). Material suppliers formulate the resin binder to meet specific flow requirements. Large molded parts of considerable depth call for long flow duration, preferably with a rapid cure; flat or solid bulk parts involving less movement utilize short flow resins. (Test methods for comparing flow rates of molding compounds are reviewed in that portion of the text dealing with quality control.)

A typical molding cycle requires from 1 to 2 min, although thin sections can be cured in as quickly as 40 sec. Massive parts such as rocket nozzles may be in the mold for 24 h. Because of the poor heat-conducting aspect of organic resins and fillers, thick sections (above 0.25 in.) are purposely molded at the lower end of the temperature range to prevent overcuring the surface. Cure time for thicker sections molded from material not preheated can extend to 10 min or more. Products having both thick and thin sections should be molded at a low temperature for the length of time required for the thickest portion to prevent overcure of thin areas and also to prevent warpage.

Opening the press   The alert press operator observes mold opening to prevent mold damage, especially during the run-in of a newly installed mold. Parts tend to stick to a cavity or force if undercured, or if mold surfaces are rough or dirty. Changing from one material to another can cause sticking until several cycles have taken place. When a mold is inadequately clamped, part sticking can be severe enough to pull either half of the mold from the platen. Molds should open slowly until free, at which time the opening stroke can be accelerated.

Examining the molded part

Overcure   Slight overcure is difficult to detect and may only be evident through laboratory tests for mechanical strength. Extreme overcure is manifested by color change, crazing, brittleness, or dull blistered surface (see end of chapter for molding defects and their causes). This is particularly true of urea, melamine, and epoxy resins where color changes in lighter shades can occur readily. Phenolics are generally more heat stable, and since they are molded in dark colors, they are more difficult to judge through visual observation. Polyesters and DAP resins generate heat as a by-product of curing, which can be high enough to cause internal or external cracking, crazing, porosity, warping, and color change.

In any event, the shading between a well-cured piece and one slightly over- or undercured can be delicate, and only through laboratory test procedures can the accuracy of the complete molding operation be verified.

In Chaps. 31, 32, and 34, which deal with testing of materials, reference is again made to the influence of molding variables on the mechanical, electrical, and chemical properties of the product.

One simple test to establish the extent of cure is to subject the part to boiling water for 10 min. Poorly cured parts may warp, fade color, or become chalky. Cutting thick sections and polishing an edge reveal undercured internal sections, either granular, soft, or as a color variance. Specific test methods applied to each type of material are described in Chap. 30, which deals with quality control.

Removing a mold

Molds can be removed hot if required. Before removal, coat all hot-molding surfaces with carnauba wax (or special proprietary, rust-preventive preparations). Avoid hydrocarbon oils or greases. They may penetrate between ejector pins and sleeves and bleed into the molded parts the next time the mold is used. Greases may carbonize on the hot mold surfaces.

Molds are removed from the press as a unit by closing the press and removing all clamps. Before final press opening, adjust any press ejection devices out of action to avoid mold damage and mistakes with a new mold setup.

Release agents

Molding compounds contain release agents to prevent parts from sticking to hot metal. When a new mold is placed in operation, it may contain minute tool marks that cause sticking until "broken in." Special break-in compounds containing high amounts of release agents are available. Products molded from these materials are discarded.

Release agents used in the normal course of molding come in many forms. Pure carnauba wax or temperature-resistant synthetic waxes are wiped directly on the mold. Excess is blown away to prevent surface smear. Aerosols containing silicone, fluorocarbon, or lecithin are popular for releasing epoxy, DAP, and alkyd resins. Silicone and other synthetic greases, stearates, and petroleum compounds are also used.

In general, release agents must not interfere with curing, must not cause crazing or discoloration, and must be easily removable when parts are to be painted, plated, or glued.

Here are some molding defects and their possible causes:

External blister   Mold temperature too high, moisture in material, insufficient cure time, incorrect degas sequence, inadequate venting, or insufficient pressure.

Internal pockets   Insufficient load, improper location of charge, mold temperature too high, incorrect degas, inadequate venting, moisture in material, or insufficient cure time and pressure.

Incomplete part   Insufficient charge, improper location of charge, press closing too slow, material preheated too long, mold cavity or force too hot, plasticity (flow grade) too low, material too old (precure), inadequate venting, piece of material stuck in mold, pressure gauge damaged.

Poor gloss   Slightly insufficient charge, excessive or incorrect release agent, poor mold polish, press closing too slow, material too old, improper location of charge, degas sequence too long, or insufficient pressure.

Orange-peel surface   Press closing too slow, mold temperature too high, or "fines" present in granular compound.

Excessive flash   Improper location of charge, press closing too fast, uneven mold temperature, plasticity (flow grade) too high, excessive weight of charge, flash on lands, platen misaligned, or cavity distorted.

Sticking to force or cavity   Insufficient or incorrect release agent, nonuniform or low temperature, lack of mold polish or scratched, flash not removed from previous cycle, ejection pins broken or inoperable, insufficient cure time, or incorrect or damaged insert.

Brittle parts   Press opening and closing too slow, mold temperatures too high, cure time too long, or material too old to densify properly.

Cracking on ejection   Sticking in one area, ejection mechanism misaligned, under-cure or overcure, or wide mold-temperature variation.

Oversized vertical dimensions   Flash on lands, insufficient pressure or cure time, improper degas sequence, poor location of charge, excessive molding charge, or moisture in material.

Undersized vertical dimensions   Slight undercharge or distorted internal lands.

Oversized horizontal dimensions   Worn mold parts, insufficient mold closing, or incorrect grade of material (shrinkage factor).

Warpage   Nonuniform mold temperature, too great a temperature variation between mold halves, insufficient pressure or cure time, or sticking on ejection.

GLOSSARY

Charge   The measurement by weight or volume of material used to load a mold at one time or for one cycle.

Draw   The depth of a cavity.

Ejector pin   A metal pin (usually cylindrical) that ejects the molded part from the cavity or off the force.

Flash   That portion of the charge which is extruded or flows from the cavity during press closing. The portion that remains attached to the molded part is referred to as FIN.

Side core   Metal component inserted in the side of a mold perpendicular to the direction of flow or ejection. Used to mold holes, slots, or other design features, or to position inserts molded in the side of a part.

REVIEW QUESTIONS

1   When are thermoplastics molded by compression?
2   Describe the compression-molding process.

3   What is a bolster?

4   What is meant by daylight opening?

5   When is a downstroke press more convenient than an upstroke type?

6   How is a compression press rated?

7   What two factors are required in order to calculate press tonnage?

8   What is kirksite? When is it used for molds?

9   What is the purpose of the following?

   (a)   adapter plate

   (b)   force

   (c)   cavity

   (d)   guide pins

10   Describe the use of flash, semipositive, and fully positive molds.

11   Sketch the proper method of clamping a mold to a press platen.

12   What is the danger in using steel to clean molds of flash, or to dislodge cured or uncured parts?

13   Explain the difference between direct and indirect heating of molds.

14   What two types of thermal energy are used to heat compression molds?

15   What temperature range is recommended for phenolic? For urea? For polyester? For silicone?

16   Why is flash necessary in compression-molded parts?

17   Why must molds be closed slowly when unheated material is used?

18   What is the purpose of breathing a mold?

19   What danger exists in opening a mold rapidly after the cure cycle is completed?

20   What is the danger in charging a mold containing cured material that has not been removed?

21   What is the purpose of cutting a section, polishing the cut edge, and examining the piece?

22   What is the purpose of a break-in compound?

23   What is the reason for using carnauba wax?

24   What causes external blisters? How are they corrected?

25   What causes an orange-peel surface?

TRANSFER-MOLDING PROCEDURES

9

Transfer molding of thermosets is similar to injection molding of thermoplastics: material is heated in a chamber outside the mold proper, then forced to flow through an orifice to fill mold cavities. The important difference is that a thermoplastic needs only to cool below its softening point to become rigid for ejection. Transfer-molding compounds are forced into a heated mold where polymerization must take place to set the shape. In this respect, transfer molding presents many of the same problems found in compression molding—namely, under- or overcure, low density, gas pockets, warping, and incorrect dimensions. Additional problems not generally associated with compression molding also need to be overcome, such as poor orientation or knitting of fibrous reinforcements, resin-rich areas, and dislocation of inserts.

Various transfer methods have evolved since the original process was introduced in 1927; these are the advantages of the process:

1   Flash is eliminated since molds are clamped before material enters the cavity. By comparison, compression molds are designed to develop flash to ensure part fill-out and density.

2   Dimensions are accurate. Material enters a cavity through a gate subject to flow directional control. By close control of temperature, pressure, and speed, parts can

be molded to extremely close tolerance. In some parts however, design is such that closer dimensions can be realized only with compression molding.

3   Delicate inserts can be molded in place. Material cannot enter a cavity unless it attains a soft, pliable condition enabling it to flow around delicate inserts or mold obstructions. Small coring pins or thin slot formers would bend or break from the pressure exerted by the press on hard compound. In compression molding, inserts are limited to heavy cross section to avoid crushing.

4   Cure is more rapid. High-transfer pressure creates frictional heat to accelerate chemical reactions. When high-speed transfer is coupled with electronic preheat, cures are extremely fast even in thick sections.

5   Designs can be intricate. Loading space is not required, and so machined cavities with intricate parting lines are less expensive than the same cavity machined as a compression mold.

All types of thermoset resins can be transfer molded by at least one of several methods. For intricate flow conditions, however, as with closely spaced inserts, high-bulk materials would not flow uniformly to provide a homogeneous product. In this case, they may be unacceptable.

TRANSFER METHODS

Integral pot method

This original concept of transfer molding, patented in 1927 as the Shaw process, utilizes the standard upstroke compression press. Fitted with a loading chamber (pot)

9-1   Electrical and electronic components transfer molded from epoxy compounds. (*Reprinted by permission of Furane Plastics, Inc.*)

above, the mold can be permanently installed or used as a hand mold. This method is analogous to placing a funnel in a bottle.

Transfer pots can be of single- or multiple-sprue design with multiple sprues feeding one cavity or each sprue a single cavity. Pots are usually round, but they can be rectangular to accommodate special preforms. Regardless of these variations, the loading methods, pressure calculations, and material transfer are undertaken in the same way.

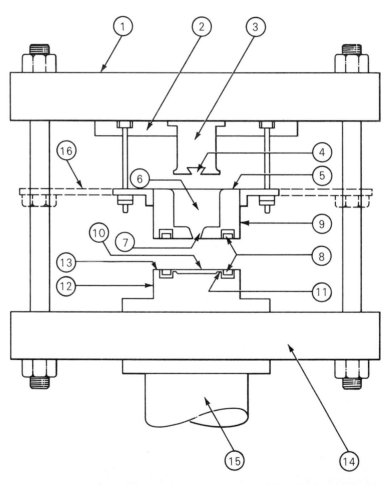

9-2 Pot-type transfer mold. (1) Upper press bolster; (2) mold-adapter (clamping) plate; (3) transfer ram; (4) sprue puller; (5) first mold-separation line; (6) transfer chamber (loading well); (7) sprue bushing; (8) cavity; (9) transfer pot; (10) runner; (11) gate; (12) mold base; (13) second mold-separation line; (14) lower press bolster; (15) upstroke ram; and (16) alternate pot-arresting arrangement.

Calculating molding pressure   A review of the method for calculating total molding pressure for compression-molded pieces shows that only one calculation is required: that of the pressure to be applied directly on the compound in the cavity, including lands, if any. In transfer molding, two opposing pressures are in effect: (1) *transfer pressure* on the total projected area of the pot, sprue, runners, and cavity (or cavities); and (2) *clamping pressure* required to keep the mold tightly closed and locked against the downward pressure exerted by flowing material.

   For example, let us take a typical transfer-molded part produced with medium-impact phenolic (cotton flock filler) as illustrated in Fig. 9-3. Transfer pressure is calculated as follows:

Part dimension 2.5 × 2 = 5 sq in., area

Number of cavities: 4

Total cavity area: 4 × 5 = 20 sq in.

Runners (four): 0.375 in. × 1.5 in. = 0.5625 sq in., area

Total runner area: 4 × 0.5625 = 2.25 sq in., area

Sprue (small dimension): 0.25 sq in., area

Total molded area: 22.50 sq in.

   Assuming a transfer pressure of 8,000 psi, then 8,000 × 22.5 sq in. indicates that a press closing pressure of 180,000 lb, or 90 tons, is required. Also, the total

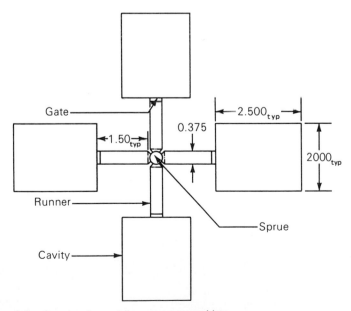

9-3   Drawing for molding-pressure problem.

mold-projected area should exceed the pot area by at least 15 percent in order to avoid flash caused by mold opening slightly. In our example, pot diameter is limited to a maximum of 5 in.

Mold temperature Mold temperatures for transfer are approximately the same as for compression molding, but they can be lower or higher by as much as 25°F, depending on flow and cure conditions. All mold parts are heated either directly or by conduction.

Transfer pressure Pressures are considerably higher than for a comparable part molded by compression (see Table 9-1). Material must flow greater distances from loading area to cavity extremities and is restricted by runners and gates. In general, pot-type molds require higher transfer pressures than other methods because of slower transfer speed developed by a compression press.

Installation Although each molder has developed his own method of building transfer molds and "floating" a transfer pot, basic procedures apply to all types:

1   Determine the size of press (tonnage) that can handle the mold, based on total molding pressure required. Platen size must be adequate to accommodate the mold, with no part of a cavity extending beyond platens.

TABLE 9-1   Typical transfer pressures, psi

Material	Pressure
Phenolic:	
General-purpose	7–10,000
Medium-impact	8–12,000
Heat-resistant	8–10,000
Glass-reinforced	10–12,000
Macerated-fabric	12–15,000
Melamine:	
Cellulose-filled	8–12,000
Mineral-filled	8–12,000
Glass-reinforced	10–15,000
Diallyl phthalate:	
Nylon-, orlon-, dacron-filled	8–10,000
Glass-reinforced	10–12,000
Silicone:	
Mineral-filled	5–8,000
Glass-reinforced	7–10,000
Polyester:	
Clay filler	1–5,000
Glass-reinforced	1–10,000
Alkyd:	
Putty	5–8,000
Granular	7–12,000

2   Minimum press daylight must provide for total mold-closed height, plus clear-
    ance to remove cull and sprue, as well as the molded parts. Sprues and culls
    are blown or tapped toward the rear of the press so that easy access is necessary.

3   As for compression molds, the entire mold is placed on the lower platen, the press
    is closed, and the mold adapter plates are clamped.

4   On the opening stroke, the transfer pot section must be adjusted to stop midway
    during press travel, while the mold continues to open, as shown in Fig. 9-2.

5   Operate the press several times to check guide-pin alignment, transfer-ram en-
    gagement, and ejection-pin movement.

6   Check temperature settings. On experimental mold setups, a good practice is to
    adjust temperatures slightly lower than maximum, using long cure times. As
    molding progresses, temperatures are increased with corresponding reduction
    in cure time.

> The sprue generally has a heavier cross section than the heaviest wall of a molded
> part and benefits little from frictional heat. If sprues are insufficiently cured, they stick in the
> bushing and must be knocked out (upward) with a brass rod. The taper is designed so
> that sprues remain on the transfer ram by means of an undercut.

INSERTS

Metal inserts can be of any form or size, usually of the pin or threaded type. They are
produced from brass, steel, aluminum, or copper, and are often plated with silver or
gold. Plastic inserts made by molding or machining are also used.

Electrical applications provide the major market for molded-in insert work.
When corrosion (caused by acids present in the molding compound) is anticipated,
inserts are precoated with a protective surface such as synthetic rubber or nylon.

Transfer molding calls for precision workmanship not only in holding dimensional tolerance but also in the spacing and orientation of an insert cluster. Inserts
must fit tightly on slender pins that hold them as compound flows around. Inserts
placed in the upper mold section require alert handling. A mold is especially vulnerable
if an insert falls out of position during closing. Pressure may *hob* the imprint of the
insert into a cavity surface and cause extensive damage.

Inserts are produced by rapid, automated methods (screw machine, stamping,

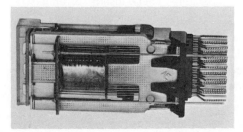

9-4   Complicated electrical device containing
expensive inserts, molded with general-
purpose phenolic. [*Reprinted by permission
of Plastics Engineering Co. (PLENCO).*]

cold-heading) and require close scrutiny before using. One defective insert in a part containing perhaps 50 others results in a defective part.

Inserts are positioned manually using tweezers or special tools, or they are tapped into place with a brass or hardwood block. Many transfer-molded parts that require delicate inserts are produced by hand molding. In this case, small molds do not require permanent press mounting. Molds are completely serviced on a bench or heated table. To speed the process, a single pot is used with two identical molds on a shuttle basis.

Molding the part   Preforms are generally used for molding, but powder is also practical. Preforms have a smaller diameter than the pot when high-frequency preheat is used, which allows for swelling. Accurate weight is important only to keep cull thickness minimal since it is waste.

To produce the part, the lower section of the mold is closed to contact the transfer pot, held there until the pot is loaded, then completely closed. Transfer of material through sprue and runners is almost immediate, so cavity filling and curing are rapid.

Molds are usually vented from cavity edge opposite the gate to mold edge by a shallow groove. If surface blisters, porosity, low density, or rough edges develop, a transfer mold can be degassed, but only after transfer is complete. Observation of travel of transfer ram into the pot indicates rate of mold filling. To degas, open press only enough to reduce gauge pressure to zero, then close to original pressure.

When cure time is completed (from 30 sec to several minutes), open the press. The sprue will break at the mold parting line because of reverse taper. As the press continues to open, the pot will be arrested by the stops, permitting the lower mold section to continue downward for part ejection. Here are the most common defects and their remedies:

Mold not filling   Reduce mold and pot temperature; increase pressure and closing speed; vent the mold; reduce preheat time; degas the mold; increase charge weight; use softer flow grade; and check for clogged sprue.

Sprue sticking   Inspect for distortion; polish if required; increase pot temperature; and increase cure time.

Parts sticking   Check mold for undercuts, distortion, or rough area; raise mold temperature; increase preheat time or preform temperature; check inserts for undersize, creating flash; and check ejector pin travel or ejector plate for dirt accumulation.

Surface blister   Predry material; check solvent content in premix compounds; reduce mold temperature; increase preform temperature; if mold is steam heated, check for condensate leak; degas mold under different conditions (sooner, later, longer); increase pressure; decrease closing speed; and add a vent or vent pin near blister area.

Internal porosity: Reduce mold temperature; reduce preform temperature; vent the mold; increase pressure; decrease closing speed; vary degas sequence; and use harder flow material.

Warped parts: Check mold for uniform heat; increase cure time; use stiffer material; change upper and lower mold heat (increase that mold section which is opposite the direction of the warp, and reduce the other. For example, if part warps upward, decrease top 10°F and increase bottom 10°F to effect even rate of cure); and preheat insert.

Flash: Check mold parting line for accumulated dirt; evaluate degas cycle or eliminate it; reduce pressure or speed of transfer; use harder flow material; increase mold temperature; and preheat to higher temperature.

Oversized parts: Excessively high pressure can cause some parts to expand by stress relief after removal.

Overcure or undercure: Because we are dealing with essentially the same resins used for compression molding, similar evaluations are made. Discoloration, streaking, fading, and density can be noted.

Additional problems: Weld areas or knit lines are more common than in compression moldings; long fibers may orient in flow direction or may block, creating resin-rich areas; inserts should be checked for spacing, length, and other factors, and for pullout; and electrical contacts should be checked for continuity.

   Molding problems also develop from improper mold design or maintenance. Runners must be short, adequate, and polished. Gates may require redesign, enlarging, or relocation. Additional vents may be required.

9-5   Twelve-ton transfer press, also used for compression molding. Note position of all gauges, switches, and controls for operator convenience. (*Reprinted by permission of Hull Corp.*)

High-speed plunger molding

High-speed plunger molding, also referred to as auxiliary ram transfer, is gradually replacing integral pot methods. Faster cycles or full automation, plus lower mold cost are the primary reasons.

Preforms are dropped through an opening in the center or to one side of the cavity block, coming to rest on the parting line of the closed mold. An independently controlled ram mounted above the press head forces the compound through runners to cavities. Auxiliary rams are fast-acting, generating more frictional heat for faster cures.

Cull, runners, and molded parts are ejected as one unit, thus eliminating the need to remove a sprue in a separate operation.

Pressure considerations   As with the integral pot type of molding, opposing hydraulic forces are in effect at time of transfer: (1) pressure required to hold the mold together to prevent flash (in tons); and (2) pressure exerted by the auxiliary ram on total projected area of cull, runners, and parts. This is the molding pressure per square inch multiplied by the projected area.

Using the example of Fig. 9-7:

Cull, 2-in. diameter:	3.14-sq in. area
Four cavities, 1.50-in. diameter:	7.07-sq in. area
Four runners, 0.1875 in. wide × 0.75 in. long:	0.56-sq in. area
Total projected area:	10.77 sq in.

9-6   Eighty-cavity mold for integrated circuits. Note cull and runner size. (*Reprinted by permission of Hull Corp.*)

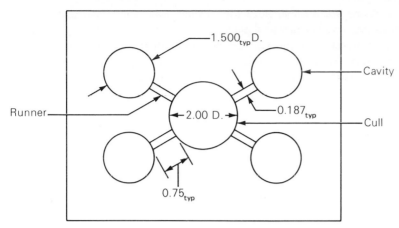

9-7   High-speed-plunger press problem.

Assuming 5,000 psi is required to fill the cavities :

5,000 psi × 10.77 sq in. = 53,850 lb

or 26.9 tons of counter force against the press clamping pressure.

Normal practice is to install a mold in a press capable of developing a clamping force from three to five times the total tonnage exerted by the auxiliary ram over the molded part. At a 3:1 ratio, the sample problem requires a minimum of 80 tons.

Theoretically, if clamping pressure exerted by the upstroke ram exactly equals hydraulic thrust of the soft, flowing compound, it should be adequate. Therefore, a slight excess of pressure upward should provide sufficient holding power. With soft flowing materials such as diallyl phthalates, premix polyesters, and some epoxies, any tendency to flash is eliminated by clamping with the 3:1 or 5:1 ratio.

Installation   Molds are fastened to a press in the same manner as a compression mold with one exception. Before closing the press, the auxiliary ram is brought down below the upper platen and the transfer hole is centered to it.

Different molds require different ram sizes, and auxiliary rams are tapped to receive the proper extension. Ejection is usually designed in the bottom half of the mold and should be checked for correct pin action.

Molding the part   To ensure complete fill and maximum density a cull thickness from 0.125 to 0.25 in. should be considered. The cull should be slightly thicker than the thickest runner so that pressure is continuous toward the cavity, with no stagnation occurring. Temperatures may be lower than for integral pot molding because of greater frictional heat.

Preforms are almost universally used with electronic preheat except when soft premix compounds can be hand packed and dropped in the chamber.

To mold the part, the mold is closed and brought to required clamp pressure.

One or more preforms are dropped into the loading well, and a separate valve is activated to move the auxiliary ram into place.

As with pot transfer, only the top half of a mold needs to be mounted when duplicate lower halves are used in a shuttle-type operation. Production can be increased by loading inserts into one mold set while the previous set cures.

Encapsulation

Chapter 28 discusses the encapsulation of electrical components using liquid resins or foams and with relatively little equipment. An outgrowth of that technique combined with transfer molding is made possible by the development of finely powdered epoxy resins with soft flow characteristics. Complicated or delicate electronic components are rapidly encapsulated by transfer molding. High-speed plunger molding is used, varying little from techniques used for normal transfer.

GLOSSARY

Cull   Waste plastic material remaining in the bottom of a transfer pot or on the parting line of an auxiliary transfer mold after runners and cavities are filled.

Gate   Section of a runner, usually reduced in size, at the entrance of a cavity.

Hob   Forcing a hardened metal object into a relatively softer metal block. (In mold construction, cavities are formed by forcing a hardened steel master pattern into a block of milder steel.)

Runner   Channel (groove) that leads from a sprue bushing to a cavity; also the material formed by the channel.

Sprue   Plastic material formed by a sprue bushing.

9-8   Ignition coil molded with liquid epoxy. Coil on left illustrates resin flow and cohesive envelopment of asbestos-insulated coil. (*Reprinted by permission of Hull Corp.*)

Sprue bushing   Tapered orifice that leads from a transfer pot directly to a cavity or to runners that feed cavities.

Transfer pot   Loading well located directly above an integral transfer mold to hold material for heating.

REVIEW QUESTIONS

1   What are the advantages of transfer molding as compared with compression molding?
2   Sketch or describe the integral pot type of transfer molding.
3   How is transfer pressure determined?
4   Which process requires the higher molding pressure, transfer or compression molding?
5   Define the following: sprue, cull, runner, and gate.
6   How is a sprue removed from a bushing if it becomes stuck?
7   What keeps a sprue attached to the transfer ram?
8   How are inserts that are to be incorporated in the molded part protected from corrosion by the plastic?
9   What type of accident can occur to the mold if inserts are not positioned properly?
10   Why must inserts be inspected before using?
11   How are inserts positioned in the proper location in a mold?
12   What is a shuttle type of molding operation?
13   Can a transfer mold be degassed? If so, how is this accomplished?
14   If a transfer mold does not fill, what are some of the causes?
15   What causes a sprue to stick in a sprue bushing?
16   What problems may arise in molding a material containing long fibrous reinforcement?
17   What other term is given to high-speed plunger molding?
18   Why should a cull be thicker than runners fed by it?

REINFORCEMENTS FOR LAMINATES

10

When two or more sheets of material are superimposed and joined by an adhesive or by fusion, the combination is called a *laminate*. This broad description includes plywood, veneered wood, automotive safety glass, and thousands of combinations of thermoplastic films used independently or laminated to paper, metal foils, wire mesh, woven fabrics, and felts. In this context, the terms *laminate*, *laminated plastics*, and *plastic laminate* have nearly the same meaning, but not quite. Thus, adjectives like rigid, flexible, and structural are added to the term to further define the type.

For our purpose, the term laminate applies to structural products formed by the combination of sheets (*webs*) and thermosetting resins. When a resin is used to join fibers, whether woven, random, or continuous, the result is a *reinforced plastic*.[1] Laminates must be considered in the broader area of reinforced plastics, which also involves molded, sprayed, or wound products.

When more than one type of reinforcement is combined in a structure, that product is called a *composite*. This includes balsa or honeycomb cores covered with a skin of rigid plastic or metal.

The greatest proportion of reinforced plastics use glass fibers, asbestos, or cotton yarns. Sheets of paper and synthetic fibers (rayon, nylon, acrylic, polyester)

[1] The abbreviation FRP is commonly used to designate thermosetting plastics reinforced with fibers of various kinds. The abbreviation RTP refers to fibrous reinforced thermoplastics.

150

are used in most of the remainder. Filaments made of graphite, carbon, metallics, and ceramic also are used as specialized reinforcement. Because glass fibers are used in the broadest scope of processes, the technology of glass is emphasized here.

HISTORY OF GLASS

Glass occurs in natural form (obsidian) as a result of volcanic activity or when lightning strikes a sandy area. Other forms of glass are known as jasper, agate, and quartz. The discovery of how to make glass is lost in antiquity although legend credits it to Phoenician mariners. The story handed down tells that the seamen built their cooking fires on beaches at night, using blocks of natron (hydrated sodium carbonate) for a fireplace. In one shoreline area, a sheet of glass formed as the fire cooled. Presumably the natron provided a flux, while ashes provided lime which lowered the melt point of the sand and produced a fused sheet.

Archeological finds show that subsequent civilizations in Mesopotamia (modern Iraq) used glass glaze as a coating for stone beads more than 14,000 years ago; Egyptians, 10,000 years later, produced glass jars and bottles for ointments.

Syrians are credited with a method of twirling molten glass on a tube and blowing it into a hollow shape. Later techniques of blowing molten glass into a hollow cavity created symmetrically molded shapes.

In 1893, visitors to Chicago's Columbian Exposition were the first to view a woman's gown, lamp shades, and other items fashioned from glass fibers. For many centuries before, filament and rod were drawn from melted glass and spirally wrapped, combined, criss-crossed, and colored for decorative effect on vases and containers. However, the glass textile advertised in 1893 was of no lasting commercial value.

Owens-Illinois Glass Company developed the first commercially feasible glass fibers (used in air filters and for insulation batts) during the period 1930 to 1934. Batts were made by collecting short lengths of coarse fibers previously drawn from a melt and attenuated by jets of air. These short lengths (called *staple filament*) were gathered, twisted, plied, and woven into fabric for use as electrical insulation (tape

10-1   Front end of truck produced for White Motor Company resists dents, corrosion, and has superior sound-insulating properties. (*Reprinted by permission of Automotive Products Div., North American Rockwell, Reinforced Plastics Operations.*)

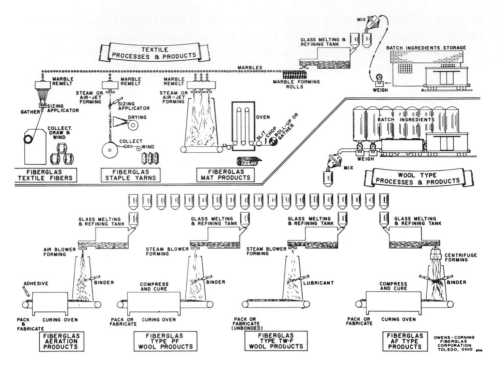

10-2 Schematic of how fiberglass is made. (*Reprinted by permission of Owens-Corning Fiberglas Corp.*)

and braided sleeving) and filter fabrics for collecting systems in steel mills, foundries, carbon black manufacturing, and chemical processing.

In 1938 Owens-Illinois and Corning Glass Works jointly formed Owens-Corning Fiberglas Corporation to develop a method of producing continuous-glass filaments. Soon afterward, fine-glass monofilaments were made available not only as a reinforcement for plastics but for any use where chemical inertness, heat, mildew, and fungus resistance were required.

Sensitive tensioning devices, plus high-speed winding units, made it possible to draw a single filament 400 miles long from 1 in.^{3} of molten glass.

Continuing research by Owens-Corning and others who entered the field has led to many glass formulations that fill a variety of uses. Glass fiber, from which all other forms of glass are developed, is also known as fiber (fibre) glass, fibrous glass, and Fiberglas.[1] *E* glass is used for most common forms of roving, chopped strands, mat, and cloth. Special purpose *C* glass (chemical) is used for such items as acid battery separators; *S* glass (high strength) is used in aerospace or pressure vessel applications to meet more exacting tensile or deflection requirements. *High modulus (M)* glass is approximately 10 times more expensive than *E* glass (electrical).

---

[1] Fiberglas is the registered trade name of Owens-Corning Fiberglas Corporation.

10-3   Fiberglas ® yarn, woven roving, chopped strand, package of roving (approx. 35 lb), and bonded mat. (*Reprinted by permission of Owens-Corning Fiberglas Corp.*)

PRODUCTION OF GLASS FILAMENTS

Individual filaments drawn through a platinum alloy bushing containing 204 holes are gathered together to form one *end*, or *basic strand*. Melt temperature, bushing orifice size, and takeoff speed determine basic monofilament diameter. Parallel and untwisted single ends can be combined to build multiple-end roving (8, 12, 20, 60, 120, 240, and 408 ends), or they can be twisted and plied for yarn.

GLASS-YARN NOMENCLATURE

Because of the virtually unlimited number of ways in which basic filaments can be combined for yarn construction, a standardized code has been adopted to aid in cloth identification. The system tells whether yarns are constructed from continuous fiber or from staple (short length). Fabric woven from staple yarn is not used for laminates requiring ultimate strength, but it is used in certain places where interlaminar shear strength is important, as in machined products.

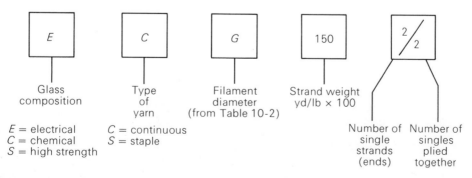

10-4   Yarn nomenclature.

Yarn nomenclature consists of three symbol groups: one alphabetical and two numerical.

In the example, the first letter indicates glass composition (see Table 10-1); the second letter tells whether the fiber is a continuous filament or staple; and the third specifies fiber diameter (see Table 10-2). The first part of the numerical coding expresses yield in 1/100 of actual yards per pound. For example, 150 = 15,000 y/lb (150 × 100) of the basic strand, which, according to Table 10-2, contains 204 individual monofilaments, each with a nominal diameter of 0.00035. By comparison, ECG 75 would mean 7,500 y/lb, since two ends of 204 monofilaments each are combined for a total of 408.

The second series of numbers, resembling a fraction, designates how single ends are twisted and plied to form yarns. The first digit indicates the number of ends twisted together. Single strands, commonly referred to as singles yarn, are designated as 1/0 when twisted. In our example, 2/2, two singles are twisted together, then two more in a separate operation. These four singles (as two groups) are then plied

TABLE 10-1 Typical glass compositions (mineral percent by weight)

Mineral	E glass	C glass	S glass	Leached glass	Quartz
Silicon dioxide	52–60	60–65	64–66	99.2 max.	99.95
Aluminum oxide	12–16	2–6	24–26	0.2–4.5	. . .
Calcium oxide	16–25	7–10	. . .	. . .	. . .
Boron oxide	8–13	2–7	. . .	0.1–0.5	. . .
Magnesium oxide	0–6	7–10	9–11	. . .	. . .
Sodium and potassium oxide	0–1	8–12	. . .	. . .	. . .

TABLE 10-2 Glass filament classification

Filament designation	Av. diameter, in.	Basic strand (× 100 = yd/lb)	No. of filaments per strand
D	0.00021	1,800	51
D	0.00021	900	102
D	0.00021	450	204
B	0.00012	450	408
E	0.00029	225	204
G	0.00035	150	204
DE	0.00025	150	408
B	0.00012	150	1,224
K	0.00051	75	204
G	0.00035	75	408
DE	0.00025	75	816
K	0.00051	37	408
G	0.00035	37	816
K	0.00051	18	408

to build the final construction, 2/2. Twisting is further designated by the number of turns per inch (tpi) ranging from one to eight or more. For example:

ECE 225 4/3 1/0.Z

where E = electrical
　　　C = continuous
　　　E = 0.00029-in. filament
　225 = 22,500 yd/lb, basic strand
　4/3 = strands total; three groups of four singles (previously twisted) plied
1/0.Z = one twist per inch, final plying done counterclockwise

　　　By providing industry with almost limitless yarn combinations, textile designers can furnish laminators with hundreds of fabric styles to solve design and cost problems.

REINFORCEMENT FROM YARN

Glass-fabric construction

Glass cloth is woven in four standard widths; 38, 44, 50, and 60 in. Narrow fabric or tape is produced in fractional widths to 12 in. Yarns of required type are gathered parallel into a beam to feed lengthwise thread (called *warp* or *ends*) to the loom.

10-5 Creels feeding warp yarns to loom. (*Reprinted by permission of the Woven Structures Div. of HITCO.*)

(See Table 10-3 at the end of this section.) Yarns carried across the width by means of a shuttle are called *fill yarns* or *picks*.

Fabric count   The number of warp and fill yarns per inch, counted separately, is called the *count*. A fabric count of 57 × 54 (or 57 by 54) means 57 ends per inch in warp direction and 54 in fill. The first number always designates yarns in length direction of a fabric.

Weaves   Weavers can design hundreds of weaves (styles) of cloth. Laminators use a relatively few styles in a wide range of weights per square yard.

PLAIN

Most common of the basic weaves, plain-weave yarns interlace in an alternating over-and-under pattern. This weave has the following features:

1   Maximum fabric stability and firmness with minimum yarn slippage.

2   Pattern provides uniform strength in two directions when yarn size and count are similar in warp and fill.

3   Fair to good porosity for resin penetration and air removal, depending on closeness of weave.

4   Makes possible lightest weight and thinnest style.

Uses: Plain weaves are widely used for coated fabrics, flat laminates, printed circuit board, narrow fabrics, tooling, and covering wood structures such as boats.

BASKET

This weave is similar to plain except that two or more warp yarns are woven as one over and under two or more fill yarns. This weave features:

1   Less stability than plain

2   More pliability

3   Flatter and stronger than an equivalent weight and count of plain weave

Uses: Similar to those for plain, with slightly better drape on mild contours.

A modification of basket weave has one warp yarn alternately weaving over and under two or more fill yarns in heavier constructions, as when roving is used in place of yarn (semibasket).

10-6   Plain-weave fabric.

TWILL

Construction of twills produces a broken or straight diagonal pattern when one or more warp yarns are woven over and under two, three, or four fill yarns. Twill has these features:

1 More pliable than plain or basket, therefore better drape

2 Better sewing characteristics

3 More difficult to wet and exclude air

Uses: Mainly filtration fabric with minor use in "sock" shapes for cylindrical laminates.

CROWFOOT SATIN

In this weave, one warp yarn is carried over three then under one fill yarn, providing a unidirectional quality. This weave has the following features:

1 High strength in one direction

2 Can incorporate higher yarn count than plain or basket

3 More pliable

4 Designed for laminating complex contours and spherical shapes

5 Provides strength in direction of applied stress

Uses: Fishing rod blanks, diving boards, skis, aircraft ducts, channel, and conduit.

LONG-SHAFT SATIN

Also called harness weave, this construction has one warp yarn weaving over four or more then under one fill yarn. Satin weaves are widely used in aircraft-quality laminates. These weaves have the following features:

1 A high degree of drape and stretch in all directions

2 High yarn density; more threads per inch

10-7 Crowfoot satin weave.

10-8 Long-shaft satin weave.

3   Less stability than plain weave

4   Less openess than most weaves

5   Difficult to remove air unless vacuum or other pressure is used

Uses: Housings, radomes, ducts, and contoured surfaces.

LENO

This weave is produced by having two parallel warp yarns twisted around each fill yarn, providing a locked effect. This weave is popular because it:

1   Reduces distortion of low-count, open-weave fabric

2   Provides heavy fabrics for rapid build-up of plies

Uses: Grinding wheels, inner core for support of thin coatings or film laminations, and tooling.

MOCK LENO

This weave is a variation of plain weave. Two or more closely spaced yarns are alternated with loosely spaced yarns in both warp and fill direction. This construction develops a textured pattern that provides:

1   Excellent mechanical bonding to the resin

2   Improved "nesting" of adjacent layers

3   Increased thickness at lower cost

4   Excellent wet-out and air removal

5   Good drape

Uses: Any application where rapid thickness build-up is required such as in tooling, boat hulls, or repair work.

10-9   Leno weave.

10-10   Unidirectional weave.

UNIDIRECTIONAL

When plain, basket, or satin styles use a higher number of comparatively strong yarns in one direction (usually warp) and fewer, lighter yarns in the opposite, the result is referred to as unidirectional fabric. Laminates for diving boards, shipping containers, or cargo compartment liners are fabricated by laying plies either in one direction or cross-ply, depending on where strength is needed. Fishing rods, vaulting poles, and other tubular products are produced from unidirectional cloth.

TRIDIRECTIONAL

Adapting special looms for use with glass fibers has produced many unique fabrics. Tridimensional fabric such as Tricon[1] permits extremely fast lay-up, helps contain low-viscosity resins on vertical surfaces, and conforms well to extreme changes of contour.

HIGH MODULUS[2]

The tendency of "glass to cut glass" is markedly reduced by lattice construction where heavy yarns are tied by light yarn. Structural yarns are not crimped, so shear factor is reduced. Uses include covers, cargo containers, and channel.

SCRIM CLOTHS

Developed primarily for adhesive-tape reinforcement, asphalt roofing, pipe-wrap membranes, facing for insulation batts, and as a stiffening core between films or foil, scrim also finds use in laminating. Depending on mesh size, scrim cloths are used for:

1   Surface ply to aid in working air from mat laminates

2   Low cost, temporary "carrier" in impregnating processes

[1] Tricon is the registered trade name of A. Wimpfheimer & Brother, Inc.
[2] High modulus is the registered trade name of J. P. Stevens & Company, Inc.

10-11   Contour-woven shape shown on check fixture and "as woven." (*Reprinted by permission of the Woven Structures Div. of HITCO.*)

TABLE 10-3 A partial list of industrial glass fabric (greige goods) specifications

Style	Count	Warp yarn	Yarn break strength, lb	Fill yarn	Yarn break strength, lb	Weave type	Weight, oz/sq yd	Thickness, in.	Fabric break strength, lb/in. Warp	Fill
104	60 × 52	ECD 900 1/0	0.5	ECD 1800 1/0	0.25	Plain	0.58	0.0010	40	× 15
102	88 × 44	ECD 900 1/0	0.5	ECD 1800 1/0	0.25	Plain	0.75	0.0015	59	× 15
112	40 × 39	ECD 450 1/2	2.2	ECD 450 1/2	2.2	Plain	2.12	0.0030	82	× 80
112/150 (1512)	40 × 39	ECD 450 1/2	2.2	ECG 150 1/0	3.0	Plain	2.60	0.0035	110	× 100
120	60 × 58	ECD 450 1/2	2.2	ECD 450 1/2	2.2	Crowfoot	3.16	0.0040	125	× 120
120/150 (1520)	60 × 52	ECD 450 1/2	2.2	ECG 150 1/0	3.0	Crowfoot	3.66	0.0043	150	× 135
57X	57 × 30	ECE 225 1/3	6.6	ECD 450 1/2	2.2	Crowfoot	5.42	0.0055	370	× 60
15057X (1557X)	57 × 30	ECG 150 1/2	6.0	ECD 450 1/2	2.2	Crowfoot	5.42	0.0055	370	× 60
126	34 × 32	ECD 450 3/2	6.6	ECD 450 3/2	6.6	Plain	5.45	0.0065	225	× 195
128	42 × 32	ECE 225 1/3	6.6	ECE 225 1/3	6.6	Plain	5.94	0.0070	250	× 200
143	49 × 30	ECE 225 3/2	13.2	ECD 450 1/2	2.2	Crowfoot	8.78	0.0090	611	× 56
150143 (1543)	49 × 30	ECG 150 2/2	12.0	ECD 450 1/2	2.2	Crowfoot	8.78	0.0090	660	× 70
7543	49 × 30	ECG 75 1/2	12.0	ECD 450 1/2	2.2	Crowfoot	8.78	0.0090	660	× 70
3743	49 × 30	ECG 37 1/0	3.0	ECD 450 1/2	2.2	Crowfoot	8.78	0.0090	660	× 60
181	57 × 54	ECE 225 1/3	6.6	ECE 225 1/3	6.6	Satin	8.92	0.0085	340	× 330
150181 (1581)	57 × 54	ECG 150 1/2	6.0	ECG 150 1/2	6.0	Satin	8.92	0.0085	340	× 330
7581	57 × 54	ECDE 75 1/0	6.0	ECDE 75 1/0	6.0	Satin	8.92	0.0085	340	× 330
162	28 × 16	ECE 225 2/5	22.0	ECE 225 2/5	22.0	Plain	12.2	0.0150	450	× 350
182	60 × 56	ECE 225 2/2	8.8	ECE 225 2/2	8.8	Satin	12.4	0.0130	440	× 400
183	54 × 48	ECE 225 3/2	13.2	ECE 225 3/2	13.2	Satin	16.75	0.0180	650	× 620
Tooling and boat-covering fabrics										
6 oz	18 × 18	ECG 75 1/2	12.0	ECG 75 1/2	12.0	Plain	6.0	0.0090	250	× 220
7.5 oz	16 × 14	ECG 75 1/3	18.0	ECG 75 1/3	18.0	Plain	7.50	0.0100	335	× 316
10 oz	16 × 14	ECG 75 2/2	24.0	ECG 75 2/2	24.0	Plain	10.00	0.0140	450	× 410
1000	16 × 14	ECG 150 4/2	24.0	ECG 150 4/2	24.0	Plain	10.00	0.0140	450	× 410

CONTOUR

To overcome inherent problems in shaping yard goods around complex sections without cutting or splicing, preshaped fabrics have been developed. Contour weave has these features:

1 Elimination of laps, seams, and nonuniform thickness

2 Reduced fabrication time

3 Uniform electrical properties in critical radar application

4 Higher in initial cost (three to five times) than comparable quantity of standard fabric.

REINFORCEMENTS FROM STRAND

Roving

All fibrous reinforcements start as basic strand. Instead of being twisted and plied into yarn, strands are gathered untwisted as they emerge from the melt and then wound ribbonlike into a roving package. Roving is designated according to number of ends gathered, from 6 to 120, with 60 end most common. Thus, a 60-end roving would be made up of 12,240 individual monofilaments. These fibers are so fine that the bundle is less than one-third the diameter of a lead pencil. (See Table 10-4.)

Woven rovings are sometimes designated by code number or style number. For example, 6060–0504 would mean 60-end roving in both warp and fill with 5 per inch in warp direction and 4 per inch in fill. Since woven rovings are 1 × 2 basket weaves with two picks weaving as one, *actual* roving count is 5 × 8 in this instance.

By comparing specific styles, it can be noted that certain series, such as 181, 1581, and 7581, are quite similar in count, weight, thickness, and strength. As shown, the finer the basic strand, the greater number of singles in the plied yarn, but the weight of the fabric is the same. However, using finer filament diameters results in a more flexible fabric, although at increased cost. A laminate required to withstand continuous flexing may utilize style 181, for example, and a product that is stationary can be made from 1581 or 7581.

TABLE 10-4 Common roving designations

No. of singles (ends)	Major use	Outstanding feature
8 or 12	Filament winding or pultrusion	Permits fast wetting Provides flat surface pattern
20	Woven roving	High bulk factor at lower cost than fabric
60	Mat, preforming, spray-up, woven roving	Suitable for chopping processes and impregnations
120	Chopped strand, milled fibers, mat	Faster production rate in processes that use 60 end

Fabricators of products made wholly or in part from roving are concerned with these four factors:

1 *Number of ends*:   For high reliability, maximum strength, and predictable service life in reinforced structures such as missile casings, storage tanks, and pressure vessels, no deviation is permitted in number of ends specified. Roving chopped for mat, winding pipe, or tanks with a lower order of strength can vary in count from $+3$ to -0.

2 *Filament diameter*:   K and G filaments are the most common rovings for chopping to make mat or other felted products such as preforms or spray-up. The coarser filament offers lower cost.

3 *Strand integrity*:   To protect fragile filaments and furnish a *coupling agent* for the resin that will be used, roving is subjected to a sizing process. Quantity and type of size influence relative softness or stiffness of the roving, and end use dictates which type is more suitable.
(*a*)   *Hard finish* (high integrity):   When little or no separation of individual strands can be tolerated as roving is put through a process, the stiffer ribbon is preferred. Manufacture of solid rod stock, tapered fishing rods, filament-wound items, corrugated sheets, or woven roving requires an absence of broken strands, fuzzing, or bunching.
(*b*)   *Soft finish* (low integrity):   Used in operations where chopping for preforms or spray-up on vertical walls requires that roving provide some "cling." If roving breaks apart in such a way as to separate some of the fine filaments from the pack, it is considered soft roving. This is desirable also when laminates or molded products contain small radii. Stiff roving would "bridge" instead of conforming to contour if sufficient pressure to force the fiber to bend were not available.

4 *Ribbonization*:   To eliminate catenary that causes production slowdown and unequal strength distribution, rovings are packaged so that each strand is under equal tension.

MAT

Glass mat is a blanket or sheet of nonwoven random lengths of chopped fibers or swirled continuous strands. Glass mat provides lower cost reinforcement than woven fabric. In general, it requires more resin to fill interstices and more positive methods for removing trapped air. Laminates made with mat are only one-third to one-half as strong as fabric laminates of comparable thickness because of lower glass content.

Chopped strand

This most common type of mat is made by depositing chopped roving onto a conveyor screen, spraying a binder to hold fibers in jackstraw arrangement, and setting the binder as the conveyor moves through an oven.

Chopped strand mats are classed by weight (0.75 to 3 oz/sq ft) and by the degree of solubility in polyester resin or styrene monomer. Mats are used primarily with polyesters, whereas fabrics are used with epoxy, phenolic, silicone, melamine, and polyesters, as well as with many thermoplastics.

Low solubility Binders of a thermosetting type usually form stiffer "boardy" mats that do not tear apart when handled, tailored, or wetted. Patterns can be die-cut or stack-cut with textile cutters. In press molding, fibers are not disrupted and pushed (*washed*) by flowing resin. Drape is sacrificed in laminated products with small compound curves.

High solubility A powdered polyester soluble in the monomeric portion of a poly-ester laminating resin aids in wet-out of the glass. For continuous saturating processes, as in corrugated or flat panel work, more rapid wetting speeds through-put rate and reduces tendency for individual fibers to be raised above the resin surface. The softer, more drapable, high-solubility type is used for laminating or low-pressure molding of housings, trays, caskets, and parts having much detail.

Swirl

Strands are laid down in a continuous flat loop or swirl pattern to provide medium-strength laminates and molded parts.

Surfacing

As the name implies, *surfacing mat* (or *veil*) is used for top layers of corrugated and flat sheets, trays, signs, and lamp shades. Ultrafine fibers are used to produce veil with a relatively smooth texture from 0.010 to 0.030 in. thick. The use of surface mat tends to conceal coarser mat fibers underneath and extends weather resistance by preventing stiffer fibers from blooming.

By controlling the amount and type of binder, mat moldability, drape, and wetting can be varied to allow for direct silk-screen or incorporating decorative effects such as leaves, butterflies, wheat, and metallized yarn.

TABLE 10-5 Types of low-solubility binders

Resin	Purpose
Amino	Provides white color so fibers stand out for decorative effect
Polyester	Provides good aging characteristics, low water absorption, compatibility
Phenolic	Lowest cost; where appearance because of dark color is not important
Acrylic	Less discoloration in ultraviolet light

Needled

These are mats produced by "dry needling." Although more expensive than others, they offer several advantages: (1) less tendency to wash under pressure; (2) excellent drape; (3) conform to combinations of thick and thin sections; and (4) can be combined by the dry-needle process with fabrics, parallel rovings, or woven roving.

Needled mats and combinations are furnished in weights from 1.50 to 10 oz/sq ft. They are used in compression molding and in vacuum-bag operations where high impact strength or intermediate mechanical properties in general are required. Examples are large tote boxes, conveyor cartons, and lug boxes.

Woven roving

In between chopped strand mat and woven cloth in price and strength-weight performance, woven roving is produced on heavy-duty rug-weaving looms in widths to 144 in. These reinforcements are used for rapid build-up of contact or other laminates or to provide backup for large structures such as boats, tanks, or tooling.

MISCELLANEOUS GLASS REINFORCEMENTS

Continuous roving, chopped in 0.25-, 0.50-, 1-, and 2-in. lengths, is purchased in bulk for reinforcing molding compounds, to mix into paste for filling sharp corners or recessed areas prior to laminating, and for wet slurry processes. Spun roving (staple) is also chopped to provide more fluffy types. These products are called *chopped strand*.

Hammer milling reduces virgin or reclaimed fibers to lengths of 1/32 to 1/8 in. for use as a filler (rather than as a strengthener) for molded, cast, or laminated products. Amounts to 70 percent of these milled fibers can be tolerated (depending on resin viscosity) to make trowel paste or adhesives. Normally amounts to 40 percent tend to reduce resin shrinkage, increase heat transfer, and improve surface hardness.

Tiny platelets of fractured sheet glass called glass flakes improve electrical and chemical values when added to a resin such as epoxy. Glass microballoons, or beads, reduce the overall weight of a filled structure and provide improvements cited for milled fiber and glass flake.

IMPROVING THE GLASS-RESIN BOND

Although noted for high-tensile strength, glass fibers are extremely fragile and abrade easily when processed. To guard against loss of strength, a starch-oil sizing is applied to filaments that are made into yarns. Sizing is applied prior to twisting, plying, and weaving as a *temporary* protection during loom operations.

Glass fabrics are used by foundries, mills, chemical plants, welding shops (for curtains), and countless other applicators besides laminators and coaters. For this reason fabrics require different types of treatments. Since they use many types of resins, laminators require specially treated fabrics to obtain the best adhesion of a particular resin.

Greige (pronounced "gray") goods

After it is loomed (loomstate), woven cloth receives no further treatment prior to coating or laminating. Textiles that still contain original sizing are suitable for laminates where strength is not paramount or for frangible uses such as armor plate or body armor. Flexible coatings (rubber or vinyl) can also be applied to greige goods with good adhesion.

Heat–cleaned fabric

Thermosetting resins do not adhere well enough to the smooth, round, glass filaments to realize maximum strength, so posttreatment with specific finishes (see Table 10-6) is normally required. Before a finish can be applied, rolls of greige goods are heat cleaned either by batch process (ovens at 575°F) or continuously at 615°F. Heat

TABLE 10-6   Finishes for glass fabrics

Code number	Chemical type	Characteristics	Thermoset-resin compatibility
111 500 516	None; heat cleaned to various levels of residual sizing	Imparts a tan color to laminates	Melamine
112	None; heat cleaned to minimum binder level, less than 0.10%	Leaves cloth natural white	Silicone
Neutral pH	None; washed to neutral (6.9–7.2)pH	Same as 112	Silicone
Volan A	Methacrylate-chromic-chloride	Imparts green hue to clear resin laminates	Polyester primarily; some epoxy and phenolic
Garan; A-172	Vinyl-silane	Imparts no color	Polyester; silicone elastomers
A-1/4 Z-6030	Acrylo-silane	Tends to stiffen fabric	Polyester
A-187 Z-6040 Y-4087	Epoxy-silane	Stability in prepreg shelf life	Epoxy; phenolic
A-1100 Y-2967 Z-6020	Amino-silane	Improved properties at elevated temperature	Epoxy; phenolic; phenyl-silane; melamine
UM Feno	Proprietary*	Improved drape	Phenolic; epoxy
S-910	Proprietary†	Clarity; fast wet-out; wet-strength retention	Polyester

*United Merchants.
†J. P. Stevens & Co., Inc.

TABLE 10-7 Comparator of roving finishes

| End use | Finish | | Polyester resin compatibility | | | | |
		OCF[a]	PPG[b]	FERRO[c]	JM[d]	GB[e]
Spray-up:						
Roving	Chrome-silane	851	539	206	71802	...
	Chrome	825	520, 522, 525, 925	160	...	FR-5, FR-13C
	FW[f]	881	...	...	...	...
Press molding:						
	Chrome-silane	851FS, 851, type 2	539F	106, 206, 226, 258	...	FR-3
Preforms, continuous roving	Silane	858	535, 537, 541	203, 206X, 219, 385	323, 7290	FR-8, FR-10
Preforms, spun roving	Chrome-silane	877	...	...	...	...
	Chrome	825, 834	508, 510	157, 159	7301	FR-1, FR-1B
	FW	881, 886	...	...	...	...
Chopped strand mat	Silane	M800, M802	AMM-1, AMM-2	GP	...	MR-1, MR-2, MR-15
Premix	Chrome-silane	851	...	...	...	...
	Chrome	863	Chrome Hard	...	...	SR-5
Sheet molding compound:	Silane	M900, M901	ACM	GP, MSGP, MSHSB, MRHSB	...	MR-9
Mat	FW	M911	...	...	...	...
Bulk molding compound: Roving	Silane	888	535, 533-13	219, 267, 307, 385	279, 393, 7201	...
Hand lay-up: Mat	Silane	M700	...	HSB, MRHSB	601, 630	MR-6, MR-18, MR-19
	FW	M710, M711	ABM	Unicomb	Plymat	MR-29, MR-40

Process	Type of finish	OCF[a]	PPG[b]	FERRO[c]	JM[d]	GB[e]
Vacuum-bag molding: Mat	Silane	...	...	GP, MSHSB, MRHSB, MSGP	...	...
	FW[f]	M710	...	...	...	...
Filament winding: Roving	Silane	888	...	307	279, 393, 7201	FR-6
	FW[f]	810, 887K, 891K, 893K	764	...	...	...
Encapsulating: Milled fibers	Cationic water	701, 709	...	HC, CATX	HDX, XDX	MF-1
Centrifugal casting: Mat	Silane	M851, M856	...	MSGP, HSB	...	MR-12
Roving	Chrome	808	508, 510	...	...	...
Mat manufacturing:	Chrome-silane	851, type 2	...	...	...	...
Roving	Silane	855-2D	56-W, 56-WF, 537, 538	203, 226, 307	7200, 7201	...
Epoxy-resin compatibility						
Filament winding:	Silane	801	...	...	...	FR-14EZ
Roving	FW[f]	392, 879, 901, 904, 836K, 859K	762	777	...	FR-14

[a] OCF = Owens-Corning Fiberglas Corporation.
[b] PPG = Pittsburgh Plate Glass Company.
[c] FERRO = Ferro Corporation.
[d] JM = Johns-Manville Corporation.
[e] GB = Gustin Bacon Company.
[f] FW = Fast wet-out.

Note: Some code numbers are occasionally replaced as new or improved finishes are developed. Consult roving suppliers for specific information.

soaking volatilizes the starch-oil size and makes fabrics more receptive to resin-compatible finishes.

Neutral pH

One exception to the use of an applied finish occurs with rigid silicone laminates. If residual alkaline elements are not removed in deionized water, the silicone cure is affected and electrical properties are impaired.

FINISHES FOR ROVING

Various finishes comparable to those employed with fabrics are applied to roving, depending on intended use (see Table 10-7). They are applied as strands emerge from the bushing.

COTTON FABRIC, PAPER, AND ASBESTOS REINFORCEMENTS

Laminates reinforced with cotton fabrics have played an important role in industry since the development of phenolic resins in 1909. Increasing use of electricity in the United States at about the same time gave impetus to the use of synthetic resin binders in cotton laminations. Previously, insulations consisted of shellac, natural rubber, bitumen, animal glues, and vulcanized fiber.

The National Electrical Manufacturers Association (NEMA) is largely responsible for specifications governing various grades of cotton-base laminates as well as glass cloth and mat types. Asbestos and paper laminates are also included in NEMA standards.

Cotton fabric grades

Grade C   This grade designates a woven cloth that weighs more than 4 oz/sq yd and has no more than 72 yarns per inch in fill direction or more than 140 yarns total in both warp and fill. Heavy cotton ducks up to 20 oz are used for laminating gear stock, bearings, machine guides, and support members. Impact strength increases as fabric weights increase, but at a sacrifice of smooth, machined edges.

Grade L   Grade L designates woven cloth that weighs 4 oz or less per square yard and has more than 72 fill yarns per inch and more than 140 yarns total in both warp and fill. Also called linen grade, this laminate is produced for applications such as small gears, pulleys, punching stock, and machined parts where smoother edges are required.

Depending on ultimate uses, electrical, mechanical, and chemical grade laminates are further classified according to type. For example, CE and LE have superior electrical values.

Paper-base grades

Laminates utilizing various types of paper (kraft, alpha cellulose, rag stock, or blends) are used when electrical properties are more important than mechanical strength, or when a lower cost material satisfies requirements.

Paper-base laminates are designated as Grades X, XX, and XXX in descending order of mechanical properties. The suffix letter P indicates a more flexible resin binder, permitting stamping or punching sheet stock to required designs (see Table 10-8).

Asbestos grades

Two forms of asbestos webs are used to make laminated products for extreme heat and flame resistance. NEMA Grade A utilizes thin paper or heavier felt; Grade AA designates asbestos woven fabric.

TABLE 10-8   NEMA properties for sheet

Grade	X	XX	XXX	XP	XXP	XXXP
Base: paper						
Resin: phenolic						
Tensile strength, psi	20,000	16,000	15,000	12,000	11,000	12, 00
Compressive strength, psi	36,000	34,000	32,000	25,000	25,000	25,000
Flexural strength, psi	25,000	15,000	13,500	14,000	14,000	12,000
Dielectric strength, V/mil						
thickness: 0.0625 in.	700	700	650	650	700	650
0.125 in.	500	500	470	470	500	470
Operating temp., °F, max.	225	250	250	250	250	250

Grade	C	CE	MC	L	LE
Base: cotton					
Resin	Phenolic	Phenolic	Melamine	Phenolic	Phenolic
Tensile strength, psi	11,200	12,000	10,000	14,000	13,500
Compressive strength, psi	37,000	39,000	38,000	35,000	37,000
Flexural strength, psi	17,000	17,000	16,000	15,000	15,000
Dielectric strength, V/mil					
0.0625 in.	200	500	300	200	500
0.125 in.	150	360	200	150	360
Operating temp., °F, max.	225	250	275	225	250

Grade	G-3	G-5	G-7	G-9	G-10	G-11
Base: glass cloth						
Resin	Phenolic	Melamine	Silicone	Melamine	Epoxy	Epoxy
Tensile strength, psi	23,000	37,000	23,000	50,000	45,000	45,000
Compressive strength, psi	50,000	70,000	45,000	75,000	60,000	60,000
Flexural strength, psi	20,000	44,000	20,000	55,000	50,000	50,000
Dielectric strength, V/mil						
0.0625 in.	700	350	400	400	500	500
0.125 in.	600	260	350	350	400	400
Operating temp., °F, max.	290	300	400	300	...	...

TABLE 10-8   NEMA properties for sheet (*Continued*)

Grade	A	AA	N-1	FR-2*	FR-3*
Base	Asbestos paper	Asbestos fabric	Nylon	Paper	Paper
Resin	Phenolic	Phenolic	Phenolic	Phenolic	Epoxy
Tensile strength, psi	10,000	12,000	8,500	12,400	12,000
Compressive strength, psi	40,000	38,000	. . .	25,000	28,000
Flexural strength, psi	13,000	18,000	10,000	12,000	20,000
Dielectric strength, V/mil					
0.0625 in.	225	. . .	600	650	600
0.125 in.	160	50	450	. . .	. . .
Operating temp.,°F, max.	275	275	165	250	300

*Burning time is not to exceed 15 sec.

Nylon fabric

Designated N-1, nylon cloth with phenolic binders finds use in applications requiring lighter weight laminates with high impact strength and high dielectric strength. It is not recommended for use at elevated temperature when dimensions are critical.

10-12   Courtroom illustrates extensive use of paper-base melamine laminates in partitions, tables, and doors. (*Reprinted by permission of Fabricon Products Div., Eagle Picher Co.*)

DECORATIVE LAMINATES

High-pressure lamination of composite layers using melamine-paper top sheets and phenolic-kraft core stock provides a wide range of decorative and functional paneling and surface applications.

Water-soluble melamine (or water-alcohol) solutions are used to saturate alpha or rag papers that progress through staging ovens to advance the resin, eliminate solvent, and control flow. Lower cost phenolic is used to saturate kraft paper to provide the base or core sheets.

Transparent melamine is used with preprinted or solid-color surface sheets or can itself be pigmented for solid colors. Hundreds of printed overlay papers simulating wood grains and marble or using special effects such as gold fleck provide wall panels, sink and counter tops, furniture, and desks with long-lasting, scorch-proof, stain- and alcohol-resistant surfacing.

LAMINATED TUBES AND RODS

Any combination of reinforcement and resin can be used to produce rod and tube stock. Small-diameter paper tubing is generally produced by spiral wrapping around steel mandrels. Fabric base tubes are produced by several methods: (1) machined

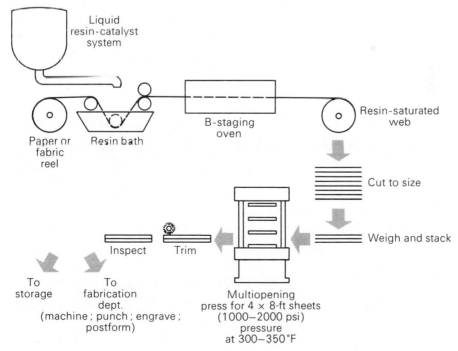

10-13   Manufacture of high-pressure laminates.

from flat stock when special diameters are required ; (2) molded in compression molds ; or (3) rolled around a mandrel.

Tubes and rods are used to produce washers, spacers, supports, rollers, and pads.

CLAD LAMINATES

Combining metal foils and sheets as part of a laminate has been practiced for many years. With the advent of printed circuits, epoxy-glass sheets with a copper-clad surface have made miniaturization of electronic assemblies feasible. Transistor radios, television, and complex machine consoles can be made smaller and more dependable.

Other composite laminates are designed for special purposes :

Stainless steel: For extreme corrosion or nonmagnetic use.

Aluminum: Engraved stock, such as that used for an office machine, where darker color of core sheets contrasts with bright or brushed aluminum surface sheet.

Silver or gold: Laminate provides backup for facing uses for optimum electrical contact.

Rubber: Both foam and solid sheets are used as mounting or isolation pads to reduce noise or vibration.

10-14   Epoxy-glass cloth laminate used as a base for printed circuit. (*Reprinted by permission of Mica Corp.*)

EXOTIC REINFORCEMENTS AND ADVANCED COMPOSITES

Constant demands for structural materials to perform under increasingly higher thermal conditions produce a never-ending search for new products. As the world entered the Space Age, new materials that were dependable at temperatures exceeding 10,000°F had to be found.

High-silica fibers

When type *E* glass in any of its forms (roving, mat, or cloth) is leached by treating with acids and re-fusing at a high temperature, a reinforcement is produced with less strength than the original glass but with the melting point increased from the original 2150 to 2950°F.

Graphite fibers

Graphite fibers are produced by several methods, depending on the starting fiber (precursor), which is usually rayon or polyacrylonitrile (a synthetic rubber).

Rayon process

1   Heat fibers at 75 to 300°F to remove as much as 10 percent residual moisture.

2   Increase temperature (using nitrogen atmosphere) from 300 to 460°F to eliminate hydroxyl (OH) groups and to form carbon-to-carbon bonds.

10-15   Carbon fabric laminated with phenolic resin is used in critical areas of rocket nozzle. (*Reprinted by permission of Aerojet General Corp.*)

3   Heat from 460 to 800°F to break down cellulosic rings and to form an almost pure carbon structure.

4   Increase temperature to 1,300°F to form carbon rings, and then follow by tensioning to orient the crystals and increase tensile strength.

Graphite fibers show great promise as reinforcement for plastics. They may be used ultimately in transportation, sporting goods, and buildings (load-bearing members).

Carbon fibers

Carbon fibers produced by methods similar to those which produce graphite, but at lower process temperatures, are used as less-expensive reinforcements for sections of solid-propellant rocket motors.

Boron filaments

Investigations of fine filaments prepared by depositing boron on a small-diameter tungsten wire and laminating with epoxy resin are being studied for helicopter rotor blades, wing flaps, and deflectors in high-speed aircraft.

Whiskers

Single crystals grown in a single crystallographic direction can attain a length to 75,000 μ from 0.5- to 10-μ diameter. The advantage of single-crystal whiskers is the reduction or total absence of flaws found in bulkier filaments.

Many inorganic elements are used to grow whiskers. Boron or silicon carbide, sapphire, aluminum oxide, and many others are being evaluated. Orienting whiskers in the direction of applied stress and obtaining uniform distribution in the resin are problems yet to be solved.

The reinforcement of synthetic resins with a widening variety of fibrous materials has resulted in the creation of a segment of the plastics industry that is devoted to the engineering and manufacture of laminated and molded products for structural, ablative, chemical, and electrical applications. Because of broad scope and constantly expanding technology, this separate division of the SPI, called the Reinforced Plastics and Composites Division, functions to promote the proper application of these

TABLE 10-9   Mechanical properties of polyester-glass laminates

Property	Woven Fabric	Random mat	Continuous or filament wound
Resin content, %	45	70	30
Tensile strength, psi	50,000	15–30,000	125,000
Flexural strength, psi	40,000	18–25,000	70,000
Impact strength, ft-lb	20–25	18–20	60–75

materials. Other professional groups, such as the Society of Aerospace Materials and Processes Engineers (SAMPE), similarly promote the use of reinforced plastics.

Although the emphasis of this chapter is placed on laminating processes, the use of all these reinforcements (particularly fiber glass) in compression, injection, blow, and rotational molding processes must also be recognized in current and future applications.

GLOSSARY

Blooming Raising of fibers above cured resin surface either during polymerization or caused by weathering and resin shrinkage.

Compressive strength Resistance to being crushed (compressed).

Dielectric strength Resistance to electrical current that tends to puncture the material.

Fiber Relatively short-length material made by chopping filament.

Filament Variety of fiber (usually solid and cylindrical) characterized by extreme length.

Flexural strength Resistance to bending.

Flux A substance that when added to a solid increases its fusibility.

Frangible Capable of fracturing easily. In armor plate, frangible layers may be superimposed with high-strength components.

TABLE 10-10 Comparative costs and properties of reinforcements

Reinforcement	Tensile strength, psi	Specific gravity, g/cc	Cost/lb
Glass:			
Roving (*E* glass)	450,000	2.49	0.35
(*S* glass)	600,000	2.50	3.00–5.00
Mat	. . .	. . .	0.55
Cloth	Varies with weave	. . .	1.00–2.00
Asbestos	310,000	3.20	1.10
Leached silica	. . .	2.50	6.00–9.00
Nylon	70,000	1.14	0.52
Graphite (filament):			
Thornel-40*	250,000	1.56	55.00–75.00
-50	280,000	1.63	. . .
-75	375,000	1.86	. . .
-100	525,000	2.0	. . .
Morganite:†			
Type I	250,000	2.0	90.00
Type II	425,000	1.8	130.00

*Registered trade name of Union Carbide Corporation.
†Registered trade name of Whittaker Corporation

Fusibility   Tendency of a substance to melt.

High modulus   Exhibiting lower order of deformation per unit of stress than comparable materials.

Interstices   Space between adjacent fibers or yarns.

Kraft   German word meaning strong. Thus, kraft paper is heavy-duty.

Microballoons   Extremely small-diameter hollow spheres.

Micron   One millionth of a meter or 0.00003937 in. Not to be confused with the American term, micro-inch, which is one-millionth of an inch.

Superimposed   Placed on top of one another.

Tensile strength   Resistance to being pulled apart.

Volatize   Cause to pass off as a vapor.

Web   Trade term applied to rolls of paper, foil, fabric, or felt.

REVIEW QUESTIONS

1   Define laminate.
2   What is meant by the term reinforced plastic?
3   Which type of glass filament is needed for roving and textile products that are used for laminating?
4   What is staple fiber?
5   Explain the term ECE 225 4/3. How many singles have been plied to produce this yarn?
6   Which term describes the lengthwise direction of a fabric? The width of a fabric?
7   Which fabric weave would you select to use in the following applications:
    (*a*)   covering a wooden row boat
    (*b*)   laminating a hemispherical shape
    (*c*)   right-angle elbow for a duct system
    (*d*)   archery bow
8   How many monofilaments are in a 60-end roving?
9   What is the purpose of applying a finish on a roving?
10   How is the weight of glass mat expressed? How does this differ from glass cloth weight?
11   What are the differences among chopped strand mat, surfacing mat, and swirl mat?
12   What is a clad laminate?
13   What are the advantages of using woven roving?
14   Which laminating processes utilize chopped strand?
15   What is the purpose of glass sizing?
16   What is meant by greige goods?
17   What is the purpose of a finish as applied to glass fabric?
18   Which finishes are recommended for polyester-resin compatibility? For epoxy? For phenolic?
19   What is the meaning of strand integrity?

20   Which organization establishes standards for laminates used in electrical applications?
21   How does Grade C cotton cloth differ from Grade L?
22   What reinforcement is used to produce a Grade XXP laminate? What does XXP mean?
23   Which type of laminate, X or XP, has the higher flexural strength?
24   What reinforcement is used to produce a G-9 laminate?
25   Sketch and explain the construction of a high-pressure decorative laminate.
26   How are laminated tubes and rods produced?
27   How are graphite fibers produced? What applications exist for graphite-reinforced laminates?
28   Where are boron fibers used?
29   What is a whisker?
30   What is a micron?

TOOLING FOR THE LAMINATING PROCESS

11

High standards of living enjoyed by certain countries are based partly on the ability of their people to develop tools that produce items at low cost. The toolmaking industry serves as a barometer for the economic climate of a nation since tools must be produced in advance of scheduled production. The level of employment in the toolmaking industry is reflected in the purchasing power of the general populace. If a business recession is expected, toolmaking is one of the areas economists analyze to predict its duration.

Toolmaking for the plastics industry is closely allied to that of the metal-working trades. Metal tools are used to produce plastic items, and plastic tooling is readily accepted in many areas of metal shaping. Most metal tools are fabricated by skilled mold makers using conventional machining practices. Tooling is also cast by foundries or fabricated from sheet metal. Other practices include electroforming, chemical milling, and electrical discharge machining.

COST

The phenomenal growth of reinforced plastics and composites for structural applications has been influenced greatly by the practicality of low-cost, easily made tooling. This does not mean that all tools used are inexpensive. A matched, tool-steel mold to produce a boat may cost $50,000, or the same boat may be made from a plaster or wood pattern costing less than $1,000. The production rate and quality of the product made on matched molds is, of course, far superior.

CLASSIFICATION

Tooling may be defined as any form, pattern, mold, or device that can be used to produce an article of value. It is classified as *temporary* or *permanent* (see Table 11-1). Temporary tools are defined as a "one-shot" or expendable unit, or as experimental prototypes for evaluation. Permanent tools are fabricated from more durable materials capable of producing the required number of articles. In the laminated plastics field, permanent tooling, or class A tools, refers to hard-surfaced, rigid materials such as reinforced epoxy or polyester, and metals.

MATERIALS

Selection of the type of material for tooling is based upon many factors, such as: (1) number of parts to be produced; (2) degree of surface finish required; (3) extent of portability (weight and size factors); (4) design and fabricating features; (5) repair or modification possibilities; (6) heat conductivity; (7) heat resistance; (8) weather resistance; (9) impact or compressive strength; (10) chemical resistance; (11) availability of skilled toolmakers; and (12) cost.

11-1 Nose cone for F104 Starfighter being filament wound on steel tool. (*Reprinted by permission of Lockheed Aircraft Corp.*)

It is difficult to determine which type of temporary tooling should be used in some cases. The rule is that the product obtained is only as good as the tool on which it was made. We shall see that in many of the processes by which reinforced plastics items are produced, the quality depends upon skill in laminating and sound knowledge of resin systems and fibrous materials. To make a simple shell structure such as a bait tank or a housing, the form could be cardboard over which a wet lay-up can be made.

Wood

Other simple forms consist of wire screening supported by wood framing or thin wood veneer shaped to contour. Hard maple or mahogany is used when difficult shapes are not easily obtained by other means. The inherent problems in using wood tools are swelling caused by moisture, the need for perfect sealing, and susceptibility to damage.

Plaster

Plaster patterns are more widely used than any other form of temporary tooling because of low cost, ease of working, and ability to provide smooth surfaces.

Casting plaster is produced from the mineral hydrated calcium sulfate ($CaSO_4 2H_2O$), which remains as a sediment when salty bodies of water evaporate.

TABLE 11-1   Classification of tooling for reinforced plastics

Temporary	Permanent
Wood:	Metals (in ascending order of durability):
Mahogany	Copper
Hard maple	Brass
Plaster:	Kirksite
Gypsum	Aluminum
Break-away (soluble)	Cast iron
Resin-impregnated	Steel:
Cement	Sprayed
Ceramic	Electroformed
Cast Plastics:	Cast
Epoxy	Fabricated plate
Phenolic	Forged
Polyurethane elastomer	Machined
Polyvinyl chloride	Laminated Plastics:
Silicone rubber	Epoxy-glass
Polysulfide rubber	Polyester-glass
Miscellaneous:	Phenolic-glass
Cardboard	Polyimide-glass
Masonite	
Acrylic sheet	
Polyethylene sheet	
ABS sheet	
Rubber	

11-2   Hollow plaster mandrel being
chipped away from laminated fuel cell
used on Poseidon missile. (*Reprinted
by permission of E. I. du Pont de
Nemours & Co.*)

The mineral is mined or quarried, then calcined (heated under controlled conditions)
to remove most of the chemically combined water. When plaster is mixed into water,
it "sets" by the formation of interlacing needlelike crystals. Chemically only 18.6 percent water is required to combine with gypsum plaster, although as much as 30
percent is required to obtain a workable consistency. After being mixed, plaster
undergoes several distinct phases that must be recognized in order to take advantage
of its unique characteristics:

First stage: Creamy, flowable consistency useful for pouring or finish coats. Entrapped air is removed during this period.

Second stage: Less mobile; can be moved by trowel or other means; will cling to
vertical surfaces or rotating mandrels.

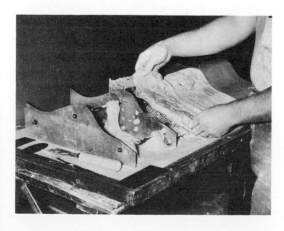

11-3   Templates are bolted together,
then tied to gypsum base, using hemp
fiber impregnated with gypsum. Expanded metal is used to span openings
between templates and provide support
for hemp. (*Reprinted by permission of
United States Gypsum.*)

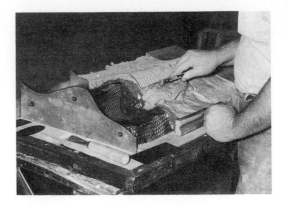

11-4 Gypsum being applied with sawtooth scraper to build surface just below template top edge. Final coat will bond mechanically and chemically to this base coat. (*Reprinted by permission of United States Gypsum.*)

Third stage: Stiff; workable only by scraping.

Fourth stage: Hard, immobile, "set" condition.

Complex contours of any size and intersecting planes are developed by a process known as the template method (see Figs. 11-3 to 11-5). Plasters must be dried to a low-moisture content before being used as a laminating tool. Drying is done in ovens at 140°F maximum or with hot air blowers when size makes ovens impractical. Failure to use thoroughly dry tooling results in poor-quality laminates. Other shapes are produced by bench-run methods or turning.

Before a plaster tool is used, surface pores should be sealed with several coats of shellac, varnish, or impregnating resin such as Plas-preg,[1] followed by several coats of paste wax. Release agents are also necessary to provide separation of laminate and tool.

Plaster tooling can be used for laminates produced by contact, vacuum-bag, or spray-up methods. They are not recommended for resins that require elevated-temperature cures except on a one-time basis since they become friable.

[1] Plas-preg is a furfural-formaldehyde resin produced by Furane Plastics, Inc.

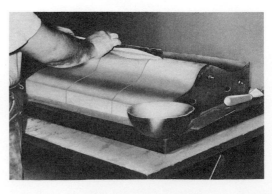

11-5 Screeding final coat of gypsum. Note screed spans three templates for accuracy. Ventilation holes in templates speed drying. (*Reprinted by permission of United States Gypsum.*)

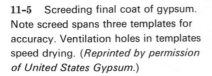

11-6 Straight bench-run method of producing profiles using a "sled." (*Reprinted by permission of United States Gypsum.*)

Water-soluble plasters are available for tooling when used as internal cores for hollow articles or for undercut sections. One such product, Paraplast,[1] is used for expendable mandrels for a wide variety of complicated ducts and similar shapes used in aircraft. After the laminate is cured, the plaster is washed out with a water jet.

A modified plaster known as Calcerite[2] is a combination of gypsum and amino resin. When mixed with water, which acts upon the latent catalyst of the amino, the resultant cure provides a denser, harder material with higher heat resistance.

Epoxy resins

The coming of epoxy resins for tooling purposes has resulted in the emergence of companies that specialize in developing epoxy tooling compounds and in advancing the art of tooling plastics. These firms, known as resin formulators, are largely responsible for the acceptance of epoxy tooling in the automotive, aircraft, marine, and general metal forming industries as well as in laminating.

Epoxy resins are more widely used as compared with polyesters, despite a much higher resin cost, for the following reasons: (1) extremely low shrinkage; (2) heat resistance to 500°F; (3) long-term dimensional stability; (4) hard surface and good

[1] Paraplast is the registered trade name of Rezolin Division, Hexcel Corporation.
[2] Calcerite is the registered trade name of Furane Plastics, Inc.

11-7 Box-turning method for cylindrical shapes. Tubing or rod is hand cranked against metal template. Cord is wrapped around rod so gypsum will adhere at first. (*Reprinted by permission of United States Gypsum.*)

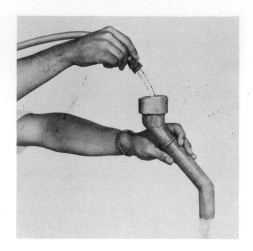

11-8   Water-soluble plaster (Paraplast) being
washed from laminated duct. (*Reprinted by
permission of Rezolin Div., Hexcel Corp.*)

abrasion resistance; (5) chemical resistance to solvents present in prepreg or used in
cleanup; and (6) high compressive and impact strength.

Epoxy tooling resins can be used to produce mass-cast tools, or they can be
combined with a variety of materials to reduce cost or weight. In some instances it
is practical to use an epoxy as the laminating surface only, backed up with less expensive polyester-glass laminations.

Large tools that must be moved from one area to another, as from the lay-up
area into an oven, are reduced in weight by several means. Materials that reduce
density by providing high bulk and light weight are incorporated into the resin and
applied behind the surface coat or after several plies of reinforcing materials are used.
These weight-saving ingredients may be in the form of porous volcanic rock particles,
vermiculite, or hollow spheres. Spheres may consist of phenolic microballoons, glass
beads, or Saran[1] spheres, which, when mixed in the resin, produce a putty that is

[1] Saran is a polyvinylidene chloride manufactured by Dow Chemical Company.

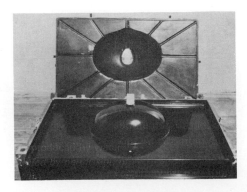

11-9   Epoxy tool for fabricating mosaic sinks.
(*Reprinted by permission of Furane Plastics,
Inc.*)

troweled into place. Resins filled with hollow, enclosed particles are known as *syntactic foams*. Density of an epoxy matrix filled with phenolic spheres can be as low as 9 pounds per cubic foot (pcf).

Whether produced by straight casting methods or from laminations, epoxy tools are poor heat conductors. Several methods are used to improve heat transfer. Aluminum powder or iron filings can be incorporated directly in the resin. Copper and aluminum screening have been used as laminations with some success. More direct approaches involve placing copper tubing near the laminating surface to conduct heating fluids or steam, or to imbed electrical heating units at strategic locations within the laminate.

Epoxy tools can be made lighter in weight if reinforced in such a manner as to provide support and prevent dimensional errors. Pipe, cardboard tubes, wood structures, and laminates can be attached to provide rigidity and a means of handling large tools.

Phenolic resins

Phenolic resins are used for tooling in laminating operations where lower impact strength and heat resistance can be sacrificed for lower cost. Phenolic resins are usually cast to shape rather than used in conjunction with glass reinforcements.

Thermoplastics

Flexible materials such as polyvinyl chloride, polyurethane, silicone rubber, and polysulfide are used to reproduce intricate surface detail. These elastomeric plastics are supported by plaster, cast or laminated plastic, or wood framing.

11-10   Installing electrical heating tapes within a laminated plastic tool. (*Reprinted by permission of Briscoe Manufacturing Co.*)

Thermoplastic sheets of acrylic, polyethylene, or other plastics can be fabricated into simple shapes and used as tools for laminating.

Curved, hemispherical, and other shapes are formed from heat-softened sheets to provide tooling for contact, spray-up, or vacuum-bag processes. Flat sheets can be fabricated into box structures or other simple shapes.

GLOSSARY

Chemical milling   Process utilizing acids to erode metal according to a predetermined pattern.

Electrical discharge machining   Removal of metal by atomic disintegration through the use of electrodes that follow a prearranged pattern.

Electroforming   Electrodeposition of metal until adequate thickness is obtained.

Polyimide   Newer class of thermosetting resins produced from BDTA (benzophenone tetra-carboxylic dianhydride) and an aromatic diamine. Applications include glass, boron, and graphite-reinforced laminates, foams, and adhesives useful up to 700°F.

Syntactic   A lightweight structure made up of hollow particles surrounded by a solid binder.

Vermiculite   A porous magnesium silicate.

REVIEW QUESTIONS

1   How important is toolmaking to the nation's economy?
2   Define tooling as it applies to the plastics industry.
3   What is a class A tool?
4   Is plaster considered a temporary or a permanent tool?
5   List several factors to be considered in selecting the type of tooling to be used for laminating.
6   What temperature is recommended for removing moisture from plaster?
7   What is water-soluble plaster used for?
8   What is a syntactic foam?
9   List several methods of providing better heat conductance in a tool.
10   How are phenolic resins used to produce tools for the laminating process?

REINFORCED-PLASTICS PROCEDURES

12

The term *reinforced plastics*, or composites, indicates that a resin is structurally or otherwise upgraded by incorporating fibers in a resin matrix. Laminates in which fibrous felts or woven goods are superimposed in either flat or contoured design belong in this category. Also included are processes in which chopped strands are laid down in jackstraw arrangement or in which continuous filaments are wound in a predetermined pattern or joined in parallel configurations. Compression and injection molding also play an important role in mass production of many reinforced plastics.

The segment of the industry directly involved in this specialized work continues

12-1 Calculator housing injection-molded from glass-reinforced polycarbonate. (*Reprinted by permission of Liquid Nitrogen Processing Corp.*)

187

to make unprecedented advancement into large-volume markets such as aerospace, transportation, storage, housing, recreation, marine, and chemical industries.

Reinforced plastics compare favorably with metals and wood for many reasons:

1 *High strength-to-weight ratio*:   Airborne components, tank trucks, and cargo containers reduce operating costs by permitting greater payloads.

2 *Greater design latitude*:   Compound curves or angles difficult to produce in metal can be more easily achieved. Extra thickness needed to strengthen certain sections can be incorporated during production.

3 *Unitized construction*:   Fitting and assembly of multiple components are eliminated by one-piece construction. Rivets, bolts, and welds are unnecessary.

4 *Ease of fabrication*:   Extremely large products can be made on comparatively inexpensive equipment. Tooling as simple as cardboard forms or wire mesh can support a laminate during cure. Low-cost plaster or plastic tools are commonly used.

5 *Ease of installation*:   Large structures can be airlifted into position and usually require lighter supporting members.

6 *Ease of repair*:   Most fractures can be repaired "on the spot" using room-temperature curing resins and a fibrous patch.

Although the choice of resin system contributes greatly to the performance of a reinforced-plastic composite, the fiber portion improves serviceability under a variety of environmental conditions:

1 *Corrosion and chemical resistance*:   Saltwater, acids, and alkalies often in a heated state are successfully combated by reinforced plastics used for piping, tanks, and handling equipment.

2 *Weather resistance*:   Reinforcing fibers enable corrugated sheet manufacturers to guarantee panels for as long as 20 years in outdoor use. Inert glass fibers minimize the tendency of resins to shrink and craze.

3 *Electrical resistance*:   Insulation values are improved when glass fibers are added.

4 *Thermal conductivity*:   The transfer of heat can be slowed or accelerated, depending on type of reinforcement. Asbestos improves thermal insulation, and aluminum screen increases conduction rate.

5 *Fatigue resistance*:   Orienting fibers in the direction of known applied stresses makes possible a number of products such as diving boards, archery bows, vaulting poles, fishing rods, and ladders.

Producing reinforced plastic also has these drawbacks:

1 Material costs are higher than comparable metal parts. Scrap factors such as trim and unused catalyzed resin must be considered.

2   Labor cost per pound of laminate is higher when hand lay-ups are made.

3   Thin sections are not as stiff as comparable metal thicknesses.

4   Product uniformity is difficult with some of the laminating procedures.

When the reinforced-plastics division of the industry is compared with other processors, it is evident that a sound knowledge of resin systems, fibers, fabrics, fillers, and catalysts, and the diverse ways these dissimilar materials can be combined is the key to success. A review of the more-automated techniques of injection, compression, extrusion, and blow molding shows that the emphasis there is more on rapid machine cycling than on materials technology. Even though some phases of reinforced plastics are suitable for mass production as, for example, continuous laminating and compression molding, most operations are time consuming and require expensive finishing steps. To complete an item may require days, weeks, and sometimes even months.

Because 90 percent of reinforced plastics involve glass roving, mat, or fabric and 85 percent of glass-resin laminates use polyester resins, we shall concern ourselves primarily with the procedures for making and evaluating fiberglass-reinforced plastic (FRP). Since phenolics, epoxies, silicones, and polyimides also find substantial application in aerospace, electrical, or chemical fields, procedures for working with these resins are included.

For convenience, FRP production techniques can be classified according to the method used.

12-2   Fiberglass-reinforced stack, 285 ft high, produced in 3 sections. Only 7 ft in diameter at the base, it tapers to a 5-ft diameter at the top. (*Reprinted by permission of Harshaw Chemical Co.*)

CONTACT METHODS

Contact laminating is also referred to as *wet lay-up*. This method is popular with the home craftsman but is also widely used in many industrial plants. Fabric or mat is saturated with liquid resin, and the lay-up is made by building layer upon layer to obtain thickness. If pressure is applied at all, it consists of rolling or wiping with a squeegee to remove trapped air.

The advantages of this method are:

1   No limit to size or shape of product.

2   Tooling can consist of any material that will hold its shape.

3   Tooling can be changed easily during experimental phases.

4   Investment in pressure devices such as a press, autoclave, or vacuum pump is not required.

5   Curing ovens are not needed.

6   Semiskilled workers can be readily trained.

The limitations are:

1   Only addition-type resins (polyester, allylic, and epoxy) can be used since condensation types require some form of pressure to avoid porous, poorly laminated structures.

2   Product uniformity within a single part, or from one part to another, is difficult to maintain. In all laminates, the relation between resin and fiber content is critical in meeting weight and design standards. Extremely high resin contents add cost and weight and reduce strength-weight ratios. If too low, fractures occur more readily, water absorption increases, and electrical and mechanical values diminish.

3   Waste factor is high. An excess of resin must be used to ensure good wetting and aid in flushing out entrapped air during rub-out.

4   Tight-weave fabrics and woven rovings are difficult to saturate with high-viscosity resins. Individual central fibers might be left "dry."

5   Draining from vertical walls is a problem, creating puddles.

LOW-PRESSURE METHODS

Pressures from a few pounds to 400 lb (per square inch of laminate) are considered low. Various techniques are used to develop pressure, depending on the size and shape of the product, quantity required, equipment available, and quality sought.

Lagging

Cylindrical objects such as fishing rods, air ducts, tubing, and elbows can be made by wrapping presaturated material over a solid removable mandrel or around a break-

12-3   Hollow plaster mandrel, lagged duct, and completed item. Polyester-glass-cloth mold is used to cast the mandrel. (*Reprinted by permission of Los Angeles Trade-Technical College.*)

away hollow core. When overwrapped tightly with cellophane, nylon, or other narrow tape, a light pressure is developed as the tape shrinks during cure. An oven cure at 200 to 300°F is required to shrink the tape. Tape is removed by wire wheel or grinding, followed by a lacquer coat to improve roughened surfaces.

Vacuum bag

The application of a vacuum to an FRP part serves to improve quality by removing air from within the lay-up before and during cure. When used with condensation-type resins, vacuum draws off gaseous by-products as they form, reducing the danger of delamination. The vacuum ensures that each layer or loose fiber is pressed tightly into place. As the air is evacuated from within the sealed assembly, atmospheric pressure exerts an external force evenly over the laminate. Most aircraft specifications require a minimum vacuum of 25 in. of mercury, although acceptable quality can be realized at as low as 20 in.[1]

To calculate the actual laminating pressure per square inch, assume a vacuum of 25 in. is drawn. Then

$$\frac{25 \text{ (vacuum-gauge reading)}}{30 \text{ (perfect vacuum or 15 psi)}}$$

$$\frac{25}{30} \times \frac{15}{1} = \frac{5}{\underset{2}{\cancel{6}}} \times \frac{\overset{5}{\cancel{15}}}{1} = \frac{25}{2} = 12.5\text{-psi laminating pressure}$$

The advantages of the vacuum-bag method are that:

1   One-piece tooling can be used, either male or female. The choice, as in wet lay-up, is governed by which surface is to be visible or present a smooth surface. For example, boats are made in a female tool, shower stalls on a male form.

2   There is virtually no limit to the size of part that can be bagged.

[1]Atmospheric pressure at sea level is stated as approximately 15 psi (actually 14.7). A column of mercury can be sustained by vacuum in a tube to a height of 30 in. (29.92) at sea level. In laminating practices, a perfect vacuum is impractical to obtain and unnecessary.

3   Void-free laminates with a high degree of transparency are possible.

4   The method applies to all resin systems.

5   Resin content can be held to reasonably close tolerance through developed skill in impregnation and rub-out of wet resins.

6   Extremely good uniformity can be obtained when used with prepreg systems.

    The limitations of this method are:

1   It is expensive because of time involved in making and testing vacuum bags. Bags are usually discarded after each cure.

2   Resins requiring heat cure limit the part size to the physical size of the oven.

3   A minor leak in the bag during cure results in poor quality or rejected part.

4   It is difficult to overcome "bridging" of the bag when small depressions, ridges, or channels are part of the design. This can be improved by using shaped wood or plastic plugs underneath the bag.

Pressure bag

Inflatable bags of silicone or other synthetic rubber sheeting are used for products that can be made in an uncomplicated female mold. Either flat rubber sheets or semitailored bags attached to a backup plate are used.

    A modification of the single-sheet pressure bag involves using a preinflated bag. Plies are laid up on the pressurized bag, after which a split mold of cast or fabricated aluminum is clamped in place. Molds can be heated by cast-in-place steam channels or externally in an oven. Marine slips and flotation docks are produced by this method. Unless a one-piece, hollow construction is required, items of this nature are more easily produced by vacuum bagging a laminate over a polyurethane or other foam core.

Vacuum impregnation

Both surfaces are smooth when this system is used, unlike wet lay-up or vacuum-bag parts. Trapped air pockets, wrinkles, and dislocation of reinforcement present problems.

12-4   Method of vacuum bagging a flat panel. Spring serves as a "bleeder" to prevent bag seal-off. (*Reprinted by permission of Los Angeles Trade-Technical College.*)

12-5 Electrically heated, water-cooled autoclaves used to fabricate aircraft assemblies. Two are 8 ft in diameter; one is 10 ft. (*Reprinted by permission pf J. P. Devine Manufacturing Co.*)

Swimming pools, boats, and similar large shapes are produced by using a dry lay-up between fitted mold halves. Low-viscosity polyesters are catalyzed to cure slowly at room temperature or in ovens.

Autoclave

Autoclaves are steel pressure vessels ranging to 20 ft or more in diameter and to 40 ft in length. Lay-ups are enclosed in special rubber bags. After evacuation of air, steam, or heated air pressure to 125 psi, exerts both heat and pressure to cure the part.

Hydroclave

Similar to an autoclave, a hydroclave uses heated fluids pressurized to 1,000 psi. As pressures increase, laminate density and interlaminar bond strength improve. Particularly useful with phenolic resins, high-reliability items such as rocket nozzles, heat shields, and reentry nose cones are routinely produced by hydroclaving.

Pultrusion

Strands of roving, drawn continuously from creels, are coated in a resin bath and pulled through a shaped orifice. Emerging from the die as solid or hollow geometric shapes, pultruded articles form the basis for such items as ladder rungs and rails, cable supports, channels, and I beams.

Filament or tape winding

The basic filament-winding process utilizes a machine that rotates a solid mandrel or collapsible core segments as resin-saturated roving or slit fabric tape is applied circumferentially or longitudinally. The tension of wrapping is generally sufficient to create good interlaminar adhesion of polyesters and epoxies. Additional pressure furnished by compression molding or autoclave is used for large rocket nozzles and similar products made with phenolic resins.

Pipe, race-track rails, missile fuel cells, and storage tanks are representative of filament-wound structures. Hollow cubical shapes also can be produced.

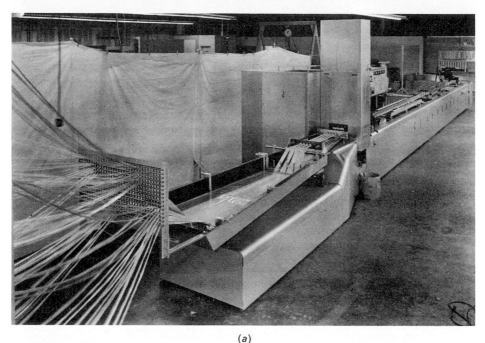

(*a*)

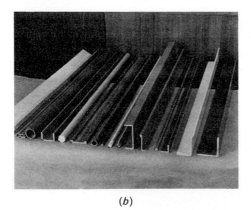

(*b*)

(*c*)

12-6 (*a*) Glastruder pultrusion machine. Roving, mat, fabric, or combinations are fed into a resin bath, then proceed to forming dies and high-frequency curing section. (*b*) and (*c*) Examples of various cross sections produced by this method. (*Reprinted by permission of Goldsworthy Engineering, Inc.*)

Centrifugal casting

Limited to round sections, centrifugal casting is used to produce tank sections and pipes of various diameters. Filament winding is more widely used where highest strength-weight ratios or high burst strengths are specified. The advantage of cast pipe is in the resin-rich inner surfaces obtained as centrifugal force pushes the heavier glass fibers to the outer diameter. Glass mat or woven roving is placed in the tubular mold, and resin is applied by means of a reciprocating dispensing unit. Curing can be effected at room or moderate temperatures.

Spray-up

The spray-up operation is actually a wet lay-up technique, using polyester resins and room-temperature catalysts. The difference is in the method of depositing glass fibers and liquid resin, compared with the use of mat or woven fabrics. The advantages inherent in using spray-up systems are the complete wetting of reinforcement and the rapidity of production cycles. The hazards lie in the fact that the total output of a company as to quality and waste rests in the hands of the equipment operator. Control of part thickness over an extended production run is entirely dependent upon the skill of the person applying the material.

Special "gun roving" produced to cut cleanly, wet rapidly, and conform to intricate contours is preferred. Specially formulated polyesters with minimum drain-off and fast gel times have been developed for spraying.

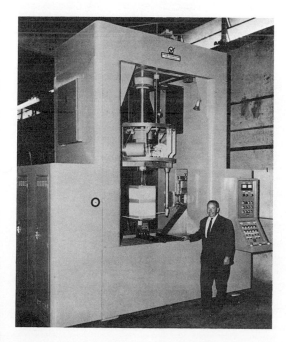

12-7 Filament-winding machine for producing cubical shapes. (*Reprinted by permission of Goldsworthy Engineering, Inc.*)

12-8 Chopper gun cuts glass roving, mixes polyester and catalyst, and coats cut fibers prior to impinging on the mold. Air is eliminated by rolling with metal disk rollers. (*Reprinted by permission of Los Angeles-Trade Technical College.*)

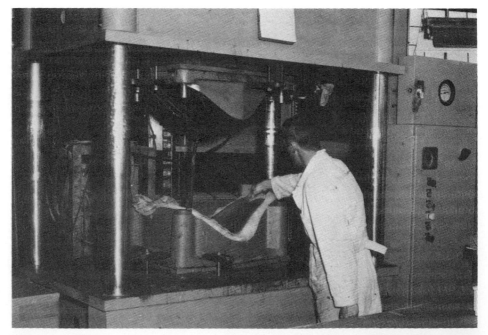

12-9 Compression molding a chair, using glass mat and liquid polyester resin. (*Reprinted by permission of Dake Corp.*)

HOT-PRESS MOLDING

The ultimate in production rate of FRP is realized by use of compression presses and matched metal molds. It is difficult to state in general terms when the expense of a matched metal mold is justified over the choice of hand lay-up, vacuum-bag, or spray methods. For simple shapes such as box, channel, or dished sections, inexpensive aluminum, cast iron, or kirksite molds are feasible and may be amortized in a run of less than 100 parts. On the other hand, a cast or machined tool steel mold may not be economically practical unless thousands of units can be sold.

When the expense of a first-class steel mold is justified, parts can be made by one of several methods, depending on other factors.

Premix

Basically, premix molding is a version of compression molding, but with sufficient difference to set it apart from conventional methods. The distinction between the two methods lies in the curing parameters and the materials (see Tables 12-1 and 12-2). Conventional molding of phenolic, amino, and other compounds in dry powder form requires molding pressures to 12,000 psi and temperatures from 275 to 360°F. Premix compounds are molded at 50 to 2,000 psi, curing at 180 to 300°F. Premix based on polyester and allylic resins is available from commercial suppliers but is more often formulated by the molder to suit his own price and end-use requirements and to reduce inventory. Only sufficient compound is batched to serve the molding department for short periods of 1 day to several weeks, depending on the stability of the resin-catalyst system.

A molder may formulate his own proprietary compounds when he:

1   Possesses the technical know-how to successfully concoct special recipes

12-10   Loading premix into compression mold. (*Reprinted by permission of Haveg Corp.*)

2   Is willing to invest in continuing research to evaluate new resin blends, catalysts, fillers, and reinforcements

3   Can invest in mixing, blending, and specialized storage facilities

4   Has adequate quality control, testing equipment, and skilled personnel

Reinforced products molded from premix often solve design or application problems that prove difficult in compression molding. Extremely thick sections or unusually large area parts are molded routinely.

Rapid cycles are made possible through choice of catalyst and promoter systems. One limitation occurs in molding thin wall (less than one-eighth in.) sections in a deep-draw or complex shape. Unless fibers are short, they may interfere with uniform flow and accumulate in one area at the expense of another.

TABLE 12-1   Typical premix formulations, recipe 1

Material	Weight (parts)
Polyester resin, promoted, low viscosity	15.00
Monomer, styrene, or vinyl toluene	0.35
Catalyst, benzoyl, lauroyl peroxide, or other	0.15
Filler, clay, talc, calcium carbonate	22.35
Pigment, titanium, iron oxide, or other	1.50
Release agent, aluminum or zinc stearate; Zelec*	0.15
Reinforcing strands, one-eighth to three-fourths in., glass, nylon, sisal	10.50
Total	50.00

*Zelec is a Du Pont product.

Based on this formula, glass content is 26.6 percent. For lighter fibers, 10.5 lb is volumetrically difficult to work mechanically into the filled resin. First, catalyst is dissolved in monomer for easier dispersion, after which other ingredients are blended in order. Intensive mixing should be avoided in order to lessen breaking glass fibers.

TABLE 12-2   Typical premix formulations, recipe 2

Material	Weight (parts)
Diallyl phthalate powder, prepolymer	62
Diallyl orthophthalate monomer	38
Tertiary butyl perbenzoate, catalyst	2
Release agent, stearates, or other	1
Filler, slate, marble dust, or clay	to 300
Pigment	to 5
Reinforcing strands, glass, nylon	to 30

Some formulators dissolve the DAP powder in acetone to reduce resin viscosity and effect higher filler and fiber loading. This can cause difficulty when molded, unless solvent is completely removed by air drying or under vacuum.

In the recipes shown in Tables 12-1 and 12-2, a high filler content with no fibers results in a molding putty. As fibers are added, filler proportions must be reduced.

DAP formulations exhibit long-term stability at room temperature, and this is extended by refrigerated storage. In working with any premix previously refrigerated, allow sealed containers to reach ambient temperature before materials are exposed to prevent condensation, which can cause blisters and porosity during molding. The higher cost of these resins increases premix costs by as much as three times that of compounds made with polyesters, which are produced for an average of 35 cents/lb. Premium products that require heat resistance to 350°F, superior chemical resistance, or specific electrical performance are molded from DAP formulations.

Preform

To ensure that fibers are located correctly within a molded part, a dry preform that conforms closely to the final molded shape must first be made. Several methods may be used in making preforms.

Directed-fiber method   A manual operation, shown in Fig. 12-11, is necessary for preforms such as boats, or housings too large to be produced automatically. Water-soluble binders hold fibers in place and prevent "washing." Their selection is based

12-11   Fiberglass preform for Corvette underbody being built by blowing chopped roving against a perforated screen fixed to a turntable. (*Reprinted by permission of Automotive Products Div., North American Rockwell.*)

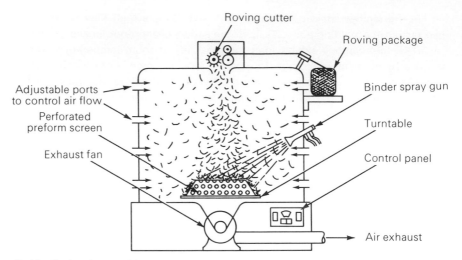

12-12   Preforming machine.

on several factors: (1) starch—low cost; (2) polyester—weather resistance and com-patibility; (3) amino resins—highlight glass fibers for prominent fiber pattern; and (4) acrylic—weather resistance.

Preforms are weighed for controlled glass content in the molded part.

Plenum chamber   Several versions of automatic preformers are used. The single-station type is useful for shapes such as chairs, or it can accommodate multiple screens for small parts.

A more expensive, multiple-station, unit rotates one set of screens into the plenum chamber while a duplicate set is in the binder curing section, and a third is removed.

When a preform arrives at the press, the press operator inspects, weighs, and makes minor fiber adjustments, then fits it over the male plug before pouring catalyzed resin over it.

Water-slurry method   Adapted from methods used to produce cellulose-pulp pre-forms later pressed into egg cartons and similar disposable paper items, this technique now serves the plastics industry.

Cut patterns   When cloth or mat is specified, patterns can be die-cut or patterned with a textile cutter. When required, glass or nylon yarn is used to stitch individual segments. Broad goods are usually preimpregnated for controlled flow and resin content. Combinations of mat and fabric plus design of thick and thin sections are made possible. For example, decorative effects such as leaves, butterflies, and metallic yarn may be incorporated.

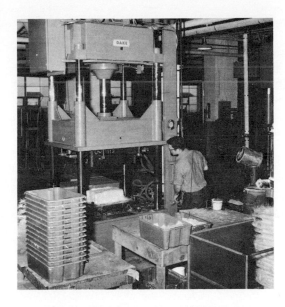

12-13 Molding tote boxes on a 75-ton press. Note preforms, completed boxes, and measuring cup for liquid polyester resin. (*Reprinted by permission of Dake Corp.*)

Polyesters predominate as the resin most widely used, but industrial items such as helmet liners are produced from phenolic or epoxy *B*-stage fabrics. Laminating resins are, by nature, very adhesive. Great care must be exercised to prevent the cured part from sticking to a mold or tool. Because they exhibit the highest adhesion

12-14 Chair preform produced by water-slurry method. (*Reprinted by permission of Cincinnati Milacron Co.*)

plus lowest shrinkage, epoxy systems are most prone to cause sticking, although all systems are troublesome.

To effect total release, parting agents (also called release agents, mold lubricants, or mold release) are required. Most parting agents are proprietary compounds and are sold under many brand names. They are based on wax, polyvinyl chloride, polyvinyl alcohol, silicones, or fluoroplastics. Application is by wiping or spraying. As a general rule, a parting agent should be pretested for compatibility with specific resins before it is used in production. A particular type may work successfully with one resin but this does not ensure similar success with another. When applied to porous tool surfaces such as plaster, a parting agent in liquid or aerosol form should be preceded with a pore sealer such as shellac.

The foregoing methods indicate the scope of the reinforced-plastics field. Each process represents an area of technology that requires specialized skills and equipment. The ambitious student desiring to further his learning in particular segments is directed to the references in Appendix B.

To develop an appreciation for the handling characteristics of FRP, several of the simpler procedures are given in detail on the following pages to serve as a learning process. Starting with a flat laminate, the same steps can be followed in covering objects such as wooden boats, polyurethane foam cores for surfboards or flotation items, or fabricating contoured structures.

SPECIFIC PROCEDURES

Making a wet lay-up

Materials

Plywood or wood, 12 × 12 × 0.25 in., or heavy cardboard: 1 piece

Cellophane, polyethylene, or vinyl film, 12 × 12 in.: 2 pieces

Glass cloth, plain open weave, 7.50 or 10 oz, 8 × 8 in.: 6 pieces

Polyester laminating resin, promoted: 1 pt

Methyl ethyl ketone peroxide (MEKP) catalyst: 1 oz

Acetone for cleanup: 1 pt

Equipment and Tools

1 paint brush, 2 in., natural bristle

1-pt paper cup or small metal container

Stir sticks and razor-type knife

Squeegee (rubber, polyethylene, or teflon). *Note*: Discarded windshield wiper or window squeegee is satisfactory.

Masking or cellophane tape, or tacks

Gram scale, eyedropper, paint roller or disk roller, all optional

Procedure: Work on a flat surface in a well-ventilated area. Cover the wood base with cellophane to provide release, and tape or tack down.

1   Estimate the amount of resin to be catalyzed to obtain a resin content of 50 to 55 percent in the lay-up.

Sample problem: Six pieces of 10-oz glass fabric weigh 3 oz. If 3 oz of resin is worked into the plies with no run-off, the laminate would have a 50-percent resin content. To allow for waste, plus an excess for rub-out, increase resin requirement to 4 oz. A typical laminating resin weighs 9.4 lb/gal, therefore 4 oz is 0.25 pt, using the volumetric method. At normal room temperature (75 to 80°F), the percentage of MEKP to allow a working time of 20 min is 1 percent. If a gram scale can be used, 1 percent will be 1.10 g.

Calculation: 28.4   (grams per ounce)

   4   (resin quantity, ounces)

113.6   (resin quantity, grams)

0.01   (percent catalyst)

1.136 (grams, catalyst)

It is advisable to use the lower 1.1 figure rather than 1.2 if scale accuracy is difficult.

Using the alternate eyedropper method, a standard medicine dropper meters 30 drops per gram of MEKP, which is readily converted.

Calculation: 1.136 (grams MEKP at 1 percent, as above)

   30 (drops per gram)

34.080 (34 drops for 113.6 g of resin)

2   Add catalyst to resin. Stir *thoroughly* for several minutes.

3   Pour or brush a thin layer of resin on the release film to cover an 8 × 8-in. area.

4   Lay down the first piece of fabric, and add a small amount of resin to the center. Work the resin into the weave, brushing or wiping from center to corner and edges. The whiteness of the glass changes to translucent green-blue as air is eliminated and wetting occurs.

5   Complete the laminate with remaining plies by repeating step 4. Daubing the resin sharply with brush bristle ends opens the weave to purge air better than using "painting strokes." If roller or squeegee is used, employ a light touch to avoid excess resin loss or cloth distortion.

6   Place the remaining piece of cellophane or film on top of the last ply and work out any remaining air. Leave the film on until resin cures to prevent oxygen absorption, which makes the surface tacky.

7   Clean tools in acetone. Wash hands in soapy water and apply hand cream if required. *Note*: Disposable plastic gloves should be worn if extensive exposure

is contemplated for future work. Some persons are allergic to certain polyesters, which precludes handling these chemicals. Catalysts should be handled with respect. If accidentally splashed in the eyes, flush with copious amounts of cold water and consult a physician.

8   Check the lay-up for gel (rubbery, unworkable semisolid) every few minutes by probing the squeezed-out resin. Check for exotherm (heat generation) after gelation by placing the back of the hand against the top.

9   When resin has gelled, prior to final hardening, trim all edges to a 6 × 6-in. laminate by cutting through the top protective film and all plies. This is the best time to trim any wet lay-up since the resin is strong enough to maintain the structure, yet soft enough to cut easily. When full cure develops, the laminate must be sawed, routed, or ground to final size.

10   Most laminates can be handled, but not machined, within a few hours. Test cure rate with a thumb nail ; a hardened resin cannot be indented. A resin may require up to seven days to polymerize fully.

11   Remove the cured laminate from the board and peel off the release films. Sticking indicates undercure, which can be improved by heating at 150 to 180°F.

12   Examine the laminate for voids, starved or resin-rich areas, waviness, delamination, and overall uniformity of thickness and appearance. White areas indicate air caused by poor wetting. Tap the laminate with a coin. A sharp sound indicates a good cure, although it is only an indication.

Decorative laminates using printed fabrics, woven burlap, metallic cloth, decals, leaves, and ferns are made by the same process. High production, of course, dictates that hot-press methods be used.

For hand lay-up, when a decorative surface is applied to one side only, it is applied first face down, then backed up with fabric, mat, or woven roving. This ensures a smooth finish on the decorative side. Printed fabrics should be checked for color fade caused by the bleaching action of peroxide catalysts, or dye smear if soluble in styrene monomer.

In working with mat, calculate resin content for 65 to 75 percent to fill voids. For sandwich composites using alternate mat and fabric layers, select an open-weave fabric and rely on resin in the mat to work through the cloth, flushing air out. Using fabric prevents loose mat strands from sticking to roller or squeegee.

Repair of laminates

Expensive laminates made incorrectly or damaged in use can usually be repaired by wet lay-up. Voids can be filled, or a small delaminated section improved, by drilling small holes into the area and injecting catalyzed resin using a disposable syringe. Fractures and tears require clamping of the sections after applying polyester resin, often accompanied by a cloth or mat patch on one or both sides to tie the joint. Major damage requires major "surgery."

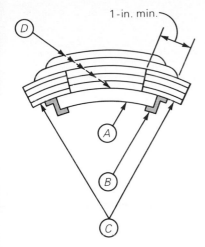

1-in. min.

12-15 Method of repairing laminated or other structure. (*a*) Backup; metal screen, expanded metal, cured laminate. Use release film on inner side. (*b*) Tape. Screws, clamps, or adhesive can be used also. (*c*) Cross section of structure to be repaired. Note tapered or feathered edge of cutout. (*d*) Successfully larger patterns cut to fit.

Procedure :

1 Cut away the entire damaged section with a saber saw, grinder, or other means.

2 Remove any paint and surface resin from both faces of the laminate for at least an inch surrounding the cutout. Try to expose raw fabric or mat fibers to get rough surfaces for bonding.

3 Wipe away all dust. Use solvent (kerosene or acetone) to remove oil or grease.

4 Cut cloth or mat patterns in successively larger patterns, as shown in Fig. 12-15.

5 Catalyze resin as for wet lay-up. Either a polyester or an epoxy can be used with the appropriate catalyst. The repair procedure also applies to secondary bonding of components. An alternate method for large or thick areas is to use a prepared block or sheets of cured laminate, as a core, surrounded by wet lay-up material.

Making a vacuum-bag laminate

Generally room-temperature catalysts are not practical because of the time involved in making and checking a bag assembly unless parts are small enough to fit into a complete envelope. It is also possible to resort to sunlight catalysts.

Materials

Polyvinyl alcohol (PVA), nylon, or Mylar[1] film : 24 × 24 in. × 0.003–0.006 in.

Paste wax, high carnauba type

Zinc chromate sealing tape : 1/2 in. wide × 8 ft

Bleeder strip (glass mat, jute, canvas) or a large spring : $\frac{1}{2}$ in. wide × 8 ft

[1] Mylar is a Du Pont polyester film.

Polyester laminating resin, promoted : 1 pt

Tertiary butyl perbenzoate (TBP) catalyst : 1 oz

Polyester gel coat resin, pigmented (optional) : 2 oz

Methyl ethyl ketone peroxide (MEKP) catalyst : 1 oz

Pigment, oil-base paste type (optional) : small tube

Fiber glass surfacing mat, 12 × 12 in. × 0.005 : 1 piece

Fiber glass bonded mat, 2 oz, 12 × 12 in. : 1 piece

Woven roving, 12 × 12 in. : 1 piece

Acetone for cleanup : 1 pt

Note : 7.50- or 10-oz glass cloth, or any other may be substituted for all or part

Equipment and Tools

Metal plate, sheet laminate, or wood if well sealed, 24 × 24 in. : 1 piece

Vacuum pump with gauge and resin trap

Air circulating oven or heat lamp

Gram scale, optional

Rubber hose, noncollapsible : 5-ft minimum

Paper cups, stir stick, squeegee, brush, eyedropper, knife, scissors

Procedure :

1   Wax center portion, 12 × 12 in., of metal plate, and buff smooth.

2   Catalyze 2 oz of gel coat at 2% MEKP, or add desired amount (usually 1 to 5 percent) of paste pigment to standard polyester laminating resin.

3   Brush a generous coat on the waxed portion. Brush in one direction, then across, using additional resin if required, and allow to gel.

4   Weigh all glass as a unit. Calculate resin content for 60 percent. If fabric is used in place of mat, reduce resin to 40 or 50 percent.

Sample problem :   Assume surfacing mat, 2-oz mat, and woven roving weigh a total of 150 g, or 5.25 oz.

Calculation :

$$150 : 40 = X : 60$$

Glass weight : percent glass = resin weight : percent resin

$$40 X = 60 × 150$$

$$40 X = 9,000$$

$$X = \frac{900}{4}$$

$$X = 225 \text{ g resin at 60 percent}$$

5   Allowing approximately 5 percent for waste, catalyze 235 to 240 g of laminating resin with 2 percent TBP. Stir *thoroughly* for several minutes.

6   Place surfacing mat directly on the gel coat and saturate with resin. Brush easily so as not to disturb application.

7   Repeat with heavier mat and woven roving. *Do not rub out!*

8   Place bleeder strips or the spring around and touching the edges of the lay-up.

9   Press chromate sealant down firmly on the unwaxed plate surface at the outside edge of bleeder, as shown in Fig. 12-4.

10 Press PVA film smoothly along one sealant strip and work it down the remaining three strips toward a corner, leaving an opening for the vacuum-hose inlet.

11 Preparation of vacuum hose:
    (*a*)   Tape a piece of mat or jute across the hose end to prevent resin sucking in or blocking the inlet with PVA film.
    (*b*)   Wrap a piece of sealant around the hose a short distance from inlet end that will seal against the sealing strip.
    (*c*)   Position hose so it draws air through a bleeder strip and seal tightly.
    (*d*)   Turn on the vacuum and note draw-down. As in wet lay-up, whiteness of fibers disappears as air is purged and fibers are wet. If vacuum holds at least 20 in., it should be difficult to raise film from the lay-up.

12 Wipe a little grease, oil, or wax on the bag and squeegee or roll the laminate, pushing air and excess resin into the bleeders. In commercial practice, special bleeder fabrics are sometimes used as the top ply. They are silicone-coated glass fabrics that peel off easily.

13 When resin has been rubbed out satisfactorily, disconnect vacuum hose from the pump inlet or the resin trap. Try to prevent air from traveling back through the hose into the laminate.

14 Place lay-up and bag assembly in the oven for 1 to 2 h at 260 to 280°F, and reconnect vacuum. Recheck suction for leaks. After laminate warms from 150 to 175°F, additional rub-out can be undertaken in the oven to eliminate more resin and to obtain higher glass ratio in the laminate.

15 Laminates can be removed hot when cured. A good way to prevent warping in an unbalanced construction or in complex design is to cool the laminate while it is still under vacuum.

16 Remove the bag, clean off sealant from the plate with acetone, remove the laminate, and saw off the bleeder (which is firmly bonded to the panel).

17 Inspection:
    (*a*)   White spots indicate a bag leak or excessive rub-out of resin.
    (*b*)   Wrinkles are created by too much pressure in rubbing out, caused by underlying layers "traveling," particularly in an all-fabric lay-up.
    (*c*)   Undercure (soft, tacky) laminate indicates miscalculation of catalyst content, weak or overage catalyst, low oven temperature, or insufficient time.

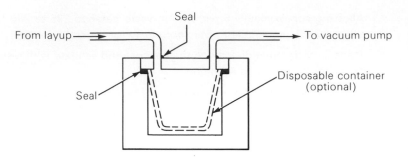

12-16   Resin trap.

18   Protecting a vacuum pump : In making large or thick laminates, some resin may
be accidentally drawn into the vacuum pump. To prevent this, resin traps are
placed in the line between the lay-up and the pump. The trap may be any small
metal container or a fairly large (5- to 30-gal) drum. If necessary, cured polyester
can be dissolved from pump components by using one of several proprietary
chemicals.[1] Hoses clogged with hardened polyester may be bent sharply to crack
the resin and then blown clean.

Bagging operations are widely used in aircraft production to obtain quality
laminates in limited quantities. Both male and female tooling are used, and, for large
assemblies, vacuum-bag films are heat sealed to obtain wide widths and provide
one-piece constructions. Skill is required to tailor a bag to conform to complex tools ;
the greatest difficulty is preventing leaks at corners.

Laminating with phenolic resins

Phenolics used in laminating consist of solid resins that must be dissolved in solvents
such as alcohol in order to apply them to the reinforcement. The solvent must be
completely evaporated from the lay-up to permit polymerization. Resin content, residual solvent, and laminate quality are virtually impossible to control by hand impregnation ; therefore wet lay-up with phenolics is not practiced. Instead, laminators
purchase preimpregnated materials (*prepregs*) from specialists (*prepreggers*). The
term prepreg is taken from the process of preimpregnating or coating fabrics, mats,
rovings, and other forms of fibrous webs.

In accordance with the user's requirements, the amounts of resin, volatile
(solvent) percentage, and flow are closely controlled, as is the degree of resin advancement (*B* staging). Dryness (or tackiness) and drape (flexibility) of the material
also are governed by these controls.

Prepreg fabrics are cut to pattern, laid up by hand, and pressurized by lagging,
vacuum-bagging, hydroclave, autoclave, and press methods, depending on the tooling and the quality of laminate desired.

The type of phenolic prepreg is chosen according to the laminating method

[1] One such chemical (solvent) is DeSolv manufactured by Ram Chemicals.

used and according to the shape of the part. For example, laminating flat sheets in a press requires a prepreg that is dry and boardlike to facilitate cutting, stacking, and pressing. The user may call for a fabric with a resin content of 40 to 45 percent, a volatile percentage of 0.5 to 1.5 percent, and a flow (resin loss under pressure) of 2 to 5 percent.

In another example, laminating a complicated duct requires a prepreg with drape and tack to facilitate wrapping around the tool. This is accomplished by increasing the volatile from 8 to 15 percent or more and the flow from 10 to 15 percent while keeping the resin content at the same 40 to 45 percent. As laminating pressures increase, e.g., vacuum-bagging to press methods (in other words, from 15 to 2,000 psi), the necessity for resin flow and volatiles decreases.

Regardless of the method used to pressurize the laminate, the polymerization temperature is similar to that used for compression molding of phenolic resins: 300 to 350°F, with undercure, porosity, blisters, and delamination the most common problems.

Laminating with polyimide resins

Polyimide resins are relatively newer types of thermosetting plastics. They have been developed for laminates, coatings, and foams that must retain a high percentage of their initial strengths at temperatures to 700°F. These resins are used in prepreg form applied to glass cloths, boron, and graphite fabrics. Laminates are prepared by methods practiced with phenolic materials. The major handling differences are:

1 Higher vacuum is required (25 to 29 in. of mercury).

2 Vacuum bags are made of heat-resistant films such as Mylar or silicone rubber.

3 Curing cycles for laminates require a rapid heat build-up from 350 to 400°F, followed by a postcure schedule of 16 or more hours beginning at 390°F and ending at 600°F, minimum, or 700°F if the laminate is to be used at that temperature.

4 Some toxicity hazards exist, particularly if DMF (dimethyl formamide) is used as the solvent for prepregging. Extremely good ventilation is required, and goggles, gloves, and caution are recommended.

Laminating with silicone resins

Rigid silicone resins are laminated by the same techniques as those used for phenolics, except that higher curing temperatures (to 375°F) are maintained. As with phenolic systems, silicones are best applied in the prepreg state, with the best laminates being made by high-pressure methods.

Laminating with epoxy resins

Epoxy resins cure by addition and therefore can be laminated by contact methods. Curing temperature is dictated by the hardener or catalyst selected and the properties desired in the end product. Epoxies can be applied by hand since many are 100 percent solids in liquid form. As described previously, epoxies can be thinned by adding reactive diluents for difficult wetting jobs.

Many laminators prefer epoxy prepregs because of the hazards involved in applying resins to reinforcements or a disinclination to become involved in the chemistry and testing of epoxy raw materials.

All prepregs must be stored in dry surroundings below 70°F in sealed containers to avoid loss of volatiles and resin advancement. As volatiles escape, resins begin to polymerize, and tack or drape is affected.

GLOSSARY

Ambient　Surrounding.

Mandrel　A solid or hollow form of metal, plaster, or other material around which reinforced plastics are laminated. Mandrels are removed after curing.

Polyvinyl alcohol (PVA)　A tough, transparent vinyl resin that, in film form, exhibits excellent hot-tensile strength and resists stretching.

REVIEW QUESTIONS

1　What advantages may be cited for reinforced plastics vs. metal?
2　List the disadvantages found in reinforced-plastic laminates.
3　What does contact laminating mean? What are its advantages? What are the disadvantages?
4　List the advantages and disadvantages of laminating by vacuum-bag methods.
5　What type of product is produced by lagging?
6　What pressure is developed if a vacuum-pump gauge reads 20 in.?
7　What is the difference between vacuum-bag and vacuum-impregnation processes?
8　Which type of product is made by filament winding?
9　What is premix?
10　What are the disadvantages and the advantages attributed to manufacturing one's own premix compounds?
11　What first aid should be given when catalyst or catalyzed resin is accidentally splashed in the eyes?
12　In making a wet lay-up, when is the best time to trim the laminate?
13　What simple tests determine if a laminate is reasonably well cured?
14　What does a white area in a laminate indicate?
15　Why will glass mat require a higher resin content than glass cloth?
16　What is the purpose of a bleeder strip in a vacuum-bag process?
17　Why is it desirable to cool a laminate under pressure after it has been cured?
18　What other devices can be used in place of bleeder strips to assist in evacuating air?
19　What is the purpose of a resin trap?

20 How can hardened resin be removed from a vacuum pump?

21 What is prepreg?

22 What factors control the degree of tack and drape that a prepreg exhibits?

23 Why can't phenolic resins be laminated by conventional contact laminating processes?

24 Silicone laminates are made by processes similar to what other thermoset resin?

25 Epoxy laminates are made by processes similar to what other thermoset resin?

DEVELOPING HEAT AND PRESSURE FOR MOLDING

13

In addition to his duty of producing quality molded parts, an important task of the molding press operator is to constantly monitor heat and pressure necessary to the operation. All major processes require close control of heat and pressure to maintain product uniformity.

Machines for molding and forming thermoplastics are self-contained; that is, their heating units and pressure systems are part of an integrated unit. This chapter is concerned mainly with methods by which compression or transfer molds are heated and pressurized. The same methods apply for the most part to producing laminates.

HEATING METHODS

Compression and transfer molding of thermosets are carried out at temperatures between 180 and 350°F—except for polyimide, which is processed to 600°F. A heating system must be capable not only of supplying heat to effect polymerization but to make up for heat losses. More thermal energy is required to offset heat losses than to plasticate and cure material. This loss comes about in two ways:

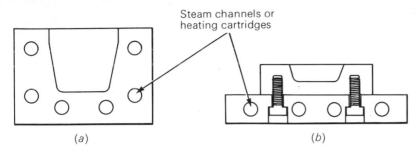

13-1 (*a*) Direct and (*b*) indirect heating from press platen or mold part.

Conduction: A great deal of heat is lost by conduction from the mold or press platens through the press frame, bolsters, and ram. Comparatively smaller amounts are consumed by conduction from mold surfaces to the molding compound.

Convection: Heat also passes off to surrounding air, which is constantly moving. Presses located near open doors are most affected.

Because of heat losses, molds must be constantly monitored for changes in temperature that may affect quality of the work. The important temperature is that of the cavity walls and the contact surfaces of the force, since this is where molding takes place. (See Chap. 8, p. 129.) Thermocouples placed at a distance from the molding surfaces help only when compensation is made for heat losses. Depending on the distance of the heat source from molding surfaces, temperature differentials can vary from 20 to 50°F.

Direct

Molds fastened to press platens are usually heated directly when cavities are deep. This ensures uniform temperature for force and cavity.

Indirect

Hand molds removed from the press each cycle, or shallow molds permanently mounted, are heated by conduction from press platens. To maintain temperature, hand molds are serviced on a bench equipped with a hot plate. Indirect heating saves cost of machining steam channels or electric cartridge holes.

HEATING UNITS

Presses and molds are heated by steam, thermal fluids, or electricity. Selection of the type of heat is guided by such factors as plant size and location, preference, and in some cases the type of molding or laminating.

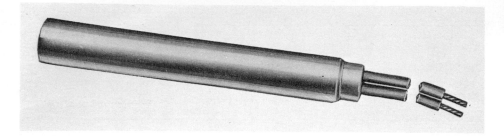

13-2   Heating cartridge. (*Reprinted by permission of Edwin L. Wiegand Co.*)

Steam heating

Heat is furnished by small boilers serving one to three presses or by one large capacity boiler operating from a central location to serve a battery of machines. Companies operating large central units install a second or standby boiler in case of shut-down.

Steam heat is preferred to electricity because heat losses are more quickly compensated, and molds can be brought to temperature more rapidly from a cold start, as when a new mold is set up. As steam courses through mold or platen channels and condenses, it is constantly replaced, maintaining constant mold temperature. The hot condensate is returned to the boiler through a closed circuit for recycling. For this reason, steam is not only channeled to prevent stagnation within a platen or mold plate, but is directed from top to bottom of platens or mold parts. Boiler pressure is controlled at individual presses by pressure regulators and indicated on a steam-pressure gauge (see Table 13-1). Generating live steam for molding requires boiler pressures of 125 to 175 psi. When generated and maintained at the proper pressure, the temperature delivered to a press is constant.

Steam heating is economically mandatory to compression-mold thermoplastics since cooling water is admitted through the same channels while the mold is kept under pressure to make the piece rigid. Electric heat combined with water cooling is not practical because of the slow reheat and space limitations in the mold or platens.

The disadvantages of steam are: (1) leaks in connections, causing rust on press or mold parts, moisture in molding compounds, or steam burns to personnel; (2) corrosion or scaling in pipes, valves, boiler tubes, and mold channels unless water is

13-3   Strip, ring, and band electrical-resistance heating units, which are fastened to exterior of molds. (*Reprinted by permission of Edwin L. Wiegand Co.*)

13-4 One 3-hp boiler serving two compression presses. (*Reprinted by permission of Mears-Kane-Ofeldt Div., S. T. Johnson Co.*)

constantly treated correctly; and (3) expense of installing boiler, valves, insulation, and standby unit, and also the expense of hiring a stationary engineer to meet code requirements for large boilers. These disadvantages, however, are far outweighed by advantages in large operations.

Thermal fluids

Some companies prefer to use thermal fluids as a heat source for molding. The method is similar to steam heating in that a boiler is required to heat the fluid, but more thermal energy can be supplied to the mold at lower pressure.

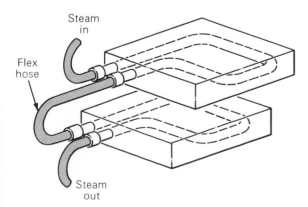

Steam
in

Flex
hose

Steam
out

13-5 Method of steam connection.

A high-pressure–hot-water system called Supertherm draws water from the bottom of a boiler and circulates it to the press at pressures of 250 to 275 psi. Temperatures are controlled by adjusting the volume of water. At such pressures the heat content of the liquid is approximately 50 times greater than steam at 170 psi. Thus smaller pipes and valves can be used, lowering the cost of the initial installation.

Other heating fluids, notably Dowtherm,[1] are used in the same manner.

Electrical heat

Most companies heat with electrical resistance units, as discussed in Chap. 8. These are most common in smaller plants and in cities where antismog regulations rule out coal- or oil-fired boilers. Electrical heating is expensive not only because of the high wattage required but because of replacement costs of burned-out heating units and the need for temperature controllers. Individual controllers for both upper and lower mold halves may cost as much as $2,000 each and must be repaired and periodically calibrated. However, molders who install electric heating systems argue that the absence of boilers, valves, and piping gives a better appearance to the molding department.

DEVELOPING PRESSURE

Various methods may be used to develop the necessary hydraulic pressure; the choice depends on the scope of the operation. Self-contained presses can be manually operated or activated from an integral pump near the press. Pumps operating at 1,000

TABLE 13-1 Steam-pressure table

Gauge pressure, psi	Absolute pressure, psi	Temperature, °F	Temperature, °C
0	14.697	212.0	100.0
20	34.71	258.8	126.0
30	44.7	274.1	134.5
40	54	286.7	141.5
50	64	297.7	147.6
65	78	311.7	155.4
75	89.59	320.0	160.0
90	104	331.2	166.2
100	113	337.8	169.9
110	123	344.1	173.4
120	134	350.1	176.7
130	144	355.6	179.8
140	154	360.9	182.7
150	164	365.9	185.5
160	174	370.6	188.1
170	184	375.3	190.7
180	194	379.6	193.1

[1] Dowtherm is the registered trade name of Dow Chemical Company.

to 2,000 psi are generally vane-type units with a booster in the system to increase pressure when required.

Some molders employ an accumulator system that can deliver hydraulic fluid under pressure to a battery of presses. This method overcomes the expense and noise of individual pumps. Individual valve controls (relief valve and volume-control units) regulate speed of press closing and final pressure for each press. Accumulators are of two types: weighted ram and inert gas.

An accumulator reduces costs of operation because only one pump is required to operate the battery of presses. The high volume of oil (or water containing rust inhibitor), plus a steady flow at constant pressure, assures that presses close smoothly and maintain required pressure.

SELF-CONTAINED VS. BATTERY SYSTEM

Whether a molding plant is operated on a system basis or functions with integrally contained presses is a matter of manufacturing policy. The choice of heating is usually an economic decision. For example, a plant with 100 self-contained presses can have

13-6 An early but outdated thermoset molding shop, with myriad of overhead pipes and valves for low-pressure press closing and high-pressure molding. Water was used as the hydraulic fluid. (*Reprinted by permission of Bipel International, Inc.*)

13-7 Modern molding shop with clean and organized arrangement. Presses are actuated by hydraulic fluid (oil) from a central supply via floor trenches. Work area is light and airy, which is conducive to production of quality parts and contributes to efficiency and morale. (*Reprinted by permission of Bipel International, Inc.*)

ten presses inoperative and still retain 90 percent of its productive capacity. Presses can be readily moved from one location to another and quickly placed in operation.

Companies utilizing a central heating plant and/or an accumulator system can operate at lower cost per press for heat and pressure. System operations require a second or backup power supply to ensure continuing production.

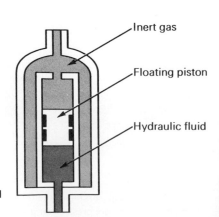

13-8 Cylindrical gas-charged accumulator.

In most businesses the plant is shut down periodically, usually to coincide with a plant-wide vacation period. During this time the molding room is cleaned, presses are repaired and painted, and hydraulic and heating systems are serviced.

GLOSSARY

Absolute pressure Pressure measured in relation to zero pressure (in other words, absolute pressure equals gauge pressure plus atmospheric pressure).

Gauge pressure Pressure measured in relation to that of the atmosphere.

Thermocouple Bimetallic heat sensor inserted in the mold or press platen.

REVIEW QUESTIONS

1 What other production duties, besides loading molds and removing parts, does a press operator perform?
2 Describe how heat is lost in a molding operation.
3 When is direct heating used? Indirect heating?
4 What is the advantage of using steam heat? Disadvantage?
5 What steam pressure is used to mold phenolic? Urea? Melamine?
6 What advantage does the use of thermal fluids exhibit as compared with steam?
7 Why do some molders prefer electric heat?
8 What is the advantage of an accumulator system as compared with individual pumps at each press?
9 Describe two types of accumulators.
10 Discuss the merits of operating a molding plant on a:
   (*a*) self-contained basis
   (*b*) battery basis

FINISHING MOLDED AND LAMINATED PARTS

14

Compression- or transfer-molded parts and laminated products always require some type of finishing before they are shipped or become part of another assembly. Modern production plants maintain a separate department where large-scale finishing operations are carried out.

FLASH REMOVAL

In many cases, only a thin flash needs to be removed from a compression-molded part to make it a completed item. When possible, press operators clean the product manually at the press by filing, scraping, or sanding.

Circular parts are trimmed on a lathe when shape permits or when flash is difficult to remove from high-impact grades. Flat parts can be die-cut to remove tough flash caused by cotton or glass fabric fillers.

Multicavity molds or fully automatic presses produce at a rate that requires bulk handling of small parts. One method of removing brittle flash from products without fragile sections is by tumbling. The products are tumbled in cylindrical wire baskets, metal drums, or octagonal wood barrels—the latter is used for heavier parts with more difficult flash-removal problems. To facilitate flash break off, hard maple or lignum vitae balls, pegs, or other shapes, sometimes in combination, are tumbled with the product.

Brittle flash can be removed from fragile products or those too large for tumbling with special equipment that supplies a high-velocity stream of crushed fruit pits, nut shells, or nylon pellets against a rotary table or belt conveyor.

BUFFING AND POLISHING

Most molded products have an acceptable surface luster provided by highly polished molds. Extrahigh finish or polishing of flash lines is accomplished on small parts by adding waxed wooden pegs or wax compounds to the tumbling barrel.

Large parts are individually buffed on muslin wheels to which cutting compounds, such as tripoli-wax, are applied. The procedure removes fine finishing marks or flash lines. The buffing wheels consist of individual muslin disks sewn together to provide a firm surface or left loose for more flexibility. Stitched wheels cut faster, generate more frictional heat, and are more adaptable to parts with regular contour. Loose buffs are used for reaching crevices or irregular surfaces and are less likely to scorch less heat-resistant materials.

Parts with extremely irregular surfaces are buffed on a wheel composed of alternating large- and small-diameter disks to provide rigidity at the arbor and extreme softness at the periphery. Wheels worn beyond usefulness are sometimes inserted as the smaller component of the assembly.

Polishing operations are similar to buffing except that soft flannel disks are used with wax compounds. Some waxes contain color to cover minute blemishes of sanding, filing, or gate cutting.

ASHING

Ashing is a process that removes material too difficult to reach by sanding. Loose muslin disks apply wet pumice to remove plastic. Ashed articles are then rinsed and dried prior to polishing.

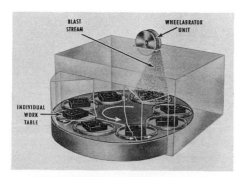

14-1   Schematic view of a table deflasher. (*Reprinted by permission of Wheelabrator Corp.*)

OTHER DEFLASHING OPERATIONS

When inserts are used, particularly in transfer-molded parts, flash must be cleaned off from metal portions. Inserts may be slotted, threaded, or have virtually any configuration. Core pins, which form blind or through holes, also are a source of flash.

With threaded inserts, flash is removed by tapping, usually on a lathe or drill press. External threads are cleaned with a thread die and smooth holes are reamed with a fluted reamer or abrasives.

In many instances, flash removal is an extremely important and painstaking procedure. Most insert work applies to electrical products where shoddy workmanship would lead to poor electrical performance.

PRESSED-IN INSERTS

When designs permit, inserts are pressed in after molding. This speeds press cycles and obviates deflashing. Inserts are pressed in by an arbor press or similar tool.

DECORATING

Painting

Molded plastics are painted for the following reasons: (1) color required is not available in the molding compound specified; (2) painting is required only on certain portions; (3) cost can be reduced by using less expensive dark compounds; (4) exact color matching of components is not always possible from one resin batch to another; and (5) information, identification, numerals, or special effects are required.

An example of a painted product is an electric iron handle molded from black phenolic and coated with epoxy in pastel colors.

Spray painting an entire product and selectively painting specific areas by using spray masks are both common procedures.

Silk-screen painting (or printing) is used for elaborate designs on flat or gently curved surfaces. In this process, a framed silk (or nylon) screen is masked to a particular design. Paint is forced by squeegee to transfer the design to the underlying surface. Multicolored, intricate designs are possible by use of a series of screens.

Raised surfaces such as lettering, trade marks, monograms, and design embellishments can be roller coated if surfaces are flat with sharp edges. Depending on the design's complexity, a blocking mask may or may not be used.

HOT-FOIL STAMPING

Molded and thermoformed products such as cosmetic containers, closures, toys, and packaging are embellished by passing a treated foil of the desired color between the plastic surface and a heated die. Color is transferred and embedded into or adhered to the article.

PLATING
Several methods of applying metals to plastics are in general use.

Electroplating
Chromium, gold, nickel, and other metals can be applied to an entire surface or selected areas of compression- or transfer-molded products. Parts are degreased to remove residual release agents, then rinsed and dried. Copper is deposited first to provide a conductive surface, followed by the surface metal. In some cases, a clear lacquer is applied to provide protection against corrosion. Plumbing fixtures, marine and aircraft hardware, and electrical components are electroplated in quantity.

Vacuum metallizing
In this process, an electrically conductive surface is not required. Parts to be metallized, usually with aluminum, are placed on racks in a vacuum chamber. Aluminum clips are placed on tungsten wires within the chamber, which is then sealed in order to produce a high vacuum. Current is applied to the tungsten to evaporate the aluminum, which then condenses in a uniform continuous coating on the molded parts. Products

14-2 Vacuum-metallizing chamber. Parts are placed on racks and are coated by vaporizing from a metal that radiates outward from energized tungsten filaments. High vacuum (one-millionth of normal air pressure) is necessary to the process. (*Reprinted by permission of Stokes Div., Pennwalt Corp.*)

may be lacquer coated prior to metallizing to provide better adhesion. Clear lacquers are generally used as an exterior coat to improve abrasion resistance of the film, which is approximately 1 or 2 mil thick.

FINISHING LAMINATED PARTS

Except for laminates made by press-molding methods, in which excess generally is trimmed by shear edges incorporated in the mold, laminated products are made over-size and trimmed to dimension.

Asbestos and fibrous glass reinforcements are extremely abrasive and require special cutting tools and techniques. High silica and quartz fibers also belong in this category. Cotton, paper, carbon, and graphite are less troublesome in maintaining sharp cutting edges.

Sawing

Glass laminates dull a band saw or circular saw blade of high-speed steel in the first 10 in. of cut. The blade may still be used for glass, but its efficiency is impaired and it is useless for any other material. Best results are achieved with carbide-tipped or diamond blades, or abrasive wheels, or by frequent replacement of standard blades.

Abrasive wheels or cutoff disks are operated at 3,000 to 6,000 rpm, depending on diameter. Disks are available in diameters to 20 in. and to 0.125 in. thick. They are extremely dangerous since they may shatter if overheated and stressed. Adequate guards and personal protection are mandatory. Wheels should be selected according to the type of cut planned. For long cuts or slotting, a coarse silicon-carbide abrasive disk results in less clogging. For slicing rod or tube stock, smaller, less coarse wheels should be chosen.

Grinding

Air-operated portable disk grinders are used for cutting laminated structures too large to handle on saws. Coarse sanding disks remove material quickly. Dry cement spread over flat surfaces will prevent loading of the disk if frictional heat softens polyester resins.

Routing

Portable or fixed routers are widely used to make cutouts, trim edges, and provide grooves. As with sawing, router bits are subject to rapid wear in machining glass or asbestos laminates, and they require frequent replacement.

Filing

Coarse rasps and files prove best for rapid removal of stock when the laminate is glass or asbestos reinforced. Finer files are used for final finishing. File pressure should be directed parallel to the laminate edge to avoid delamination.

A dust hazard arises when laminates and molded products are finished by the methods outlined. Suitable exhaust systems for sanding, buffing, filing, sawing, and related operations are more necessary in the use of plastics than with metal or wood because chemicals in the resin binders can cause allergic reactions in some people and because of the irritating nature of glass, silica, and asbestos fibers, or inorganic fillers such as mica. Dust particles made up of these fillers abrade the skin, particularly when deposited under cuffs, neckbands, rings, and wrist watches.

Respirators are highly recommended when grinding, routing, or machine-filing glass or asbestos laminates.

GLOSSARY

Ashing   A finishing process used to produce a satin finish by applying the part to a rotating muslin wheel loaded with wet pumice.

Tumbling   A finishing operation for small articles by which flash, fin, and gates are removed and/or surfaces are polished by rotating them together in a barrel. Wood pegs, sawdust, and polishing compounds are added to facilitate the process.

REVIEW QUESTIONS

1   Describe several methods of removing flash from molded parts.
2   What does buffing molded parts accomplish?
3   When are loose (unstitched) buffing wheels recommended?
4   What is ashing?
5   How is flash removed from threaded inserts?
6   When is it advisable to paint a molded product?
7   Describe the procedure for electroplating plastics.
8   What danger exists in using an abrasive cutoff wheel?
9   What type of machining is accomplished with a router?
10   Why should filing pressure be directed parallel to the laminate edge?

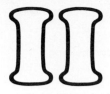

THERMOPLASTIC MATERIALS
AND PROCESSES

Section II is devoted to a study of the manufacture of thermoplastic resins, major processing methods, and the properties and applications of individual materials.

A thermoplastic material is defined as one that can be repeatedly softened when heated and becomes firm when cooled. Thermoplastic polymers are comparable with candle wax or cooking grease in their reaction to temperature changes. Unlike wax or grease, they do not liquefy to pourable consistency. Instead, they attain a state of high melt viscosity, requiring suitable pressure to mold or shape them. There are several exceptions, notably certain nylons and vinyls. In most cases, decomposition occurs before liquefaction.

Compared with the thermosetting class of resins, the thermoplastic class is much larger, both in number of types and in volume used. The number of companies engaged in processing thermoplastics is far greater, and products are more diversified. *Modern Plastics Encyclopedia* (1970 edition)[1] lists 9 distinct thermosetting resins and 20 major types of thermoplastics. Because some thermoplastics of a particular type are sufficiently different from each other, the list could be more than 30.

The technological advance of thermoplastics since 1940 can be attributed to the following:

1   Introduction and acceptance of numerous new resins having superior or unique properties, particularly heat resistance and chemical inertness

2   Development of machines to produce high volume on an automated or continuous basis

3   Incorporation of fibrous glass to upgrade standard or general-purpose resins

[1] This is published annually by McGraw-Hill, New York.

4   Refined or new processing methods such as rotational and blow molding

5   Improved or new techniques in producing foamed articles

6   Increasing scarcity or cost of competing materials (hardwoods, zinc, brass, copper)

7   Acceptance of thermoplastics (based on their merits) by the general public, architects, engineers, designers, building trade unions, and by other industries

These aspects are going to be discussed in detail for each thermoplastic, in the succeeding chapters. To form a basis for understanding these aspects, it is necessary to obtain an insight into the structure of each resin and the effects of their molecular formation on moldability and final properties.

Thermosetting resins, it may be recalled, during processing undergo a chemical reaction that results in a thermally stable, three-dimensional molecular network. Molding compounds and laminating and casting resins are cured by catalysis, often accompanied by heat and pressure. In each instance, conversion to final form involves a chemical *and* a physical change. Processing involves close control over batch-to-batch uniformity, storage, shelf life, preparation of materials, and reaction time.

With the exception of foams, thermoplastics should be considered chemically complete, requiring only softening by heat to help change their physical condition from pellet or sheet to the desired shape. Polymerization is completed by the resin manufacturer, who supplies a polymer in ready-to-use form.

Thermoplastics are readily softened (and in most instances easily dissolved) because they are not polymerized into three-dimensional molecules; instead, they are made up of linear chains that may be parallel, coiled, interconnected slightly, or tangled. A few resins exhibit a crystalline structure in which repeating groups of molecules assume a more precise spatial pattern. These crystalline plastics have sharper melting points and lower melt viscosities than noncrystalline (amorphous) types. Adding to the complexity of structure are resins that contain both amorphous and crystalline regions, giving rise to a wide range of different grades of one particular type.

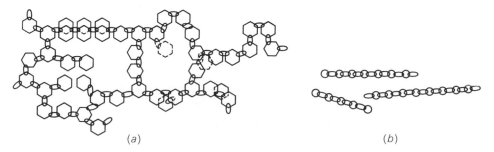

(*a*)   (*b*)

II-1   (*a*) Three-dimensional thermosetting structure; (*b*) linear or two-dimensional thermoplastic.

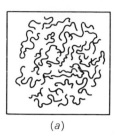 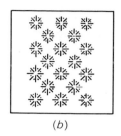 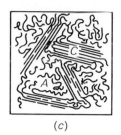

(a)   *(b)*   *(c)*

II-2   *(a)* Amorphous thermoplastic. Molecular chains are of varying length and have no orientation. Examples: polystyrene and polyvinyl chloride. *(b)* Crystalline thermoplastic. Closely packed and orderly arrangement of molecules. Examples: polyamides and polyvinylidene chloride. *(c)* Amorpho-crystalline thermoplastic. Areas of crystallinity or linearity *(C)* surrounded or intercepted by amorphous regions *(A)*. Example: polyethylene.

     Irregular molecules can literally be "made to behave" by pulling them in one direction so they line up in more orderly rows. Nylon fishline, apparel fibers, and extruded sheet are made many times stronger by orienting their molecular structure. Sheets of material can be heated, formed to a contour, and then cooled. Stresses created by forcing molecules to assume a new position are responsible for a sheet returning to its original flat shape if reheated. This phenomena is called *elastic memory*. It permits reworking sheet material when forming operations are not successful.

     In another instance, the rate at which a polymer is cooled to control size of *spherulites* (crystal aggregates) influences final properties and clarity of film or molded products.

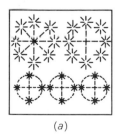 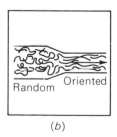 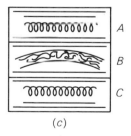

(a)   *(b)*   *(c)*

II-3   *(a)* Spherulitic plastic. Rearrangement of molecules into a crystalline structure during processing. Example: polypropylene. *(b)* Orientation of molecules. Stretching while cooling during processing (or stretching during testing) aligns molecules parallel to cold-drawing or test-load direction. Examples: nylon monofilament, polystyrene extruded sheet, and polyvinyl chloride extruded tubing. *(c)* Elastic memory. *A* is original molecular structure; *B* is rearrangement by heating, forming, and cooling; and *C* is recovery to original shape by subsequent reheating. Examples: cast acrylic and cast films such as cellulose acetate.

$$-M-M-M-$$

Monomer →(heat; pressure / catalyst)→ Polymer

Example: Ethylene ⟶ Polyethylene

II-4 Monomer polymerization.

Monomer #1
+
Monomer #2 ⟩ Copolymerization ⟶ Copolymer

Vinyl chloride
Example: + ⟩ Copolymerization ⟶ Polyvinyl chloride-acetate copolymer
Vinyl acetate

II-5 Simultaneous polymerization of two monomers.

All thermoplastics have one feature in common. They contain molecules of extremely great length; these molecules are often referred to as giant or macromolecules. To produce giant molecules, the chemist must first develop a monomer capable of combining with other similar units to form a polymer.[1] Monomers may exist in gaseous or liquid form, derived from petroleum, natural gas, coal, air, and water. As they combine to form long chains, the resin becomes a solid, or polymer. The reaction that achieves this chain growth is called *polymerization*. Thus, the prefix *poly* indicates the *polymer*. For example, *ethylene* (a gas) is converted to the solid, *polyethylene*.

Other thermoplastics are brittle or difficult to melt and process. To overcome these limitations or to provide variations of basic resins, copolymers are made by simultaneously polymerizing two monomers.

To differentiate between copolymers and polymers, a single polymer is referred to as a homopolymer. Further proliferation of thermoplastic resin types introduces terpolymers and alloys.

Resin manufacturers can produce an almost limitless variety of grades of one particular polymer by the method used to effect polymerization and choice of catalysts. More than one resin manufacturer has the capability of producing a particular polymer, and specifications concerning melt temperature, flow, and processing characteristics may be identical. However, mixing similar polymers of two different suppliers may be

[1] Mer is from the Greek word meaning part or unit. Mono-mer means one mer or complete chemical unit; poly-mer is many combined monomeric units.

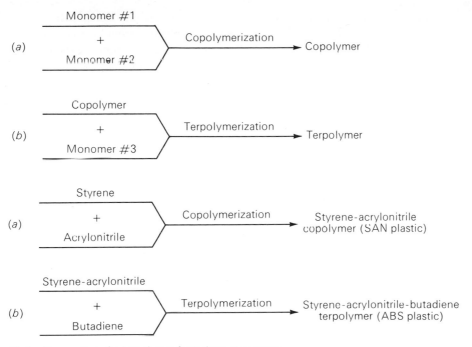

II-6   Preparation of a terpolymer from three monomers.

disastrous unless experimentation by the molder or extruder offers proof to the contrary. Rules to follow include:

1   Avoid mixing one grade of material with another grade of the same material unless previous experience shows the mixture is practical. Stresses can be induced that may cause product failure.

2   Avoid blending the same grade of material obtained from two different suppliers unless warranted by experience. Specifications are developed under a fixed set of test conditions seldom duplicated in processing. Thus each material in the blend can react differently during a molding operation.

METHODS OF POLYMERIZATION

The method of converting a monomer to its polymeric form depends largely on monomeric state, ease of production, and degree of molecular chain growth required.

Bulk

This method is used when monomer conversion produces small or controllable exotherm and to produce resins of high-molecular weight that yield a useful polymer not attainable by other methods.

Bulk polymerization can be carried out on a continuous or batch basis. Liquid monomers are charged directly into a reactor with catalysts, where heat converts the monomer to a viscous liquid. The monomer acts as a suspension agent as the polymer develops. When reacted to the desired degree, unreacted monomer is evaporated off and the polymer washed and dried. Other steps may involve extrusion-compounding before pelletizing for shipment. Polystyrene and acrylic polymers are produced from liquid monomers.

Gaseous monomers are passed through heated tubular reactors that convert the gas to polymer as they pass through a catalyst grid or bed. Polymers form as chains grow and combine. Subsequent cooling produces the solid form. An example of continuous bulk polymerization of a gaseous monomer is polyethylene.

Solution

Solution processes are similar to the bulk process in which a monomer (gas or liquid) is dissolved in a suitable solvent with a reaction initiator during polymerization. The less-viscous solution aids in heat dissipation and prevents runaway reactions. Solution polymerization makes polymers that are used in solution form, such as paper coating, floor wax (polyvinyl chloride, acrylic), or latex paint. Some copolymers (impact polystyrene) and homopolymers (nylon) are produced in the form of molding pellets.

Expensive recovery steps, such as precipitating polymer in a nonsolvent, stripping unreacted monomer, filtering and drying, and retrieving spent solvent, create economic barriers to more widespread use of this method.

Suspension

Suspending a monomer in water forms a slurry that can be easily stirred and pumped through a succession of polymerization steps. Constant agitation keeps the polymer in suspension. Polymer may form as tiny beads or hemispheres that are readily separated from the suspension fluid by spray drying. Suspension methods also result in fine powders of very high molecular weight, such as polyvinyl chloride.

Emulsion

Similar to suspension methods, emulsions yield smaller particle sizes that are used as latex paints, paper or textile impregnants, or adhesives. Emulsifiers such as soaps maintain resin particles within the water-polymer mixture to prevent settling and agglomeration. Typical resins produced in this manner are acrylic and vinyl latex paints.

When polymers in the form of solids or gums are desired, the emulsion is coagulated through removal of water and soaps. This is done by spraying over heated drums, then scraping particles off with blades. Solid polymers in pellet form often do not have the crystal clarity obtainable by bulk or solution methods because traces of emulsifier contribute to haze.

Combinations of Methods

In some situations, it is more convenient to start polymerization with one method and complete it by another. For instance, in preparing some types of nylon, reactants are introduced in a water suspension, where polymerization results in nylon salt as water evaporates. The salt is further polymerized in bulk by either a continuous or batch process.

POLYOLEFIN PLASTICS

15

The term *olefin* is applied to a group of unsaturated hydrocarbons extracted from petroleum or natural gas. Examples are ethylene, propylene, and butylene. Commercially useful products from oil or gas are referred to as *petrochemicals*.

HISTORICAL DEVELOPMENT

The fact that ethylene (a gas) can be reacted with a catalyst to form a polymeric substance has been known for many decades. The first polymers were unsuited to compete in the market because the products formed were like wax or grease. However, in 1933, British researchers at Imperial Chemical Industries, Ltd. developed a method of obtaining a polymer of sufficiently high molecular weight to provide acceptable physical properties and processing reliability. The British discovery was a culmination of experiments on the effects of extremely high pressures on polymerization. Their efforts were greatly aided by the invention of compressors capable of delivering pressures above 30,000 psi. The timely appearance of a useful polyethylene was a vital factor in perfecting radar insulation during World War II.

Though numerous other thermoplastics attained commercial importance prior to polyethylene, the diversity of types and grades, plus low cost and unique properties of this synthetic resin led to its becoming the first to reach the billion-pounds-per-year level (in 1959). Applications continued to increase, so that in 1968 U.S. production was 4 billion lb. Much of the expanded usage results from development of new polymerization techniques and new methods of catalyzation.

CHEMICAL NATURE OF POLYETHYLENE

Polyethylene can now be produced by many processes, each resulting in resins having different properties and thus different end uses. High-purity ethylene, obtained as a by-product of petroleum refining or from natural-gas fractionating, is the starting monomer. Ethylene is a gas that boils at −104°C. Although it is a simple hydrocarbon molecule made up of four hydrogen atoms attached to two carbon atoms, scientific efforts to promote chain growth to produce giant molecules were thwarted for 50 years.

The structure of ethylene can be illustrated as:

$$
\begin{array}{cc}
\text{H} & \text{H} \\
| & | \\
-\text{C} & = \text{C}- \\
| & | \\
\text{H} & \text{H}
\end{array}
\quad \longrightarrow \quad
\begin{array}{cccccc}
\text{H} & \text{H} & \text{H} & \text{H} & \text{H} & \text{H} \\
| & | & | & | & | & | \\
-\text{C} & -\text{C} & -\text{C} & -\text{C} & -\text{C} & -\text{C}- \\
| & | & | & | & | & | \\
\text{H} & \text{H} & \text{H} & \text{H} & \text{H} & \text{H}
\end{array}
$$

Ethylene ⟶ Polyethylene
$(H_2C = CH_2)$

The presence of the double bond $C = C$ indicates a sharing of electron forces between adjacent carbon atoms. When conditions of temperature, pressure, and catalysts are correct, the double bond can be "opened," and an electron can attach to carbon or hydrogen atoms of other ethylene units to form long chains having a carbon backbone.

The polymer thus formed is a mixture of ethylenic groups containing chains of extreme length (hundreds of thousands of units) interspersed with short chains containing as few as 10 units. Polyethylene chains are not flat or two dimensional as shown. Instead, they are made up of coiled, irregular molecules containing side branches, much like tree branches that spread in all directions from the main trunk.

The degree of branching influences properties such as density, melt temperature, softness, rigidity, transparency, and oxidation resistance. Side branching is also accompanied by sharing of some double bonds with others in an adjacent chain. This is undesirable since it provides reactive sites for oxygen to attach. Oxidation, increased

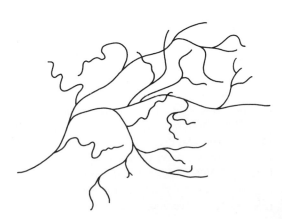

15-1   Graphic representation of polyethylene.

by overheating either in processing or in service, results in embrittlement of poly-ethylenes used for electrical insulation, shorter life expectancy in applications such as pipe, and processing difficulties such as "fisheyes" in film.

In addition to the branched chains, varying amounts of molecules are lined up in parallel fashion like a truckload of neatly stacked lumber. These molecules are linear or crystalline, as opposed to the random amorphous molecules in branched chains. Thus, branching can be compared with a spilled load of lumber.

All polyethylenes contain both amorphous and crystalline chains in varying proportions. A totally amorphous resin would be soft and greasy; a totally linear one would be hard, brittle, and incapable of being processed by standard melt techniques. Control of the degree of orientation and its location within the resin mass provides a great variety of grades of polyethylene.

MANUFACTURING METHODS

Several basic methods are now used to produce polyolefins. In 1954, polymerization of ethylene by a process utilizing low pressure and reduced temperatures was an-nounced almost concurrently by Phillips Petroleum Company in the U.S. and Professor Karl Ziegler in Germany. Oxygen and peroxide catalysts used in high-pressure poly-merization were replaced by Marlex catalysts (chromium oxide on a silica or alumina support) or by Ziegler catalysts (titanium tetrachloride with triethyl aluminum). The net result was a polyethylene resin having greater numbers of linear molecules of controlled chain length and distribution. Closer packing of molecules yields resins with higher density. Thus polyethylenes exhibiting this characteristic came to be known as linear or high-density types.

Since production method influences the type of polymer obtained, a comparison of the basic types is in order. Low-density polyethylene (produced by the high-pressure process) and high-density polyethylene (produced by the low-pressure process) contain both linear and branched molecules. The difference is a matter of proportion, chain length, and distribution. Low-density resins contain from 40 to 60

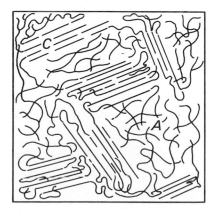

15-2   Crystalline (linear) regions (*C*) and amorphous (random) molecules (*A*).

percent linear crystallites; high-density grades contain 70 to 95 percent. Density differences between all grades of resin are very slight, but these minor differences influence molding or film-forming processes and the serviceability of the product. For purposes of classification, polyethylenes are grouped into three density categories:

Type	Weight, g/cc*	Density
1	0.910–0.925	Low
2	0.926–0.940	Medium
3	0.941–0.965	High or linear

*Products made from any type would float since all are lighter than water.

In addition to low resin cost (12 to 27 cents/lb), low density enables the molder or film producer to obtain more productive yield from polyethylene than from any other resin, except polypropylene and polyisobutylene.

Linear polyethylene exhibits greater heat resistance, stiffness, and chemical inertness than lower density types. These properties suggest its use for items that require sterilization, or for tool boxes, lab ware, and sporting goods. Low-density resins are softer, more flexible, and less heat-resistant. Products such as snap-on lids, squeeze bottles, and packaging film utilize these inherent properties.

DETERMINATION OF MOLECULAR WEIGHT (MW)

Still another factor in differentiating polyethylenes is their molecular weight. It has been shown that as polymerization takes place, ethylene molecules do not align in orderly rows. Chains do not grow to uniform length, either in the linear phase or the

15-3   Clear, glossy wrap is extruded from medium-density polyethylene. (*Reprinted by permission of U.S. Industrial Chemicals Corp.*)

branched condition. Branched areas consist of both long and short chains, from a few molecules to many thousands. In comparing resins, molecular weight can be taken as a number value or an average.

The atoms that make up chemical elements and molecules each have been assigned arbitrary atomic weights: carbon 12; hydrogen 1; oxygen 16; nitrogen 14; lead 207.2, and so on. An ethylene molecule containing two carbon and four hydrogen atoms has a molecular weight of 28, the sum of the weights of the six atoms:

Carbon $2 \times 12 = 24$

Hydrogen $4 \times 1 = \dfrac{4}{28}$

If one chain contains 1,000 ethylene units, the molecular weight is 28,000. Another chain may have only 100 units, or molecular weight of 2,800. The average molecular weight of the two adjacent chains is then

$$\dfrac{\begin{array}{r}28,000\\2,800\end{array}}{30,800}$$

$30,800 \div 2 = 15,400$ av. MW

The number average in molecular weight can then be defined as the total weight of all molecules divided by the number of molecules.

High-molecular-weight resins exhibit better properties and a wider range of uses when structural reliability is a requirement. Pipe and pipe fittings, wire insulation, and high-strength film are good examples. It is possible to produce a low-molecular-weight resin having high or medium density and vice versa. When the polymer contains a high proportion of linear molecules of short chain length, combined with amorphous areas of short branches, the total weight is divisible by a greater number of individual molecules, thus reducing the average molecular weight. However, the short branches permit closer grouping of all molecules, resulting in higher density.

15-4   Rigid containers blown from high-density types. Open-mouth jars for mayonnaise, mustard, and other food and a 1-gal gasoline can attest to chemical resistance of polyethylene. (*Reprinted by permission of U.S. Industrial Chemicals Corp.*)

MOLECULAR-WEIGHT DISTRIBUTION

To determine the percentage of short, medium, and long chains in a particular polymer, chemists use laboratory techniques such as precipitating the different fractions out of the polymer and also separating and recovering each portion for comparison. A resin having a wide range of molecular weights, for example, very short to extremely long, is classified as having a broad distribution. Resins with a majority of molecular weights near the average are classified as having narrow distribution.

Resins of narrow distribution are used for thin-walled, molded products where warping may be a problem, and in paper or textile coating where uniform flow and penetration are required. Broad distribution grades are widely used in blow-molded containers and tough packaging films.

MELT-INDEX DETERMINATION

Perhaps the simplest way of classifying a polyolefin is by determining its melt index (MI number),[1] a useful but potentially misleading evaluation of the flow characteristic. Under controlled conditions of heat (374°F) and pressure (2,160-g weight), the weight in grams of resin extruded from a small orifice in 10 min is stated as the MI number. Melt index measures the ability of a molten material to flow and fill out a mold or form a blown, hollow shape uniformly. Resins are furnished in a range of 0.25 to 25 or more.

A high MI number means that the resistance to processing pressure is minimal and viscosity is low. This is important to the injection molder plagued by mold-flashing problems or to the blow molder who must control the formation of blown objects. Often a processing problem can be overcome by changing to a resin with a lower MI without sacrificing cost or properties. In general, resins with high-average-molecular weight exhibit a low MI rating.

In practice, it is unwise to mix resins classified as having the same MI but produced by different companies, unless experimentally it has proven feasible. The MI value is obtained under a precisely fixed set of laboratory conditions, which cannot be duplicated in production machines. For instance, processing temperatures for injection molding may be considerably higher than 374°F, but in extrusion machines they are much lower. Each resin of the mixture then behaves differently as to melt viscosity and flow and cooling rate, resulting in a poor-quality product. Shrinkage differences, warping, and similar problems may result. For example, note the following:

Company	MI obtained at 374°F, 2,160-g piston weight	Melt viscosity at 425°F, machine temperature	Molding defects
A	5	Slightly lower	Excessive flash; warpage
B	5	Greatly reduced	Inconsistent size; locked-in stress

[1] Details of the melt-index determination are found in Sec. III, which deals with quality control.

Great care must also be exercised in control of reground scrap to avoid mixing resins not properly identified (see Table 15-1).

PROPERTIES OF POLYETHYLENE

Polyethylene continues to lead all other plastics in volume of resin used annually throughout the world. It is not a high-strength material, nor does it exhibit high heat resistance, so its popularity must be attributed more to a combination of other assets.

Low cost

Ethylene is abundantly available at low cost and is polymerizable at high-production rates. Low-pressure types command higher prices.

Wide range of resin types and colors

A large number of resin producers have entered the polyolefin field. Some firms offer more than 50 grades and have the ability to "tailor" additional types for specific, new applications.

TABLE 15-1   A general summary of the effects of molecular structure on resin properties

Properties	As density increases	As MI decreases or av. MW increases	As MW distribution narrows
Thermal:			
Softening point	Great increase	Slight increase	Slight increase
Viscosity at melt temperature	Increase	Increase	Slight increase
Flow resistance to pressure	Increase	Increase	Increase
Low-temperature flexibility	Decrease	Increase	Increase
Mechanical:			
Stiffness	Increase	Increase	Slight increase
Tensile strength	Increase	Increase	Slight increase
Torsional strength	Increase	Increase	Slight increase
Impact strength	Decrease	Slight increase	Difficult to define
Scratch resistance	Great increase	Slight increase	No effect
Chemical*:			
Stress crack resistance†	Decrease	Decrease	Increase
Permeability	Decrease	Slight increase	Decrease
General resistance to chemicals at room temperature	Increase	Increase	No effect
Electrical:			
Insulation values	Slight increase	No change	No change
Processing characteristics:			
Production rate	Increase	Increase	No change
Shrinkage	Increase	Increase	No effect
Warpage	Slight increase	Increase	Decrease
Gloss	Increase	Decrease	No effect

*Although laboratory tests may indicate satisfactory performance, molding conditions may induce stresses that can render a product unfit for use with a particular reagent.
†Environmental stress crack resistance is discussed more completely in Sec. III.

The natural state of polyethylene is translucent white, similar to paraffin wax and tending toward transparency as thickness is reduced. Any degree of pigmentation is possible.

Resistance to specific reagents

In general, each thermoplastic is susceptible to attack or completely soluble in specific chemicals. Attack may be mild or severe; solvation can be partial or complete. As a class, polyolefins are among the most chemical resistant of thermoplastics, excelled only by fluorocarbons. Among the chemicals that swell, weaken, or discolor polyethylene are acetone and related ketones; oils like linseed and camphor, heptane, toluene, chlorinated hydrocarbons; and oxidizing acids (nitric, sulfuric).

Linear types are more inert to the action of reagents, although deterioration of all grades is accelerated at elevated temperatures. All types have high resistance to water absorption and low permeability to moisture vapor. Packaging of bakery goods, produce, and a variety of dry household chemicals provides a steady outlet for polyfilm and polyethylene-coated paper. Heavier gauges, to 40 mil, are used to cover freshly poured concrete, as greenhouse cover, drop cloths, for soil erosion and weed control in ponds, reservoirs, and irrigation canals, and to retain ground moisture for seedlings.

15-5 Water-storage basin being lined with polyfilm, which is sealed on-site with portable thermal sealing unit. [*Reprinted by permission of Leister Co. (Switzerland) and B. R. White Co.*]

Environmental stress cracking   Blending incompatible grades, using improper mold-ing techniques, or selecting of an unmodified grade of resin may lead to failure when exposed to chemicals that ordinarily have no effect. Such environmental stress cracking (ESC)[1] can result from chemical vapors or direct contact. Minute fissures in amorphous sections expand into larger cracks, thus causing failure. Special formu-lations containing small amounts of butene or isobutylene decrease stress cracking.

Electrical insulation

As stated, polyethylene exhibits excellent insulation values over a wide frequency range. This feature, combined with water and chemical resistance, plus a natural flexibility, makes polyethylene a logical choice for wire and cable primary or jacketing applications, e.g., underwater cable, communication networks, and tree wire.

Weathering properties

Unmodified polyethylenes are not recommended for outdoor applications. Ultraviolet light accelerates oxidation, causing embrittlement. Incorporation of small amounts of carbon black significantly increases the service life of film or molded items.

Flammability

Polyethylene is slow burning. Once ignited, it continues to burn with a characteristic blue flame. An odor of paraffin may be noticed, which serves to identify the resin.

JOINING POLYOLEFINS

Polyolefins are difficult to join to each other or to dissimilar materials with adhesives, and solvents are impractical. Their waxy surface and chemical resistance preclude dissolving mating surfaces, a common practice with most thermoplastics. Special adhesives based on epoxy-polysulfide, polybutadiene, or silicone-oxylene provide intermediate bond strengths.

Fortunately, the characteristic of polyolefins to melt easily without decomposi-tion permits the joining of two surfaces by fusion. Two mating pieces are joined by melting on a hot plate and clamping until cool. Automatic thermal welding units are used for large production of molded items.

Either flat or thermoformed sheet stock is welded into a variety of products by using a torch and polyethylene welding rod. Chemical tanks, sinks, fume hoods, and similar structures are fabricated by welding.

Flexibility and toughness of polyethylene permit incorporation of metal con-nections in many applications. Spring clips, snap rings, and rivets can be used with stiffer grades of polyethylene, but mechanical fasteners that exert pressure may create a condition known as creep in the softer grades, thus loosening the assembly. Some

[1]Detailed test procedures for ESC are discussed in Chap. 34.

designs utilize the elongation of soft grades to provide stretch over a rigid component. Examples are packaging and refrigerator food containers.

DECORATING MOLDED PRODUCTS, FILM, AND SHEET

Parts requiring painting or printing call for postmolding treatment amounting to controlled oxidation of surface molecules. Molded parts are conveyed through gas flames or past electrical discharge coils, as in the case of film, which is treated as it is produced. Slight absorption of oxygen improves adhesion of printing inks.

A test for adequacy of treatment is to wet treated surfaces with water. Water should cling and coalesce. Adherence of printing or decorating is tested by pressing a strip of pressure-sensitive tape against the product, then pulling the tape away.

APPLICATIONS OF POLYETHYLENE

Packaging

Film and sheeting provide the largest single outlet for polyolefins. Garment and produce bags, trash-can liners, construction and freight transportation uses, soft goods wrap, and thousands of other uses account for several billion pounds annually, mainly low-density types.

Injection-molded articles

Molded items exemplify the versatility of polyethylene. Its wide range of flow characteristics and good mold-filling ability permit molders to manufacture products of virtually any size and configuration. Typical applications of molded polyethylene include baskets, buckets, hospital disposables, chair glides, pipe fittings, valves, chemical lab ware, pump parts, flashlights, sporting goods, food containers, toys, gaskets, grommets, tool boxes, hair curlers, and trash containers.

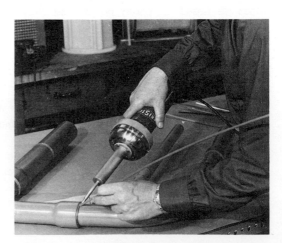

15-6 Heating polyethylene rod with hot air to weld a bead. [*Reprinted by permission of Leister Co. (Switzerland) and B. R. White Co.*]

Blow-molded articles

Squeeze bottles of softer grades and more rigid bottles of medium- or high-density resins are a steadily growing market for polyethylene. Because of their more pleasing and functional design, improved surface detail, color, and breakage resistance they are replacing many glass and paper containers. Toys, flashlights, canteens, and gasoline cans are typical of blow-molded items.

Extrusion-molded articles

Wire and cable coating, pipe and conduit for plumbing, profile shapes such as table edging, channel, and weather strip are all extruded from appropriate grades.

Thermoformed articles

The more-rigid grades are used for items such as trays, tote boxes, and containers. Low-density grades are formed into flexible covers.

Rotational molding

Hollow parts too large or complex for blow molding or injection are produced by rotational molding. Drums, carboys, tanks, and smaller items are produced from finely pulverized powder by rotating closed molds on two axes in a high-temperature oven where melting and fusion occur.

Miscellaneous uses

Low- or medium-density resins are used to upgrade other products such as: (1) fortifiers for waxes, like scuff-resistant floor wax; (2) coatings for paper, cardboard, and aluminum foil, often in conjunction with paraffin; (3) rug backing for automotive flooring; (4) monofilament for rope and carpeting; and (5) hot-melt adhesives.

SPECIAL TYPES OF POLYETHYLENE

Ultrahigh-molecular-weight resin

Special grades, known as UHMW, have molecular weights ranging from 1.5 to 3 million at densities of 0.93 to 0.94. Melt index is zero, so molded items must be produced by compression molding; extrusion requires special equipment. The product is a hard, tough, antifriction material with good fatigue resistance, often used to replace hardwoods, leather, bronze, and hard rubber in chemical plants, or in machinery such as textile looms in which items can be molded to shape or machined from slabs.

Cross-linked Polyethylene

Polyethylene becomes cross linked and thermosetting when treated with organic peroxides or electron beam radiation. It cannot be reprocessed afterward. Chemical cross-linking is accomplished by incorporating peroxides with the molding material.

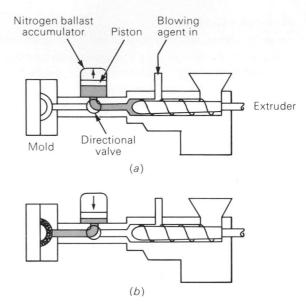

15-7 Injection-foam process for polyethylene. (*a*) Prefilling the accumulator; (*b*) mold filling and expansion. (*Based on materials supplied by Modern Plastics, Inc.*)

Extruded wire insulation and formed products thus have improved resistance to abrasion, weathering, and chemicals.

Foamed Polyethylene

Although not limited to polyolefins, foamed articles can be made by a process similar to conventional injection molding or extrusion. Figure 15-7 illustrates a process developed by Union Carbide Company. Foamed products having 50 percent lower density than solid polymers are used in such applications as crates, floats, agricultural field boxes, arm rests, and fan shrouds.

ETHYLENE COPOLYMERS

Ethylene-vinyl acetate (EVA)

By copolymerizing ethylene with vinyl acetate[1] in amounts to 50 percent, improvement can be noted in such uses as heavy-duty film. Clarity, tear-resistance gains, and electronic heat seals are improved. Bulk chemicals, peat moss bales, and plastic molding pellets are packaged in EVA film. Molded items include syringes, shoe soles, and truck mud flaps.

[1]Vinyl acetate homopolymer is discussed in Chap. 17.

Ethylene–ethyl acrylate (EEA)

A copolymer of ethylene and ethyl acrylate is extruded, injected, and blow molded for products requiring greater heat stability, stress crack resistance, and performance at subzero temperatures. Sterilizable hospital sundries, transparent flexible tubing, and squeeze toys are representative products.

CHEMICAL NATURE OF POLYPROPYLENE

The successful development of linear polyethylene led industry to investigate the possibility of forming a polymer from propylene, a refinery off-gas used mainly for producing high-octane gasoline. As in early efforts to form a useful polymer from ethylene, the first polymerized propylene was low in crystallinity and nonuniform. Finally, in Italy in 1954 Professor G. Natta reported that *stereospecific* catalysts produced useful polypropylene with ordered crystalline chain formations. A stereospecific catalyst is one that causes monomeric units to promote chain growth in a particular spatial position with respect to chain segments previously formed. The resulting chains are spiral and can intertwine to form a thermoplastic with a high degree of toughness, flex-fatigue resistance, and other desirable properties.

If we recall that polyethylenes form irregular chains, with some linear areas, we can compare the regularity of chain formation of polypropylene. It is important to recognize that improper heating and cooling of polypropylene change the properties of the extruded or molded product by changing crystallite size.

Propylene, a gas, with the structural formula $CH_3 — HC = CH_2$ can be polymerized under conditions similar to those used for high-density polyethylene but with some important differences.

If we visualize polypropylene segments as placed around a wheel hub having three spokes pointing out at 120° angles from the edge, we can have one spoke pointing toward us, the second pointing down and away from us, and the third directly away from us from the opposite edge. With the hub representing the carbon atom, the methyl group CH_3 attaches to one spoke and hydrogens to the remaining two. On the next monomeric unit, the methyl group is attached to a spoke 120 deg apart; in the third unit it attaches to the one remaining spatial position. This config-

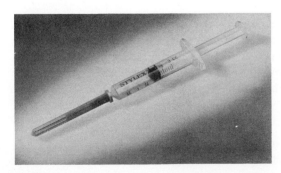

15-8 Disposable syringe barrel, plunger, and needle protector injection molded from polypropylene. (*Reprinted by permission of Pharmaseal.*)

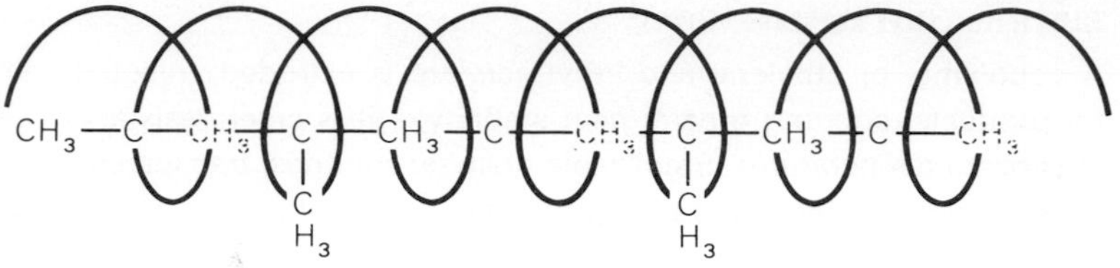

15-9 Isotactic polypropylene.

uration can be represented as shown in Fig. 15-9. The term given to this molecular alignment is *isotactic*, and commercially it is the most important type.

When methyl groups arrange in an up-and-down random pattern, propylene is called *atactic*:

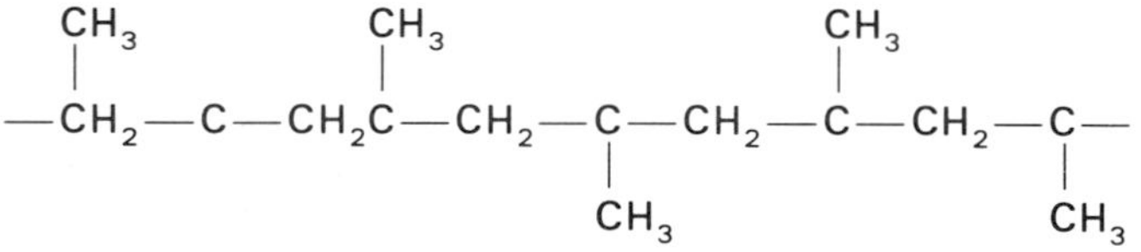

$$-CH_2-\underset{\underset{CH_3}{|}}{C}-CH_2C-CH_2-\underset{\underset{CH_3}{|}}{\overset{\overset{CH_3}{|}}{C}}-CH_2-\underset{\underset{CH_3}{|}}{C}-CH_2-\underset{\underset{CH_3}{|}}{\overset{\overset{CH_3}{|}}{C}}-$$

Polymers having different percentages of both isotactic and atactic blocks have different properties. A totally atactic polymer would be amorphous, and thus soft and weak. Most commercial grades contain only minor amounts of atactic segments. As with polyethylene, processors should avoid mixing different grades of resins because of differences in manufacture.

Another form of polypropylene occurs as a stereoblock polymer:

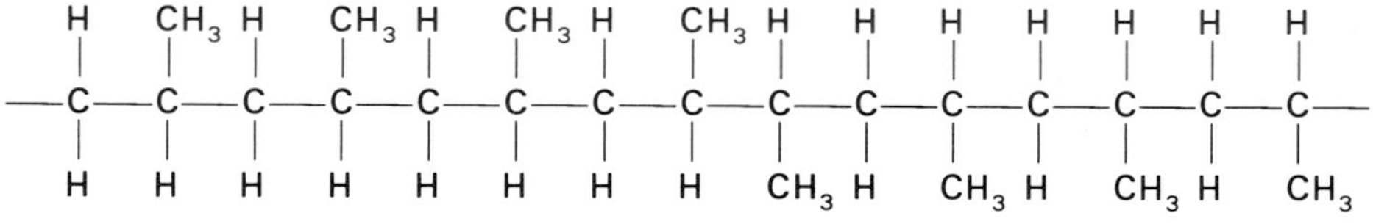

In this configuration, isotactic chains of indeterminate lengths alternate in opposite positions. Polymers formed in this manner are stretchable but recover rapidly.

PROPERTIES OF POLYPROPYLENE

Polypropylenes are not available in as many diversified types of polyethylene, mainly because of the balance of properties inherent in fewer grades. General-purpose grades are slightly higher in cost than linear polyethylene. However, specific gravity is lower, 0.905, making it competitive in many film and molded applications. Color selection is unlimited, the natural state being translucent white, with high clarity in thin sections.

Resistance to chemicals

No solvent affects polypropylene at room temperature; aromatic and chlorinated solvents soften or swell the polymer at elevated temperatures. Strong oxidizing acids (e.g., fuming nitric) slowly attack the resin; environmental stress cracking is not a problem.

Electrical properties

Polypropylene is an excellent insulator and is used in many molded products, as well as in foil form for winding coils and transformers. Stiffer than polyethylene, it is not used for wire and cable insulation even though it has superior abrasion and heat resistance.

Weathering properties

Standard grades have shorter life expectancy when exposed to the outdoors. Discoloration, color fade, and crazing occur in products not stabilized with antioxidants or carbon black. Furniture webbing, marine tow ropes, and indoor-outdoor carpeting produced from polypropylene-extruded filament are examples of increasing usage for outdoor purposes.

Flammability

Polypropylenes are classed as slow burning and can be identified by an odor of crude oil. Flame-retardant grades are available for specific electrical applications.

Mechanical properties

Commercial grades of polypropylene can be classified as tough and as having good impact resistance. Such molded products as automotive scuff pads, arm rests, wheel housings, and fan shrouds demonstrate these properties. Other applications include vacuum-sweeper housings, washing-machine agitators, and a wide variety of toys.

Polypropylene becomes more brittle than many other thermoplastics at subzero temperatures unless modified with rubber or other monomers.

POLYALLOMERS

If propylene and ethylene are copolymerized, an ethylene-propylene copolymer, or polyallomer, is produced. Improvements in low-temperature impact strength, heat-distortion temperature, and abrasion resistance are possible through selection of applicable grades.

GLOSSARY

Amorphous   Without form as compared with crystalline. Amorphous solids have no sharp melting point.

Aromatic solvent   Solvent containing a hydrocarbon ring compound such as benzene.

Creep   Dimensional change occurring in a material under load taken as a function of time after the initial elastic deformation has taken place.

Crystalline   Regular geometric arrangement of particles of matter. A few synthetic resins are totally crystalline, and a few contain some areas of crystallinity. Crystallinity denotes compact, uniform molecular-chain formation in the polymer. At processing temperatures, crystalline structures become amorphous, then recrystallize as they cool.

Fisheyes   Small globular mass within a material, usually attributed to incomplete blending.

Permeability   Passage of a liquid or its vapor through a barrier without physically or chemically affecting the barrier material.

Torsional strength   Resistance to twisting.

REVIEW QUESTIONS

1   Explain what an unsaturated hydrocarbon is, and give several examples.
2   What effect does excessive oxidation have on the properties of polyethylene?
3   What is the difference between amorphous and crystalline polyethylene?
4   Explain the three classes of polyethylene resins.
5   How is the average molecular weight of polyethylene determined?
6   Of what value is melt index rating?
7   What processing problems may be created by mixing two resins with the same MI number?
8   List the properties of polyethylene that contribute to its wide use.
9   What additive renders polyethylene suitable for outdoor use?
10   What difficulties are encountered in attempting to join polyethylene by solvent bonding?
11   What practical method is used to join polyethylene pipe sections?
12   What is creep?
13   What process is used to render polyethylene surfaces paintable?
14   What is the largest market for polyethylene?
15   Describe the properties and uses for UHMW polyethylene.
16   How can thermoplastic polyethylene be made thermosetting? What are the reasons for doing this?
17   What is the difference between isotactic and atactic polypropylene?
18   List several outstanding properties of polypropylene.
19   What characteristic odor is emitted by polypropylene when burned? How does its odor compare with that of polyethylene?
20   What is a polyallomer?

POLYSTYRENE AND ABS PLASTICS

16

During the early period of injection molding, polystyrene resin was an important factor in the development of that process. Molders readily accepted the material because of lower cost, lighter weight, and good electrical properties as compared with cellulosics and vinyls then available. Its crystal clarity, high gloss, and outstanding resistance to strong acids made polystyrene a contender in uses where these qualities were needed. Housewares, containers, and radio and clock housings were some of the early, large-volume, molded items. Polystyrene still maintains a leading place as a high-volume resin, outranked only by polyethylene.

PREPARATION OF POLYSTYRENE

Styrene monomer is a clear liquid (boiling point of 293°F) that converts slowly to the solid polymer upon aging or by ultraviolet light unless inhibited with hydroquinone or catechol. Styrene, derived from coal, has been described in laboratory references since 1839, but a method of economic conversion to a stable polymer was to elude the industry for 100 years.

Production of synthetic rubber during World War II required tremendous quantities of styrene monomer, an important chemical for Buna-S rubber used for tires. As a result of U.S. government developmental funding, chemical firms such as Dow,

16-1 The crystal clarity of polystyrene is illustrated in this flash cube. (*Reprinted by permission of Branson Sonic Power.*)

Koppers, Monsanto, and Union Carbide achieved breakthroughs in producing styrene monomer of high purity and moderate price. Since 1945, many improved and modified polystyrene resins have appeared.

Polystyrene is commercially prepared by reacting benzene and ethylene in a gaseous phase to obtain ethyl benzene, which is dehydrogenated at 600°C. The resultant styrene monomer (vinyl benzene) is then polymerized to the transparent solid thermoplastic, polystyrene.

16-2 Dow Chemical Company's styrene complex at Midland, Mich. (*Reprinted by permission of Dow Chemical Co.*)

GRADES OF POLYSTYRENE

Many grades of resin are available to suit a variety of end uses. Styrene is also copolymerized or blended with other thermoplastics to improve specific properties or modified with additives to aid processing.

General purpose

Housewares, packaging, toys, wall tile, knobs, and novelties are molded in a wide choice of solid and mottled colors. General-purpose grades are brittle, making a sharp, metallic sound when dropped.

Light stabilized

Styrene copolymerized with methyl methacrylate is used for extruded sheet or molded panels for reflective or diffusion lighting applications.

Heat resistant

A copolymer of styrene with methyl styrene permits use in continuous service at 150 to 210°F. Appliance parts and small radios are examples.

16–3   Cups molded of impact styrene get handles welded to bodies by ultrasonic welding. (*Reprinted by permission of Branson Sonic Power.*)

Impact

A wide range of impact types, with medium to high shock resistance, is formulated by blending or copolymerizing styrene with butadiene or other rubber. Transparency is sacrificed, although a wide range of color is possible. Impact grades are used for refrigerator parts such as thermoformed door liners and inner shells, defroster trays, and light shields. Injection-molded products have thousands of uses, from toys, containers, and housings to appliance components.

Chemical resistance

A copolymer of styrene and acrylonitrile improves chemical resistance. Called SAN types, these resins are molded, extruded, and thermoformed into products requiring resistance to acids, alkalies, mineral oils, and detergents.

PROPERTIES OF POLYSTYRENE

Specific gravity

Polystyrenes vary from 1.04 to 1.14 g/cc, depending upon composition.

Mechanical properties

General-purpose grades exhibit higher tensile, compressive, and flexural strength than impact types. Parts molded from general-purpose resins require annealing if they are to be machined. Cracking is avoided by annealing in air at 140 to 170°F or in hot water. Stress relief requires from several hours to several days.

Chemical resistance

Polystyrene is soluble in many hydrocarbons, including its monomer. Bonding is rapidly accomplished with benzene, methyl ethyl ketone, toluene, and dioxane. Solvent vapors may cause crazing of unannealed parts. Acetone, kerosene, and heptane, although not true solvents, induce stresses or reduce mechanical properties. Water absorption of polystyrenes is exceptionally low, less than 0.25 percent, suggesting uses where dimensional and electrical stability are required.

Abrasion resistance

Polystyrene has a hard surface in comparison with most thermoplastics but is readily scratched. Dust collects on molded and formed products because of a static charge incurred in manufacture or stacking. Dry wiping only adds to surface static and increases the possibility of scratching. Resin producers incorporate antistatic agents that migrate to the surface and provide a slightly hydrophilic layer that dissipates electrostatic charges. Molders dip or spray products, using proprietary solutions for temporary static elimination.

Transparency

General-purpose, light-stabilized, and heat-resistant grades are highly transparent, transmitting as much as 90 percent visible light. Unstabilized types tend to yellow. Polystyrene is not recommended for outdoor weathering.

Electrical properties

Dielectric strength of general-purpose grades is exceptionally high (700 V/mil). Negligible water absorption permits use of polystyrene foil as spacers, slot liners, and coverings in many electrical circuits and components. Power factor, surface, and volume resistivity[1] remain fairly constant over a wide frequency range, permitting use in capacitors, coils, and radar assemblies.

Heat resistance

Polystyrenes are dependable in environments to 210°F for short-term, no-load conditions. Normally a maximum service temperature is 160°F continuous. All grades burn slowly, emitting a black, sooty smoke and a characteristic sweet odor.

EXPANDABLE POLYSTYRENE

Because of their unique combination of low cost, almost zero water absorption, rigidity, and low thermal conductivity, slabs of cellular polystyrene are finding growing markets as insulation for cold-storage warehouses, building foundations, and marine use. Other uses include flotation, display bases, novelties, and carvable blocks for sculptures.

Slabs are produced by extruding the polymer through an orifice to form logs 24 in. in diameter or boards. A volatile blowing agent such as methyl chloride is forced into the molten polymer prior to the time the melt reaches the die lips. As polymer emerges, it expands and cools to form a closed cell foam.

Expandable beads 0.012 to 0.090 in. in diameter are produced by polymerizing styrene in a water suspension, to which a volatile gas such as pentane is added. As monomer solidifies, the volatile is surrounded by polymer to form spherical, gas-filled beads. When heated to approximately 200°F, the blowing agent expands. Beads simultaneously fuse together because of the heat-induced tackiness of surface and the pressure of escaping gas.

An examination of any item, such as a hot-drink disposable cup, discloses that the original spheres were distended to form a closely knit structure, with each bead separate. Beads can be precolored to provide variations in product appearance, although natural white is used in the majority of products. Beads are purchased in fiber drums or in rail-car lots. Storage presents a fire hazard because of gas content, static charge, and flammability of polystyrene.

[1] These properties are discussed in Chap. 32.

Prior to use, beads require a preexpansion step carried out on a continuous or batch basis. The purpose is to expand the bead wall as much as 40 times the original surface volume. If this is not done, beads will expand poorly during processing. As expanded spheres cool, the internal blowing agent condenses, creating partial vacuum. The beads are then stored for a brief period to "season," during which time air permeates cell walls until equilibrium is attained with the gas.

Large blocks of expanded polystyrene are produced in a steam chest measuring $12 \times 8 \times 8$ ft, on a 30-min cycle. Parts with standard or irregular shapes are mass produced by a process similar to injection molding. Ice chests, packaging, flotation items, insulated cups, and lunch boxes represent a fair cross section of molded-bead application. Aluminum molds containing both steam-injecting and water-cooling features are required.

Beads are introduced and steamed to soften and force the spheres together. Water cooling extracts the heat used to fuse the beads. Cycles range from less than one to several minutes, depending upon part thickness. Thin, extruded sheets of foam are thermoformed into egg cartons, meat trays, and fruit separators, which compete with pulp-paper items. They have the advantage because of their cleaner appearance, color, and printability. Also, meat juices are not absorbed as with other packaging.

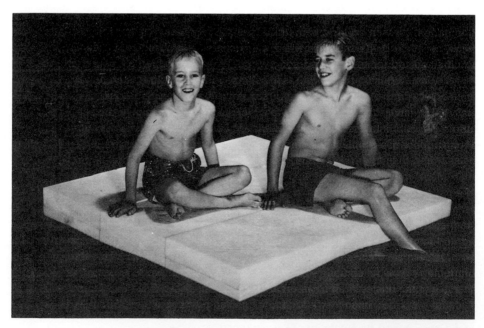

16–4 Flotation was one of the earliest uses for expanded polystyrene. One cubic foot of Styrofoam weighs 2 lb and has a buoyancy of 55 lb. (*Reprinted by permission of Dow Chemical Co.*)

ABS PLASTICS

As an outgrowth of developmental work that produced styrene-modified synthetic rubber, rubber-modified styrenes appeared at the end of World War II. In this manufacturing area, the plastics and rubber industries are closely allied.

The original synthetic rubber was an elastic material known as GR-S (Government Rubber, styrene) made up of 75 percent butadiene, 25 percent styrene. Compounded with carbon black and antioxidants, this rubber is widely used to manufacture passenger car tire tread stock. Buna-S, as it is now identified, is characterized by good abrasion resistance, low-temperature impact strength, and general toughness.

The war-induced shortage of natural rubber also spurred investigation of another synthetic elastomer for tires; acrylonitrile. Subsequently, nitrile rubber was to become more important as an apparel fiber marketed as Acrilan,[1] Creslan,[2] Dynel,[3] and Orlon.[4] Heavier monofilaments are employed for carpeting. The major uses for acrylonitrile rubber, other than its major application as a fiber, are hose, floor mats, printing rollers, and shoe parts, where its outstanding oil and grease resistance is used to advantage.

Copolymerized with styrene, acrylonitrile provides chemical resistance and improved impact strength. Perhaps the ultimate in toughness, weather resistance and other desirable properties are obtained in the production of the terpolymer, acrylonitrile-butadiene-styrene (ABS) plastics.

The copolymerization of styrene with either butadiene or acrylonitrile, or terpolymerization using all three, form resin intermediates that provide the widest range of properties extant in any single family of thermoplastics. Several methods of reacting the major components can be used to produce either SAN or ABS types:

Dry blending: Banbury mixers hot-mix polystyrene with the rubber portions, after which the blend is more intimately mixed by extruder and chopped into pellets. The term *impact polystyrene* is applied to this type. Rubber particles form individual areas surrounded by more brittle styrene. Impact strength is improved by a factor of 10 as the rubber absorbs most of the kinetic energy.

Graft polymerization: Previously copolymerized impact grade is dissolved in styrene monomer, which is then reacted in a kettle. As monomer polymerizes, some of its chains fuse to the rubber components, thus the term *graft copolymer*. High-impact polystyrene is produced by this method.

Suspension polymerization: Styrene and acrylonitrile monomer are diffused in water and copolymerized, then extruded and pelletized. Heat- and chemical-resistant types are produced.

Solvent polymerization: Styrene monomer is dissolved in a suitable solvent such as ethyl benzene together with acrylonitrile. The solution is passed through a reactor to

[1] Acrilan is the registered trade name of Chemstrand Corporation.
[2] Creslan is the registered trade name of American Cyanamid Company.
[3] Dynel is the registered trade name of Union Carbide Corporation.
[4] Orlon is the registered trade name of E. I. du Pont de Nemours & Company.

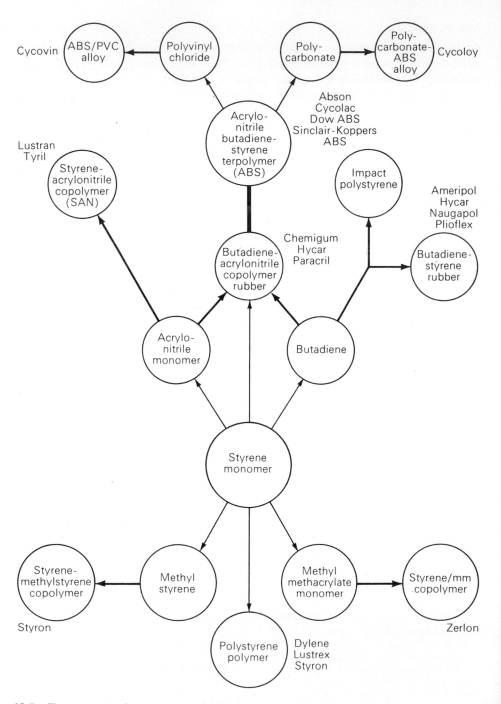

16–5   Thermoplastics from styrene and other monomers.

achieve a polymer yield to 60 percent. Unreacted monomers and solvent are removed through a vented extruder and recycled.

Resins can be tailored to satisfy a wide range of requirements by varying the ratio of styrene to rubber. ABS plastics provide a balance of properties, with tensile strength, impact resistance, and scratch resistance most outstanding. All resins are opaque, available from light ivory to dark colors. Standard grades are used in outdoor applications such as lawn-mower housings, golf carts, marine hardware, automotive instrument panels, and steering wheels. Special types suitable for electroplating are used for automotive trim, instrument clusters, and plumbing hardware.

The impact strength and ductility of ABS make it useful for such applications as luggage, fishing-tackle boxes and reels, and refrigerator parts since the material retains toughness down to −40°F.

Surface textures embossed into calendered sheets retain a measure of elastic memory permitting deep-drawn articles to be formed without losing their texture. Imitation leather, pebble grain, and hair-cell textures provide interesting surface treatments for luggage, cosmetic cases, and other thermoformed shapes.

Success in applying ABS in a wide variety of environmental conditions has prompted some manufacturers to install extensive thermoforming equipment in recent years. Canoes, small boats, and camper and trailer sections formed from large calen-

16-6 Banbury mixer; 200-lb-batch capacity. (*Reprinted by permission of Stewart Bolling and Co.*)

16-7   Snowmobile hood proves toughness and durability of ABS. (*Reprinted by permission of Marbon Div., Borg Warner Corp.*)

dered sheet are now commonplace. Sections are joined by solvent bonding (15 percent resin solids dissolved in methyl ethyl ketone) or ultrasonic welding. Of interest to car makers is an experimental car body thermoformed in two sections and ultrasonically welded. The vehicle cannot corrode, requires no painting, and is easily repaired. ABS resins are slow burning, emitting an odor of rubber. Alloys of ABS and polyvinyl chloride are self-extinguishing (see Table 16-1).

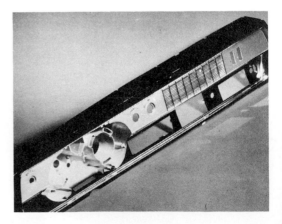

16-8   Automobile instrument panel injection molded from ABS. (*Reprinted by permission of Marbon Div., Borg Warner Corp.*)

TABLE 16-1   Properties comparison chart

General-purpose polystyrene				
Property	Regular	Lubricated	Heat resistant	Light stable
Specific gravity	1.04	1.05	1.04	1.04
Tensile strength, psi	6,000	6,000	7,000	6,000
Tensile modulus, psi	460,000	500,000	500,000	500,000
Impact strength, ft-lb	0.45	0.45	0.45	0.45
Elongation, percent	1.8	1.5	2.3	1.5
Heat-deflection temperature, °F	195	176	213	176
Clarity	. . .	Excellent	. . .	. . .
Gloss	. . .	Excellent	. . .	. . .

	Styrene-butadiene		Styrene-acrylonitrile	
Property	Impact	High impact	Heat resistant	High modulus
Specific gravity	1.06	1.04	1.05	1.08
Tensile strength, psi	4,600	4,500	6,700	10,000
Tensile modulus, psi	410,000	338,000	410,000	550,000
Impact strength, ft-lb	2.1	3.6	1.5	0.4
Elongation, percent	30.0	56.0	5.5	3.5
Heat-deflection temperature, °F	178	195	210	215
Clarity	Good	Opaque	Translucent	. . .
Gloss	Good	. . .	Excellent	. . .

Acrylonitrile-butadiene-styrene (ABS)			
Property	Medium impact	High impact	Heat resistant
Specific gravity	1.05	1.03	1.05
Tensile strength, psi	7,700	5,000	7,000
Tensile modulus, psi	350,000	240,000	330,000
Impact strength, ft-lb	3.4	8.0	5.0
Elongation, percent	3.0	50.0	30.0
Heat-deflection temperature, °F	200	185	225
Clarity	. . .	Opaque	. . .
Gloss	. . .	Fair to excellent	. . .

GLOSSARY

Acrylonitrile   Synthetic monomer produced from propylene and ammonia.

Annealing   Process whereby stresses are relieved by warming a thermoplastic at a temperature below its softening point.

Butadiene   Petroleum hydrocarbon obtained from butane.

Kinetic energy   Energy of motion.

Proprietary   A proprietary product is one in which the ingredients are not disclosed, or they are protected by patent.

REVIEW QUESTIONS

1   What other industry contributed to the development of polystyrene for the plastics industry?
2   From what major chemicals is styrene monomer derived? What is the source of these hydrocarbons?
3   Prepare a list of different grades of polystyrene and give examples of products made from each type.
4   What is the purpose of annealing? How is this carried out?
5   Name several solvents for polystyrene.
6   Why is a knowledge of specific solvents important?
7   What are the outstanding properties of polystyrene? What are some disadvantages to be recognized?
8   How can polystyrene be identified by sound? By burning?
9   List several applications for expanded polystyrene beads.
10   What hazards exist in handling and storing expandable beads?
11   What is the purpose of preexpanding beads prior to molding?
12   Which copolymer does the abbreviation SAN identify?
13   Which monomers are used to produce an ABS terpolymer?
14   What is a terpolymer?
15   What is meant by elastic memory?
16   What two nonmechanical methods are used to join ABS?
17   What odor aids in identifying ABS materials?

VINYL PLASTICS

17

Vinyl resins make up the third largest group of commercially significant thermo-plastics. The term *vinyl* applied to the unsaturated molecule $CH_2 = CH$ can be used to describe many linear resins that polymerize from a simple organic monomer. Thus monomers having the same general formula $R - CH = CH_2$ can polymerize to build long chains containing as many as several hundred monomer molecules.

$$-CH-CH_2-CH-CH_2-CH-CH_2-$$
$$\quad\ \ |\qquad\qquad |\qquad\qquad |$$
$$\quad\ \ R\qquad\qquad R\qquad\qquad R$$

In the general formula above, the most important feature is the long hydro-carbon chain. With R as a substituent or side chain recurring at intervals, it should be apparent that a wide variety of vinyl resins can be produced depending upon the degree or type of substitution. (See the section on vinyl reactions at the end of this chapter).

Vinyl reactions may be more fully understood by a graphic representation, as shown in Fig. 17-1. Note that ethylene, from petroleum or natural gas, and acetylene, from coal, are the important raw materials for vinyl plastics.

POLYVINYL CHLORIDE (PVC)

The reaction by which acetylene and chlorine formed vinyl chloride was recorded in France over 130 years ago, and resinous substances based on similar experiments

263

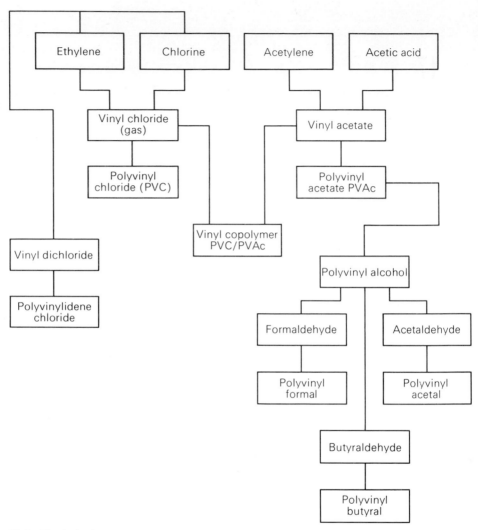

17-1 Vinyl plastics.

were periodically reported by chemists in other countries. The polymer polyvinyl chloride was then, and is today, difficult to dissolve in organic solvents and presents problems in processing unless modified.

These problems arise because the softening temperature of PVC is so close to its decomposition point. Being very inert chemically, PVC resists solvating in common solvents, and methods had to be found to provide solution coatings and processable resins.

During World War II, research into polyvinyls was accelerated when supplies of natural rubber were cut off from both Germany and the United States. German efforts to find a substitute for rubber in military vehicle tires, waterproof fabrics, and wire insulation led to the development of many high-quality vinyl compositions. In the United States, research chemists at B. F. Goodrich, Union Carbide, and Firestone worked to extend stockpiles of natural rubber until a satisfactory replacement could be found.

As work to develop synthetic rubber was progressing, the development of vinyls for use as coatings for tents, tarpaulins, rainwear, liferafts, and many other military applications enabled these resins to gain a firm foothold in peacetime markets, which it has never relinquished.

The key to processing vinyls, particularly PVC, was found in the incorporation of *plasticizers*. A plasticizer may be considered as a high-boiling solvent that has certain effects on the resin; that is, a plasticizer acts as an external molecular lubricant, permitting greater freedom of chain movement.

We shall see later that the successful application of the first plastic, cellulose nitrate (in 1869) occurred because of the incorporation of camphor as a plasticizer. Polyvinyl chloride was originally made processable by addition of tricresyl phosphate as a plasticizer.

Plasticizer technology represents an area of organic chemistry that today is responsible for an annual production of more than 1 billion lb. Hundreds of plasticizers are now available, with 75 percent used to modify properties of vinyl plastics. It is not unusual for companies engaged in manufacturing coated fabrics or formulating adhesives to maintain a file of 5,000 formulas based on combinations of plasticizers and vinyl resins.

Selection of the proper plasticizers is based on nine important factors. Plasticizers should *not*: (1) migrate or exude under service conditions; (2) contribute to flammability; (3) be toxic; (4) be expensive; (5) result in bleeding of color; (6) cause blocking of coated fabrics; (7) detract from, but should aid in preparation of solutions; (8) reduce chemical resistance or increase water absorption; nor (9) be extractable by sunlight.

The use of plasticizers has a marked effect on processing and on the final properties of the product.

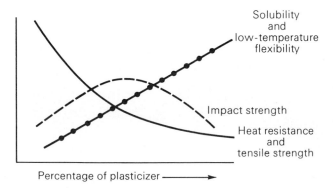

Percentage of plasticizer ⟶

Solubility and low-temperature flexibility

Impact strength

Heat resistance and tensile strength

17-2   Effect of plasticizer on properties.

Preparation

Vinyl chloride may be obtained by any of several methods, depending upon geographic location. Germany, because of its proximity to large coal deposits, utilizes acetylene and hydrogen chloride or hydrochloric acid to produce the gaseous monomer. This process was originally used in the United States until the availability of ethylene derived from petrochemicals provided a method of producing vinyl chloride.

Ethylene, obtained by thermal cracking of petroleum, or from the propane fraction of natural gas, is more widely used in this country and abroad because it is important as a starting chemical for polyethylene, polystyrene, acrylics, and other high-volume plastics.

Processing and application

Calendering   Calendering is a method for producing film and sheet for a variety of familiar end uses, primarily upholstery materials and waterproof fabrics such as rainwear, shower curtains, and baby pants. Calendered sheets may be transparent, pigmented to any desired shade, embossed, or printed. They are laminated to knitted cotton for upholstery, or applied as facings for nylon or other wide mesh fabrics for tear-resistant tarpaulins. Flexibility, abrasion resistance, resistance to household chemicals, and self-extinguishing properties, plus low cost and excellent heat sealability are reasons for its popularity.

17-3   Four-roll inverted L calender. (*Reprinted by permission of Stewart Bolling and Co.*)

Extrusion   Extruded PVC meets many stringent electrical codes when used as wire insulation. Formulations vary, depending on operating temperatures and whether or not ozone, grease, oil, or ultraviolet may contribute to deterioration. Other uses are extruded medical and chemical tubing, garden hose, refrigerator door gaskets, and vacuum-sweeper bumper strips. In these applications, water and chemical resistance, or the ability to recover when deformed, are paramount.

Injection molding   Injection-molded parts are manufactured for use in environments where the above-mentioned properties are required. Depending on formulation, resins may be extremely flexible, semirigid, or rigid. Common examples are bicycle handlebar grips, atomizer-type closures, pipe fittings, and snap-on devices.

Blow molding   Blow-molded containers are beginning to appear in quantity largely because the FDA recently sanctioned certain plasticizers as nontoxic. Transparency, toughness, and crack resistance are properties that enable PVC to compete with other thermoplastics in packaging food.

Thermoforming   Thermoformed PVC in corrugated or other design competes favorably with polyester and acrylic for applications in awnings, glazing, and sky-lights. It is also recommended for light-diffuser panels because of its inherent flame resistance.

Coating   Fabrics utilizing glass cloth, nylon, cotton, and other webs are coated with PVC on one or both sides for an unlimited number of industrial, commercial, and decorative uses: school and theater drapery, oil well covers, tents, awnings, clothing for personnel working in chemical industries, and for flexible ducting.

17-4   Cord sets, previously insulated with an extrusion coating of PVC, being inserted into injection machine to have plug molded on. (*Reprinted by permission of Pacific Electricord Co.*)

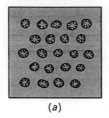   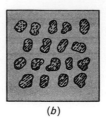   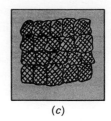

(a)   (b)   (c)

17-5   Fusion of PVC Plastisol. (a) Vinyl resin suspended in and partially solvated by plasticizer; (b) vinyl particles swelled by heating; and (c) plasticizer completely dispersed within resin structure.

Lead-filled, vinyl-coated fabrics[1] are used for sound dampening in a host of applications, from jet aircraft to typewriters, and even as x-ray barriers.

PLASTISOLS

Polyvinyl chloride can be plasticized with a high proportion of plasticizer into a viscous solution called a *plastisol*, or *fusing vinyl*. When heated from 330 to 450°F, the plasticizer forms a more intimate part of the structure, and the liquid becomes firm. The degree of hardness is determined by a Shore durometer, which measures indentation at room temperature.

Plastisols resemble corn starch in water: unless heated, the starch remains as discrete particles. When heated, the particles expand and coalesce as water penetrates, and a gel is produced. The two components cannot be separated. In the same way, vinyl and plasticizer fuse under heat to produce plastisol.

Plastisols are used for dipcoating metal products such as baskets for dishwashers, electroplating racks, tool handles, and chemical ware. They are cast into hollow molds for doll parts, novelties, and a variety of industrial products. Fabric coaters employ plastisols where heavy build-ups, single-surface coating, or different formulations or colors are required on opposite sides.

FOAMED VINYLS

Foamed vinyls are produced as either open- or closed-cell structures, depending upon the method used. To produce open-cell foam, a gas or chemical blowing agent is combined with the plastisol, which is then spread on a conveyor belt or on release paper and carried through a fusing oven. Closed-cell foam is produced by confining the material in molds during heating, which retains the blowing agent in nonconnecting cells. Closed-cell foams are used in flotation devices and shaped cushions. Open-cell foams are used for rug backing, cold-weather apparel, sporting goods, automotive arm rests, visors, and dashboard covering.

[1] For example, Coustifab, a trade name of Cordo Chemical Division, Ferro Corporation.

ORGANOSOLS

Dissolving a vinyl in a solvent with small amounts of plasticizers results in an *organosol*. Such solutions are important in textile coating for better penetration into close-weave fabrics, or as top coats for paper and cardboard. Because of lower viscosity, organosols can be extended with fillers such as clays or carbon black to modify properties. Notebook and book covers, lamp shades, and bottle-cap liners are examples of organosol-coated products.

Rigid PVC   Even with plasticization, vinyl resins are subject to degradation when overheated. This is a function of the heat history of the resin during a particular process. For example, calendering exposes the resin to less heat and mechanical shearing than does extrusion or screw injection. Also there is no opportunity for material to "hang up" on calendering equipment. Overheating results in dechlorination of PVC, thus liberating hydrogen chloride. The result is severe corrosion of machine-heating cylinders, screws, and dies. Degradation is autocatalytic in that a small amount of decomposed material starts what amounts to a chain reaction to discolor and decompose quantities of resin.

17-6   Hollow plastic head produced in electroformed mold by slush casting; basketball produced by rotocasting; change mat made by straight casting; bottle-cap crowns cast and pressure molded from FDA grade of PVC; nylon fly line; PVC coating applied by die wiping; and plier handles hot dip coated. (*Reprinted by permission of Chemical Products Corp.*)

Because PVC cannot be processed unless plasticized, rigid types were not available until stabilizers were developed to absorb or accept the hydrogen chloride released, and thus prevent runaway reaction. Stabilizers based on organo-tin compounds make possible the extrusion of rigid pipe for conveying chemicals, irrigation and potable water, and tubing for electrical conduit. Injection-molded items include pipe fittings, pump parts, and chemical laboratory equipment.

Properties

An extensive comparison of flexible and rigid vinyls would fill several volumes. Most high-volume processors compound formulations to suit particular equipment and product requirements. The competitive nature of vinyl processing indicates increased in-plant compounding to reduce cost and develop superior proprietary articles. Technicians who understand polymer compounding and product evaluation will be in increased demand in the future.

Specific gravity   The average density of PVC ranges from 1.32 to 1.45 but may be reduced to 1.16 for highly plasticized elastomers and increased to 2.00 for mineral-filled compositions.

Impact strength   This is applicable to rigid PVC only and varies between 0.5 and 18.0 ft-lb tested under standard conditions.

Heat resistance   Most vinyl resins are not recommended for continuous exposure above 160°F unless specially formulated to meet Underwriters Laboratory requirements, as in wire-coating grades for appliances, machine tools, radio, TV, and transformer conductors. Low-temperature flexibility is possible to −65°F.

Chemical resistance   Polyvinyl chloride resists most chemicals and solvents and is extremely water repellent. Depending on formulation, elastomers are soluble in acetone, toluene, methyl ethyl ketone (MEK), tetrahydrofuran (THF), methyl isobutyl ketone (MIBK), dimethylformamide (DMF), and xylene.

Optical qualities   PVC is available in extremely clear film, sheet, and pellet forms.

Weather resistance   Properly compounded and processed articles have a long service life. Draperies hung in New York City schools have retained their flexibility and color for 15 years, and outdoor cable insulation stabilized with dark pigments and antioxidants has provided reliable service for over 20 years in outdoor use. However, clear flexible vinyls used, for example, as rear windows in convertible cars gradually turn yellow, then brown to black. They also become brittle, all within the span of a few years. Expected to replace wood, stucco, and brick in home building, rigid vinyl siding is assumed to last indefinitely.

Flame resistance   Because they contain chlorine, PVC resins are self-extinguishing. When burned, they emit a pungent odor and a smoky flame, with charring of the polymer.

Fungus resistance   Some formulations are susceptible to bacterial growth. Coated fabrics, film, and sheeting to be used in tropical areas require fungicides and selected plasticizers.

POLYVINYL ACETATE (PVAc)

Although it is chemically similar to PVC, polyvinyl acetate is quite different in physical properties. It is transparent and resists ultraviolet light but is brittle at low temperatures. It is seldom used alone for molding since it becomes tacky at elevated temperature.

PVAc is formed by the reaction between acetylene and acetic acid, which is carried out in either a solvent solution or water emulsion. Water dispersions are sold as the familiar "white glues" and are also used for sizing agents in textile processes. Coated on paper, these resins provide grease- and water-proof characteristics.

Polyvinyl acetate is soluble in many organic solvents and is compatible with nitrocellulose for lacquers. Filled with wood flour, the solution is sold as patching compounds for furniture and woodwork. Toughness and adhesion are evidenced by the fact that the patch can be sanded, nailed, or sawed.

In addition to its use as adhesives for paper tape, bread wrap, bookbinding, and wood, PVAc is the major ingredient in emulsion-type paints. These are low cost, odorless, and fast drying, and tools can be cleaned by water rinsing, which makes them popular with the do-it-yourself home owner.

POLYVINYL CHLORIDE-ACETATE

Polyvinyl chloride and polyvinyl acetate are at opposite extremes in physical and chemical properties; the first is flexible, chemically inert, and has poor adhesion characteristics, but the other is hard, soluble, and easily tackified. It would appear that a combination of the two homopolymers would yield a series of materials with the best properties of each. Unfortunately mixtures of PVC and PVAc are not practical. Instead, the monomers of each are copolymerized to produce the vinyl copolymer.

PVC-PVAc resins generally contain 85 percent or more vinyl chloride, giving them the characteristic strength, water and chemical resistance, and flame-proof quality of polyvinyl chloride. The vinyl acetate portion of the chain makes the copolymer more soluble in organic solvents and contributes to moldability at lower temperatures. Increasing the proportion of vinyl acetate improves abrasion resistance. Formulations of copolymer and asbestos account for the largest single market in floor tile and linoleum. These products are produced by calendering.

As with PVC, the vinyl copolymer decomposes rapidly when overheated in processing. Iron oxide and zinc oxide may act as decomposition catalysts, and their use as pigments should be avoided. Widely used as a lubricant for polystyrene, zinc stearate can cause violent decomposition of both PVC and copolymer.

POLYVINYLIDENE CHLORIDE (PVDC)

Thermal cracking of trichlorethane yields vinylidene chloride that, when polymerized by itself, produces a strong, tough resin with a high softening point but low solubility. It may be classed as a specialty resin that finds use as a thin film for food wrap and as an extruded monofilament for auto upholstery, furniture webbing, and similar applications.

OTHER SPECIALTY VINYL RESINS

Polyvinyl alcohol (PVA)

Vinyl alcohol cannot exist in monomeric form since it undergoes a molecular re-arrangement to form acetaldehyde. The polymer must therefore be made by acid or alkaline hydrolysis of polyvinyl acetate. PVA is water soluble to the degree that hydroxyl (OH) groups are present in the chain. This also depends on the degree of polymerization, which is, in turn, dependent on that of the original polyvinyl acetate. PVA resins are mainly used as adhesives that can be remoistened, as grease-resistance coatings, and as sizings for loom operations.

Of particular interest to laminators and molders of polyester and phenolic structures is PVA film produced in thicknesses to 6 mil. Plasticized with glycerine or glycols, the film exhibits no stretch under temperatures used to cure the laminate. In fact, some shrinkage and embrittlement occur because of the loss of plasticizer, with the result that the film is used on a one-time basis for vacuum-bag or lagging operations. It is readily heat sealed with hot-bar sealers, more effective seals being made by first wiping the seam with a moistened sponge.

PVA solutions are available for brush or spray release agents that form a tough, wrinkle-free, continuous coating on tooling surfaces. For laminates requiring painting or bonding, their advantage is that they can be easily removed with water.

Scrap PVA film can be put into solution in hot water and tinted with water-soluble dyes such as confectioner's colors (often available in a glycerine base). Small amounts of methyl alcohol can be added if desired to provide a fast-drying release agent for laminating tools. PVA films are also used to package soaps, swimming pool chemicals, and dry bleaches. The package is simply tossed, unopened, into water.

Limited quantities of polyvinyl alcohol are used as molding or extrusion com-pounds. Its excellent resistance to gasoline, hot oils, and greases allows its use as tubing and hose to convey these materials.

Polyvinyl acetals

When polyvinyl alcohol is reacted with an aldehyde, an acetal is produced, which is not to be confused with polyacetal (a more recent development), discussed in Chap. 19.

Polyvinyl acetal Polyvinyl acetal is produced with acetaldehyde. Adhesive formulations based on the acetal alone, or in combination with phenolic or epoxy-phenolic resins, represent the largest single use for this acetal. As a dry-film adhesive unsupported or calendered on a carrier fabric, it is used to bond metal to metal and to bond honeycomb structures.

Polyvinyl formal Formed from PVA and formaldehyde, polyvinyl formal is used for coating fine magnet wire and to provide high-peel-strength adhesives when modified with epoxy systems.

17-7 U.S. Senator Alan Bible looks on as U.S. Senator Peter Dominick wields sledge hammer against laminated security (safety) glass with polyvinyl butyral interlayer. (*Reprinted by permission of Monsanto Co.*)

Polyvinyl butyral This acetal, formed from butyraldehyde and PVA, is used as the thermoplastic interlayer for automotive safety glass because of its toughness, flexibility, and outstanding adhesion to glass.

Polyvinyl butyral also finds application as can coatings, waterproofing natural and synthetic textiles, and as a vacuum-bag film where conformity to contour is required.

TYPICAL VINYL REACTIONS
Polyvinyl chloride

1 Ethylene + chlorine → Ethylene dichloride (by chlorination of ethylene)

2 Ethylene dichloride → Vinyl chloride + hydrogen chloride (heat)

3 Vinyl chloride → polymer (catalyst)

4 $HC{\equiv}CH$ + HCl → Vinyl chloride (from acetylene)
   Acetylene   Hydrochloric acid

Polyvinyl acetate

1 $HC{\equiv}CH$ + Acetic acid → Vinyl acetate
   Acetylene

$$
\begin{array}{cc}
& \text{O} \quad \text{H} \\
\text{H} \quad \text{O}-\text{C}-\text{C}-\text{H} \\
\text{2} \quad \text{C}=\text{C} \qquad \text{H} \\
\text{H} \quad \text{H}
\end{array}
\quad\longrightarrow\quad
\left[
\begin{array}{c}
\text{H} \qquad\qquad \text{H} \\
\text{O}=\text{C}-\text{C}-\text{H} \quad \text{O}=\text{C}-\text{C}-\text{H} \\
\text{H} \qquad\qquad \text{H} \\
\text{H} \quad \text{O} \qquad \text{H} \quad \text{H} \qquad \text{O} \\
\text{C}-\text{C} \quad\quad \text{C}-\text{C} \quad\quad \text{C} \\
\text{H} \quad \text{H} \qquad \text{H} \quad \text{H} \qquad \text{H}
\end{array}
\right]_{n}
$$

Vinyl acetate → Polyvinyl acetate

Polyvinyl chloride-acetate

$$
\begin{array}{c}
& & & \text{H} \\
\text{H} \quad \text{H} \quad \text{H} \quad \text{O}-\text{C}-\text{C}-\text{H} \\
\text{C}-\text{C}+\text{C}=\text{C} \qquad\quad \text{H} \\
\text{H} \quad \text{Cl} \quad \text{H} \quad \text{H}
\end{array}
\quad\rightarrow\quad
\left[
\begin{array}{c}
\text{H} \ \text{H} \ \text{H} \ \text{H} \ \text{H} \qquad \text{H} \qquad \text{H} \ \text{H} \\
\text{C}-\text{C}-\text{C}-\text{C}-\text{C} \quad \text{C} \quad \text{C}-\text{C} \\
\text{H} \ \text{Cl} \ \text{H} \ \text{Cl} \ \text{H} \qquad \text{O} \qquad \text{H} \ \text{Cl} \\
\qquad\qquad\qquad\qquad\qquad \text{H} \\
\qquad\qquad\qquad\qquad \text{O}=\text{C}-\text{C}-\text{H} \\
\qquad\qquad\qquad\qquad\qquad \text{H}
\end{array}
\right]_{n}
$$

Vinyl chloride   Vinyl acetate

Polyvinyl chloride-acetate copolymer

Polyvinylidene chloride

1 $C_2H_2 + 2Cl_2 \longrightarrow HCl_2C-CCl_2H$
   Acetylene + chlorine   Tetrachloroethane

2 $HCl_2C-CCl_2H \longrightarrow HClC-CCl_2H + Cl$
   Trichloroethane   Chlorine

3 $HClC-CCl_2H \longrightarrow$
   $$
   \begin{array}{cc}
   \text{H} & \text{Cl} \\
   \text{C}-\text{C} & + \text{Cl} \\
   \text{H} & \text{Cl}
   \end{array}
   $$
   Vinyl dichloride   Chlorine

4
$$
\begin{array}{cc}
\text{H} & \text{Cl} \\
\text{C}-\text{C} \\
\text{H} & \text{Cl}
\end{array}
\quad\longrightarrow\quad
\left[
\begin{array}{c}
\text{H} \ \text{Cl} \ \text{H} \ \text{Cl} \ \text{H} \ \text{Cl} \ \text{H} \ \text{Cl} \\
\text{C}-\text{C}-\text{C}-\text{C}-\text{C}-\text{C}-\text{C}-\text{C} \\
\text{H} \ \text{Cl} \ \text{H} \ \text{Cl} \ \text{H} \ \text{Cl} \ \text{H} \ \text{Cl}
\end{array}
\right]_{n}
$$

Vinyl dichloride   Polyvinylidene chloride

Polyvinyl alcohol

Polyvinyl acetate — acid or alkali → Polyvinyl alcohol

Polyvinyl acetals

1　Polyvinyl alcohol + Formaldehyde → Polyvinyl formal

2　Polyvinyl alcohol + Acetaldehyde → Polyvinyl acetal

3　Polyvinyl alcohol + Butyraldehyde → Polyvinyl butyral

GLOSSARY

Blocking   Minor adhesion between contacting layers of film or sheet occurring during storage or use.

Embossed   Raised texture on the surface of a material.

Ozone   Molecular oxygen containing three atoms of oxygen, as compared with ordinary oxygen, which has only two atoms. Ozone is a powerful oxidizing agent.

Solvation   Process of swelling (or putting into solution) a resin by the action of a solvent or a plasticizer.

REVIEW QUESTIONS

1   What is the function of a chemical plasticizer?
2   What are the factors that determine the selection of the proper plasticizer?
3   List several uses for PVC film or sheet.
4   List several commercial applications for PVC-coated fabrics.
5   How is a plastisol prepared?
6   List some applications for fusible vinyls.
7   What is the difference between a plastisol and an organosol?
8   What are the hazards if PVC is overheated?
9   What are some outstanding properties of PVC?
10   Why is PVC self-extinguishing?
11   What major markets are served by polyvinyl acetate?
12   What is the largest single market for the vinyl copolymer?
13   What is the danger in using zinc stearate as a lubricant in vinyl processing?
14   How do laminators of reinforced plastics use PVA film?
15   Which acetal is used as the interlayer in automotive and other types of safety glass? What is the underlying principle involved that prevents the glass from flying upon impact?

FLUOROCARBON PLASTICS

18

The family of fluorocarbons (or fluoroplastics) is unique in its properties, applications, and in some methods of processing (see Table 18-1 at end of chapter). The term *fluorocarbon* is used to describe resins containing only the elements *fluorine* and *carbon*. A *fluoroplastic* refers to resins in which other elements replace part of the fluorine.

Fluorocarbons appeared about 1930 as a refrigerant trade named Freon.[1] Other uses such as aerosol propellants and blowing agents for urethane foams have since created an increasing market for these materials.

In 1938 du Pont researchers succeeded in forming the first polymeric resin based on fluorocarbons, although an accidental circumstance contributed to its formation. According to the story, an alert technician noted a marked pressure drop in a cylinder containing the gas, tetrafluoroethylene, which had been stored in dry ice. Yet the cylinder weighed the same, indicating that there had been no leakage. A spontaneous polymerization of the gas had taken place, forming the polymer polytetrafluoroethylene (Teflon TFE), a white powder.

Many other fluoroplastics have been developed since, but the original TFE type still remains as the prototype for chemical resistance, electrical properties, and usefulness both at high and subzero temperatures.

[1] Freon is the registered trade name of E. I. du Pont de Nemours & Company.

POLYTETRAFLUOROETHYLENE (TFE)

The source of fluorine (also called fluorite or calcium fluoride) is the mineral fluorspar, which is mined in Illinois, Kentucky, New Hampshire, and New Mexico.

Processing and applications

The homopolymer is obtained as a dry powder by reacting fluorspar (CaF_2) with sulfuric acid and chloroform to yield the fluorocarbon, chloro-difluoromethane (Freon 22). This gas is subjected to high temperature to remove hydrogen chloride and produce the monomer tetrafluoroethylene. Final polymerization is effected with water containing a peroxide catalyst.

TFE cannot be processed by conventional molding methods. The properties that make it so desirable for high-temperature use and its extremely high resistance to solvents prevent melting, plasticizing, or dissolving to any practical extent. TFE is highly crystalline, with a melt viscosity too high to permit flow in injection machines or screw extruders.

Products of limited size and shape are made by preforming the powder to approximate shape under high pressure with or without heat. Preforms are then transferred to an oven or salt bath for fusing at approximately 620°F, followed by cooling. This process, called sintering, is similar to that used for powdered metal. Products can be further processed in molds (coining) or machined to precise tolerances. Gaskets, seals, O rings, and packings are produced in quantity by these methods.

Rods and tubes are produced by a process known as ram extrusion. Powder is introduced near the top of a long vertical cylinder, where it is compacted by a reciprocating ram. Alternate feeding and ramming force the powder through heat zones where it coalesces, cools, and emerges as rod or tubing. Size is limited to 2-in. maximum diameter. Larger tubes and rods or billets are made by the molding process described earlier, as is sheet up to 2 sq ft.

Properties

TFE is characterized by a waxy "feel," more slippery than polyethylene. It is nonflammable, with no odor or flame when burned. Products are opaque ivory in color, becoming translucent in thin sections.

The closely packed fluoride-carbon structure makes TFE one of the heaviest unfilled resins, with specific gravity of 2.13 to 2.2. Heat resistance of TFE is outstanding, leading to uses in areas where most thermoplastic materials would decompose. Being inert to all solvents and practically all acids and alkalies, TFE is used in industrial applications such as tank and pipe linings, filter media, and numerous valve and pump components.

The antistick quality of TFE makes it useful for covering conveyor rollers, chutes, and drum rollers in bakeries, fabric coating, and other industries handling sticky materials. Rolling pins, frying pans, and cooking utensils are modern examples of the antistick, heat resistance, and chemical inertness of TFE.

Diaphragms, seals and O rings in aircraft, space hardware, and cryogenic systems are made from TFE because it remains flexible even at −450°F.

Electrical properties of TFE resin are excellent both in normal conditions and in humid environment. TFE resin excels most thermoplastics under severe arcing conditions and exhibits high-insulation values at high operating temperatures. Computer circuits, thermocouples, oven lead-in wire, and coaxial cable are effectively insulated.

POLYCHLOROTRIFLUOROETHYLENE (CTFE)

Replacing a fluorine atom with chlorine results in the formation of the fluoroplastic CTFE, which is significantly different from TFE. This thermoplastic is extruded and injection-molded on conventional equipment and is also compression-molded. Like TFE, it is characterized by chemical inertness, high-temperature resistance, and low-temperature flexibility.

Because of the high cost of CTFE ($6 to $8/lb), some molders keep a special injection cylinder on hand and make a changeover to avoid purging and contaminating this expensive resin. CTFE is transparent in thickness to 0.125 in., becoming more translucent as sections increase. In transparent sheets, it is used for windows in infrared tracking missiles, flexible printed circuits, and instrument faces.

18-1 Polished metal surfaces of rolls on a papermaking machine are protected from pitting by a 20-mil covering of Teflon. (*Reprinted by permission of E. I. du Pont de Nemours & Co.*)

Extruded over wire, CTFE is used in computer circuits, chemical processing equipment, and other applications where thin-wall insulation, corrosion resistance, or low-temperature flexibility is required. Tubing, expanded while emerging from the die, can be shrunk after cooling to half the original diameter with a hot air blower. This process is used to protect electronic terminals, connectors, and other components against shock, loosening, moisture, and dirt.

FLUORINATED ETHYLENE PROPYLENE COPOLYMER (FEP)

A copolymer of tetrafluoroethylene and hexafluoropropene called FEP is melt processable by injection and extrusion methods. Like other fluoroplastics, it is a useful resin where chemical and heat resistance justify its higher cost. FEP is used in applications similar to those used for TFE, when the lower processing costs justify its use.

POLYVINYL FLUORIDE (PVF)

Acetylene gas reacted with hydrogen fluoride produces a tough, transparent polymer useful for applications where moderate heat and chemical resistance are required. PVF is an excellent film-forming resin and is cast or extruded for heat-sealable packaging for chemicals, electrical wrapping tape, and parting agent for laminates.

POLYVINYLIDENE FLUORIDE (PVF2)

Direct heating of difluorochloroethane, in turn produced from hydrogen fluoride and methyl chloroform, produces vinylidene fluoride. Its transparent polymer is capable

18-2   Flexible circuit made of Teflon FEP used in airborne countermeasures receiver. (*Reprinted by permission of E. I. du Pont de Nemours & Co.*)

18-3 World's largest structure (129 million ft), the Apollo Moon Rocket Assembly Building, has reinforced plastic windows protected by a thin film of Tedlar (polyvinyl fluoride). (*Reprinted by permission of Filon Div., Vistron Corp.*)

TABLE 18-1 Fluorocarbon comparison chart

Property	PTFE	PCTFE	FEP	PVF	PVF$_2$
Specific gravity	2.15	2.12	2.15	1.39	1.76
Tensile strength, psi	2,500–3,500	4,600	2,500–3,500	. . .	7,000
Elongation, percent	2–300	120	300	. . .	300
Water absorption, 24 h	0.00	0.00	0.01	. . .	0.04
Max. service temperature, °F	500	390	400	. . .	300
Min. low-temperature serviceability, °F	−450	−400	−425	. . .	−80
Dielectric strength, V/mil	3,200 (1 mil)	3,700 (1 mil)	6,500 (1 mil)	. . .	1,280 (8 mil)
Chemical resistance	Excellent	Excellent	Excellent	. . .	Soluble in ketones, and esters such as dimethyl phthalate; otherwise excellent
Cost per cubic inch			45–60 cents		

of wide color acceptance for wire coating, flexible tubing, and coatings for exterior siding.

TYPICAL FLUOROCARBON REACTIONS

TFE

1 $CaF_2 + H_2SO_4 \longrightarrow 2HF + CaSO_4$
   Fluorspar   Sulfuric   Hydrogen Calcium
               acid   fluoride sulfate

2 $2HF + CHCl_3 \longrightarrow CHClF_2 + 2HCl$
   Hydrogen Chloroform   Chloro   Hydrochloric
   fluoride   difluoromethane acid

3 $2CHClF_2 \longrightarrow CF_2CF_2 + 2HCl$
                           Tetrafluoro-
                           ethylene

4 $CF_2CF_2 \longrightarrow$

$$\left[\begin{array}{cc} F & F \\ | & | \\ C\!-\!C \\ | & | \\ F & F \end{array} \right]_n$$

Polytetrafluoroethylene

CTFE

1 $CCl_3CCl_3 \xrightarrow[\text{catalyst}]{\text{HF}} CClFCF_2$
   Hexachloroethane   Chlorotrifluoroethylene

2 $CClFCF_2 \longrightarrow$

$$\left[\begin{array}{cc} F & Cl \\ | & | \\ C\!-\!C \\ | & | \\ F & F \end{array} \right]_n$$

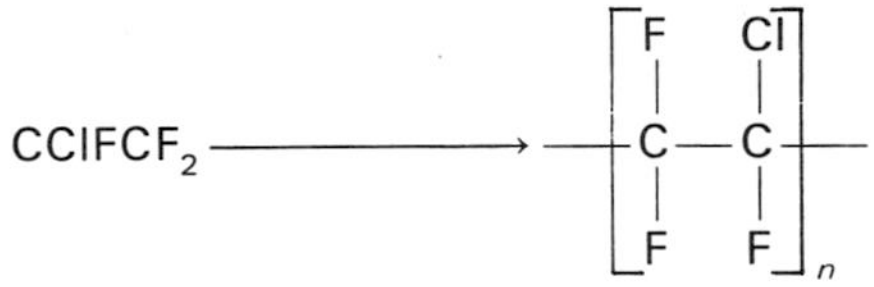

Polymonochlorotrifluoroethylene

FEP

$CF_2CF_2 + CF_3CFCF_2 \longrightarrow \left[A\!-\!B\!-\!A\!-\!B\!-\!A\!-\!B \right]_n$

  (A)   (B)   FEP copolymer

Tetrafluoro- Hexafluoropropene
ethylene   (HFP)
(TFE)

PVF

1   $HC\equiv CH$ + $\quad$ HF $\qquad\xrightarrow[\text{catalyst}]{\text{barium chloride}}\qquad$ CH_3CHF_3
 $\quad$ Acetylene $\qquad$ Hydrogen fluoride $\qquad\qquad\qquad$ Ethylidene fluoride

2   CH_3CHF_3 $\xrightarrow{\hspace{4cm}}$ CH_2CHF $\quad$ + HF_2
 $\qquad\qquad\qquad\qquad\qquad\qquad\qquad$ Vinyl fluoride $\quad$ Hydrogen fluoride

3   CH_2CHF $\xrightarrow[\text{pressure}]{\text{peroxide catalyst}}$

$$\left[\begin{array}{cc} H & H \\ | & | \\ C - C \\ | & | \\ H & F \end{array}\right]_n$$

PVF

PVF$_2$

1   CH_3CClF_2 $\xrightarrow{900-3000°F}$ CH_2CF_2 + HCl
 $\quad$ Chlorodifluoroethane $\qquad\qquad$ Vinylidene $\quad$ Hydrochloric
 $\qquad\qquad\qquad\qquad\qquad\qquad$ fluoride $\qquad$ acid

2   CH_2CF_2 $\xrightarrow{\text{catalyst}}$

$$\left[\begin{array}{cccccc} H & F & H & F & H & F \\ | & | & | & | & | & | \\ C - C - C - C - C - C \\ | & | & | & | & | & | \\ H & F & H & F & H & F \end{array}\right]_n$$

Polyvinylidene fluoride

GLOSSARY

Coining   A process by which thick metals are cold-pressed to form shapes and designs; for example, coins and medallions.

Sintering   A fusing process by which powdered particles are made to adhere to each other.

REVIEW QUESTIONS

1   What is the difference between a fluorocarbon and a fluoroplastic?
2   Why can't TFE be processed by conventional injection or extrusion methods?
3   By what processing methods are products made from TFE?
4   What are the outstanding characteristics of TFE?
5   How does CTFE compare with TFE in molecular structure?
6   What common household articles are coated with Teflon for antistick qualities?
7   Which fluorinated resin is useful over the widest temperature range?
8   How does specific gravity of fluoroplastics compare with other chemical-resistant thermoplastics such as polyethylene and PVC?

POLYAMIDE, POLYCARBONATE, AND ACETAL RESINS

19

Nylon is a household word throughout the world. Nearly everyone is familiar with nylon fibers used for wearing apparel, brush bristles, and carpet. Less widely known are the commercial and industrial uses of nylon molding resins, coatings, adhesives, and films.

POLYAMIDES

Polyamide is the chemical term used to describe linear polymers in which the structural units are connected by *amide* groups.

$$CH_2-CH_2-\overset{\displaystyle O}{\overset{\displaystyle \|}{C}}-\overset{\displaystyle H}{\overset{\displaystyle |}{N}}-(CH_2)_6-\overset{\displaystyle H}{\overset{\displaystyle |}{N}}-\overset{\displaystyle O}{\overset{\displaystyle \|}{C}}-CH_2-CH_2$$

Nylon 6,6

Research on polyamides started in 1927, primarily by du Pont, and culminated in the issuance of Patent Nos. 2,071,250 and 2,071,251 in 1937.

The first polyamide, produced in continuous filament form, was enthusiastically

Patented Feb. 16, 1937   **2,071,250**

UNITED STATES PATENT OFFICE

2,071,250

LINEAR CONDENSATION POLYMERS

Wallace H. Carothers, Pennsbury Township, Chester County, Pa., assignor to E. I. du Pont de Nemours & Company, Wilmington, Del., a corporation of Delaware

Application July 3, 1931, Serial No. 548,701

28 Claims. (Cl. 260—106)

This invention relates to the preparation of high molecular weight linear superpolymers having unusual and valuable properties, and more particularly it relates to the production of fibres from synthetic materials, and to a new method of propagating chemical reactions.

Linear condensation polymers of various types, particularly linear polyesters, have been described in the scientific and patent literature (Carothers and Arvin, J. Am. Chem. Soc. 51, 2560 (1929); Carothers and Van Natta, ibid., 52, 314 (1930); Lycan and Adams, ibid., 51, 3450 (1929), and in the applications of Wallace H. Carothers, Serial Numbers 382,843 now Patent No. 2,012,267 and 406,721 now Patent No. 1,995,291 filed August 1, 1929 and November 12, 1929, respectively). Through the application of the principles set forth in these citations, linear polyesters of moderately high molecular weight have been obtained. Thus, if ethylene glycol and succinic acid in equivalent amounts are heated together in a closed container one obtains a mixture in the form of a liquid or pasty mass which is partly composed of polymeric ethylene succinate of low molecular weight together with water, unchanged succinic acid, and unchanged glycol. If the same reactants are heated in a distilling flask arranged so that the water can distill out of the reaction mixture as fast as it is formed, practically all the succinic acid and glycol are finally used up and the product consists of polymeric ethylene succinate having a molecular weight of about 1500. If the heating of this product is continued in the distilling flask under vacuum its molecular weight finally rises to about 3000. At this stage an apparent limit has been reached, and so far as I am aware no linear condensation polymers having the same unique properties and having an average molecular weight as high as my new compounds have ever been prepared hitherto.

The synthetic linear condensation superpolymers produced in accordance with the present invention are suitable for the production of artificial fibres which are pliable, strong, and elastic and which show a high degree of orientation along the fibre axis. In this respect they resemble cellulose and silk which, as recent researches have proved, are also linear superpolymers. (Meyer, Biochemische Zeitschrift, 214, 253–281 (1929)). So far as I am aware, no synthetic material has hitherto been prepared which is capable of being formed into fibres showing appreciable strength and pliability, definite orientation along the fibre axis, and high elastic recovery in the manner characteristic of the pres-

ent invention. It is true that Staudinger has frequently emphasized the probable structural analogy between polyoxymethylene and cellulose, and he has shown (Z. Krist. 70, 193 (1929)) that it is possible to obtain polyoxymethylene in the form of oriented fibres, but these fibres are only a few millimetres in length and they are very fragile. It is true also that threads or filaments can be drawn from any tough thermoplastic resin, and British Patent 303,867 (French equivalent 667,077) discloses a process for making artificial silk in which a condensation product of a polyhydric alcohol and a polybasic acid or its anhydride is employed as a raw material. British Patent 305,468 discloses a process for making synthetic fibres from a urea-formaldehyde resin. But there is nothing in the disclosures of these references to indicate that the filaments or fibres are sufficiently strong or pliable to have any utility, and insofar as I am able to ascertain, filaments or fibres produced in accordance with the disclosures of these patents do not have any useful degree of pliability, strength, or elasticity.

Before considering in detail the objects of the invention and the methods for their attainment, it will be advantageous, for a better understanding of the present invention involving the production of linear condensation superpolymers, to refer to certain definitions and considerations involved in the production of the known linear condensation polymers.

I use the term condensation to name any reaction that occurs with the formation of new bonds between atoms not already joined and proceeds with the elimination of elements (H_2, N_2, etc.) or of simple molecules (H_2O, C_2H_5OH, HCl, etc.). Examples are: esterification,

$$R—COOH + HO—R' \rightarrow R—CO—O—R' + H_2O;$$

amide formation,

$$R—CO—OC_2H_5 + NH_2—R' \rightarrow$$
$$R—CO—NH—R' + C_2H_5OH;$$

ether formation,

$$R—OH + HO—R \rightarrow R—O—R' + H_2O;$$

anhydride formation,

$$2R—COOH \rightarrow (R—CO)_2O + H_2O.$$

Condensation polymers are compounds formed by the mutual condensation of a number of (functionally) similar molecules to form a single molecule.

Linear polymers are compounds whose molecules are long chains built up from repeating

19-1   U.S. Patent No. 2,071,250 for linear condensation polymers.

2 2,071,250

units. This type of structure may be symbolized by the general formula

$$... -A-A-A-A-A-A-A-A-A. ...$$

The unit or radical —A—, I call the structural unit of the polymer. Depending upon the nature of the starting materials the units of a given chain may be alike or different.

Linear condensation polymers are linear polymers formed by reactions of condensation as defined above. Examples are:

(a) Polyesters formed by the intermolecular self-esterification of hydroxy acids,

HO—R—COOH+HO—R—COOH→
HO—R—CO—O—R—COOH+H₂O

2HO—R—CO—O—R—COOH→HO—R—
CO—O—R—CO—O—R—CO—O—R—
COOH+H₂O, etc.;

(b) Polyesters formed by the action of dibasic acids on dihydric alcohols;

n HOOC—R—COOH+(n+1)HO—R'—
OH→HO—R'—O—(CO—R—CO—O—
R'—O)ₙH+H₂O, etc.;

(c) Polyamides derived from esters of amino acids,

H₂N—R—COOR'+H₂N—R—COOR'+
H₂N—R—COOR'→H₂N—R—CO—NH—R—
CO—NH—R—CO—OR'+2R'OH, etc.;

(d) Polyethers formed by the intermolecular self-etherification of glycols,

HO—R—OH+HO—R—OH+HO—R—OH+
etc.,→HO—R—O—R—O—R—
O—R—O— etc.,+H₂O

(e) Polyanhydrides derived from dibasic acids,

HOOR—COOH+HOOC—R—
COOH+etc.→HOOC—R—CO—O—OC—
R—CO—O—OC—R—CO—etc.,+H₂O

It may be observed that each of these starting compounds contains two (and only two) functional groups capable of participating in the condensation. I call such compounds bifunctional compounds, and their reactions bifunctional reactions. My invention is concerned not with polyfunctional condensations generally, but only with bifunctional condensations. This restriction is necessary since the presence of more than two functional groups in any of the reacting compounds introduces the possibility of developing a three-dimensional polymeric structure; and this involves a complication with which my invention is not concerned.

It is a characteristic feature of bifunctional condensations, such as those exemplified in the general equations (a), (b), (c), (d) and (e), that they present the formal possibility of producing molecules of infinite length. Thus the self-esterification of ten molecules of hydroxy acid, HO—R—CO—OH would lead to the formation of the polyester,

HO—R—CO—(O—R—CO—)₈—O—R—CO—OH,

and this, since its molecule still bears the terminal groups which were responsible for the initial reaction, is (potentially) capable of reacting with itself to produce a new molecule twice as long. A continuation of this progressive coupling would finally yield a molecule of infinite length.

In practice there are several factors that may be expected to interrupt this progressive coupling before the molecules have grown to any very great length. The following may be mentioned:

(1) Reaction may be intramolecular at some stage (i. e., it may lead to the formation of a ring).

(2) The terminal functional groups responsible for the progressive coupling may be lost or mutilated through side reactions

(3) Mechanical factors such as solubility and kinetic effects may come into play.

I have devoted considerable study to reactions of the type defined above as bifunctional condensations and have discovered the following facts: Such reactions are usually exclusively intermolecular at every stage, and the terminal groups responsible for the initial reaction are still present at the ends of the product molecule. Interruption of the progressive coupling through intramolecular reaction (ring formation) occurs generally only through the formation of 5-atom rings, less frequently through the formation of rings of 6 atoms, and rarely through the formation of rings of 7 or 3 atoms. The loss or mutilation of terminal groups through side reaction occurs only when patently inappropriate experimental conditions are adopted. Thus in most cases there is nothing theoretically to preclude the possibility of producing exceedingly long molecules in bifunctional condensations.

It is well known that reactions such as esterification, ester interchange, amide formation, etherification, anhydride formation or acetal formation are reversible reactions, and that such reactions can be forced to completion by the separation of the reaction products as they are formed. But from the facts outlined above it follows that if the reversible reaction is a bifunctional condensation, the degree of completeness of the reaction will regulate the size of the molecule in the polymeric product; the more nearly complete the reaction, the longer the molecule; and the reaction can be absolutely complete only when the product molecule is infinitely long. According to the present invention reactions of this type are brought to a degree of completeness heretofore unknown.

This invention has as an object the preparation of linear condensation superpolymers, and by this I mean linear condensation polymers capable of being formed into useful fibres. A second object is the manufacture of synthetic fibres. A third object resides in a method of propagating reversible chemical reactions involving the simultaneous formation of volatile and non-volatile products.

The first of these objects, briefly expressed, is accomplished by subjecting the linear polymers resulting from reversible bifunctional condensations to the action of heat under conditions which particularly facilitate the removal of any possible volatile reaction products. The second of these objects is accomplished by spinning or drawing filaments from a synthetic linear condensation superpolymer. The third object is accomplished by utilizing a molecular still, for removing the volatile product or products of a reaction where their effective vapor pressure is too low to be removed by the usual distillation technique and where the non-volatile product is substantially completely non-volatile.

It may be observed that the results described herein furnish a very satisfactory verification of my theory. It is to be understood, however, that I do not desire the claims to be limited thereby inasmuch as the theory has been presented in detail merely to make clear the nature of the invention and especially to explain the terminology used in describing it.

received by the public for hosiery and undergarments as a replacement for natural silk. Marketing of the original products was accompanied by glowing advertisements proclaiming "the new miracle material from coal, air, and water." Figuratively this was true, for indeed the early polyamide had its origins in these common materials. As we saw in our earlier study of phenolic resin, coal is a source of phenol; chlorine is obtained by electrolysis of brine; and oxygen is present in air.

The trade name Nylon was adopted, but today the term *nylon* is used generically to describe any polyamide capable of forming polymers. The types of nylons in general use are so numerous that a numbering system to describe them has evolved, in which single or double digits denote the number of carbon atoms in the parent chemicals:

Nylon type	Chemical	Structure	Source
6,6	Adipic acid	$HOOC\ (CH_2)_4\ COOH$	Phenol
	+		
	Hexamethylene diamine	$NH_2\ (CH_2)_6\ NH_2$	Phenol
6	Amino caproic acid	$NH_2\ (CH_2)_5\ CO$	Phenol
6,10	Hexamethylene diamine		
	+		
	Sebacic acid	$HOOC\ (CH_2)_8\ COOH$	Castor oil
11	Amino undecanoic acid	$NH_2\ (CH_2)_{10}\ COOH$	Castor oil
12	Amino dodeanic acid	$NH_2\ (CH_2)_{11}\ COOH$	Butadiene

Various other nylons (types 3, 5, 7, 8, and 9) have been under investigation or are used in Europe, Japan, and the U.S.S.R.; type 13 is being researched in the United States. In addition, it is possible to copolymerize or terpolymerize different nylons to obtain intermediates.

Preparation and applications

Nylon 6,6, the first resin introduced, is prepared by a condensation reaction. It is still widely used despite possible economic advantages resulting from manufacture of other polymers, particularly nylon 6. Because of its many successful applications, it is considered the standard with which all other nylons are compared.

Nylon 6,10 is prepared in much the same manner as 6,6, or both may be copolymerized to modify the end product. Heavy-duty brush bristles such as those used for street-sweeping equipment are extruded from nylon 6,10 because of better dimensional stability and lower water absorption.

Nylon 6 can be prepared on a continuous basis by heating the anhydride of amino caproic acid (caprolactam) in the presence of a catalyst. Controlled amounts of monomer remaining in the final polymer act as internal plasticizers, enabling the manufacturer to offer many grades with varying degrees of stiffness and toughness. As with other plastics, varying amounts of external plasticizers (graphite, molybdenum

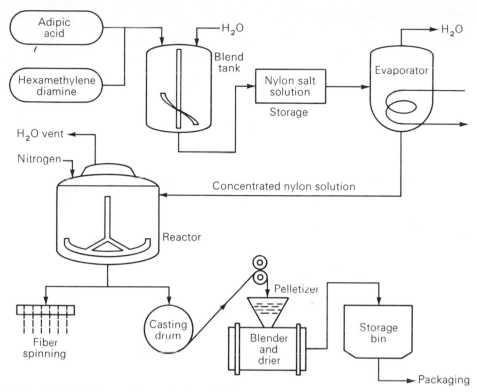

19-2 Simplified diagram for production of nylon.

disulfide, or fluorocarbon) can be added for applications where antifriction is required, or lubrication is difficult.

Nylon 11 is injection molded and extruded but is considered more as a specialty resin with commercial significance as a coating for metal products where resistance to chemicals and toughness are required. Applied by electrostatic spray or dip methods, nylon 11 proves a durable protection for tubular garden furniture and bicycle and lawn-mower parts.

Properties

Nylons, as a family of thermoplastics, possess a unique combination of physical and chemical properties that enables them to fulfill special applications (see Table 19-1).

Transparency The natural state of nylon molding and extrusion resins is translucent beige or off-white. Films can be made from formulas that are transparent. Extruded films find applications in prepared food packaging where foods are boiled in the package. Films meeting FDA standards are also used to package cheese and luncheon meats.

TABLE 19-1   Properties of injection-molded nylon

Property	Type 6	Type 6,6	Type 6,10	Type 11
Specific gravity	1.13	1.12–1.15	1.07–1.10	1.04–1.08
Tensile strength, psi	9–12,000	7–11,800	8–9,000	7–9,000
Flexural strength, psi	11–16,000	12–17,000	10,000	7–8,000
Flex after 16-h boil	3,000	4,000	4,000	4,000
Impact strength, ft-lb	0.2–4.2	0.6–3.3	0.8–2.6	1.4–4.4
Heat-distortion temperature, °F				
at 66 psi	365	360	300	155
Melt point, °F	425	490–530	400–420	375–380
Water absorption, %:				
25 days at 95% relative humidity	2.0	1.7	1.0	0.7
After 2-h boil	5.75	3.5	1.5	1.0
After 16-h boil	9.0	7.5	3.0	2.0
Arc resistance	Melt without carbonizing			
Flame resistance	Self-extinguishing			

Antidrag and antifriction properties   Extruded over wire, nylon 6,10 and others provide abrasion and cut-through resistance, and they can be pulled through complicated conduit paths because of their inherent antidrag property. Antifriction, a property common to all nylons, is responsible for their selection for gears, rollers, cams, door-latch components, drapery slides, and many other moving or bearing parts. Nylons replace bronze, brass, and zinc alloys in automotive cable units, business-machine gear trains, and food-processing equipment.

19-3   Nylon conveyor rollers replace metal bearings in food-processing plant. Lubrication is not required, thus ensuring cleanliness. (*Reprinted by permission of The Polymer Corp.*)

Perhaps nylon can best be characterized as tough, resilient, and stable, with good fatigue resistance. For these reasons, it is a logical candidate for injection-molded structural parts such as electric razors, kitchen appliances, and valve bodies.

Heat resistance   Most grades are self-extinguishing and impart an odor of burning wool. Heat resistance based on deflection under load is from 150 to 360°F, with types 6 and 6,6 exhibiting the higher value. Continuously exposed to dry heat, nylons embrittle at 250°F.

Moisture absorption   All nylons absorb moisture to a degree, depending on formulation, reaching an equilibrium somewhere between 0.20 and 2.5 percent. Selection of the proper formulation and design tolerances becomes critical when moving parts are required to operate in a humid environment. Nylon pellets, as used for injection, extrusion, and blow molding, are hygroscopic and must be kept to a moisture level below 0.2 percent for trouble-free processing. Resins that absorb moisture in storage must be dried prior to processing to avoid steam development.

Chemical resistance   Nylons are generally resistant to many chemicals, notably gaso line, liquid ammonia, acetone, benzene, and organic acids. They are attacked, as evidenced by swelling and loss of strength, by chlorine and peroxide bleaches, nitrobenzene, hot phenol, and hot formamide. Nylons are not recommended for extended exposure to ultraviolet light, hot water, or alcohols. They are resistant to moth larvae, fungus, and mildew.

Melting point   Unlike most thermoplastics, nylons are highly crystalline, with sharply defined melting points. When the melt point, which is higher than for most resins, is reached, they are extremely fluid and free flowing. To compensate for nylons' low viscosity when melted, injection machines are equipped with special devices to prevent loss of material; molds must be built with strict precision to prevent escape of molten polymer when high molding pressure is applied. Extrusion units and blow-molding equipment must also be modified to handle the sharp change from solid pellet to low-viscosity fluid, as will be explained later.

Nylons degrade rapidly when held too long in the heating chambers of extruders or injection machines. Parts are discolored and lose much of their mechanical strength or chemical resistance.

Nylon fiber and filament applications

The use of nylon as a clothing fiber was reviewed previously. The sheer quality obtainable from fine fiber and filament is caused by high-tensile strength, with elasticity, imparted by a process called cold-drawing. As filaments are drawn from the melt, they are stretched as they cool. This orients the crystalline structure to align the molecules along the fiber axis.

Heavier filaments are used for insect screening, surgical thread, parachute fabric and line, tents, tarpaulins, and carpet. Nylon woven fabrics coated with polyvinyl chloride are used for rainwear, aircraft covers, and chemical-resistant clothing. Used with polyesters, nylon fabrics provide lightweight laminates for aircraft-canopy attachments and other applications. Chopped nylon monofilament is used to reinforce diallyl phthalate molding compounds for compression- and transfer-molded items.

Miscellaneous uses

Nylon flake dissolved in phenolic resin improves flexibility of laminated fishing rods. Polyamide adhesives based on thermoplastic formulations are used as hot melts to improve sealing of food-wrap films such as cellophane, aluminum foil, and acetate, or to provide "dot-seals" for paperboard cartons. Thermosetting varieties often compounded with epoxies provide excellent bonds for metal-to-metal, metal-to-plastic, and are used for seam-filling metal components.

POLYCARBONATES

In the United States, polycarbonate resins are produced by General Electric Company[1] and Mobay Chemical Company.[2] Beginning in 1960, a joint venture participated in by Monsanto Company in this country and Farbenfabriken-Bayer of Germany (hence Mobay) resulted in the perfection of polycarbonate. Though polycarbonate materials have been reported in chemical literature for more than half a century, they were brought to commercial importance only through the independent efforts of these two companies.

Processing

Polycarbonate resins are shipped in sealed containers to avoid moisture absorption or adsorption. Moisture pickup greater than 0.03 percent prior to actual molding causes frothing with attendant porosity, bubbles, or sink marks. Parts that may be visually acceptable show poor impact resistance. To prevent moisture degradation, the following steps should be taken:

1   Preheat and maintain machine hoppers at 250°F. Keep hoppers covered.

2   Preheat pellets in their original, unopened containers at 250°F (maximum 260°F). Open-tray oven drying or closed-drawer cabinet drying is successful only with dehumidified air.

3   Allow to 4 h for heat to penetrate a 25-lb container.

4   Place only as much material in the hopper as required for the day's production, if small.

5   Reseal unused portions.

[1] Lexan is the registered trade name of General Electric Company.
[2] Merlon is the registered trade name of Mobay Chemical Company.

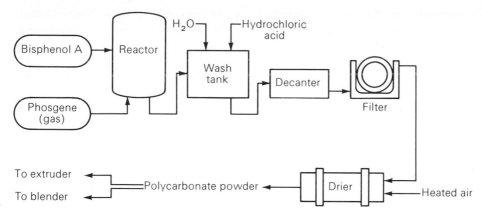

19-4 Simplified diagram for production of polycarbonate.

6   Remove unused material from hopper and store in sealed containers.

7   With plunger-type injection units, scrap can be fed back as produced. Otherwise scrap requires drying after regrind.

Properties

Polycarbonate is furnished as pellets for injection, extrusion, and blow molding, and as sheet, rod, tube, ball, and slab stock shapes. Film and sheet are produced by solvent casting and melt extrusion, respectively.

The natural state of polycarbonate is clear, with more than 80 colors available. Items fabricated from thermoformed sheet stock or by injection molding rank high among the engineering-type thermoplastics. Metal products and those formerly made from other plastic (both thermoset and thermoplastic) are continually challenged by polycarbonates. There are valid reasons for this trend:

Ductility   The ability of polycarbonates to be nailed, sawed, punched, cold-drawn, sheared, drilled, and riveted without cracking is without equal.

Toughness   Impact strength is rated at 50 times greater than federal requirements for safety glass. Street-lamp globes, helmets, factory and school windows, and machine guards utilize impact resistance.

Dimensional stability   Polycarbonates retain mechanical value over a wide temperature range from -215 to $+250°F$.

Heat resistance   Standard grades are thermally dependable to 250°F; glass-filled types are useful to 275°F. Housings for high-wattage lamps and aircraft instrument dials are molded of transparent grades. Thicknesses over 50 mil do not support combustion. For this reason, aircraft ducts, shrouds, and junction boxes formerly made from polyester-glass laminates are replaced by thermoformed polycarbonate at a reduction in cost and weight.

Chemical resistance   The National Sanitation Foundation and FDA list polycarbonate as safe for potable water, food processing, and packaging. Refrigerator parts, food-vending machines, and household appliances provide many applications. Stain tests indicate excellent resistance against catsup, coffee, mustard, tea, wine, and vinegar, among others.

Electrical resistance   Tested at room temperature and after immersion in boiling water, polycarbonate's initially high insulation values are substantially unchanged.

Machining qualities   Machine tools used for metal working can be used for threading, turning, boring, milling, and shaping. Being ductile and heat resistant, polycarbonates do not soften and clog machine tools.

Joining   Ethylene dichloride, methylene chloride, or their mixtures provide fast, excellent bonds. Dissimilar materials can be cemented with epoxy, polyester, poly-urethane, or synthetic rubber adhesives.

Limitations

Chemical resistance   Ketones, chlorinated hydrocarbons, strong acids, and alkalies have a deleterious effect. Some detergents, ammonia compounds, and high-octane gasoline attack the resin.

Weathering   Some yellowing occurs, along with a reduction in surface gloss, because of ultraviolet absorption.

19-5   Carburetor body of 1-hp engine used to power chain saws and lawn mowers is strong and dimensionally stable at elevated temperature; it resists attack by oil and gaso-line. Material is 30% glass fortified. (*Reprinted by permission of Liquid Nitrogen Processing Corp.*)

ACETAL

Another engineering-type thermoplastic, termed acetal, became available in 1960. The resin is referred to as a linear polymer of formaldehyde. From earlier chapters on phenolic-, urea-, and melamine-formaldehyde, we have seen that formaldehyde is a gaseous substance, HCHO. Earlier trials to polymerize the gas to stable solid were unsuccessful in that the polymer would readily revert to the vapor phase when attempts were made to process it.

The homopolymer Delrin[1] is identified as polyoxymethylene, in which methyl (CH_2) groups are linked together by an oxygen atom.

$$
\underset{\text{Formaldehyde}}{H - \overset{\displaystyle \overset{O}{\|}}{C} - H} \longrightarrow \underset{\text{Polyoxymethylene}}{- \overset{\displaystyle \overset{H}{|}}{\underset{\displaystyle \underset{H}{|}}{C}} \quad \overset{\displaystyle \overset{O}{\diagup\diagdown}}{\underset{\displaystyle \underset{H}{|}}{\overset{\displaystyle \overset{H}{|}}{C}}} \quad \overset{\displaystyle \overset{O}{\diagup}}{\underset{\displaystyle \underset{H}{|}}{\overset{\displaystyle \overset{H}{|}}{C}} -}
$$

The Celanese resin Celcon[2] is reportedly based on trioxane and a minor constituent, ethylene oxide.

Acetals compete with nylons and polycarbonates in many structural applications such as hardware components, gears, bearings, business-machine assemblies, and housings. Like nylon and polycarbonate, stiffness and dimensional stability are improved by incorporating glass fibers, and lubricity is enhanced by adding fluorocarbons.

A review of advantages and limitations provides a basis of comparison between these competitive materials.

Processing

No special problems are encountered when acetals are injection molded. Molds previously designed for another thermoplastic may require modification of runners and gates. Raising the temperature of the polymer does not appreciably lower the melt viscosity. Therefore, mold filling is not as dependent on temperature-pressure relations as it is on correct mold design.

As we shall learn in the sections dealing with injection molding and extrusion, thermoplastics soften more as temperatures rise, though not in a straight-line fashion. Even though acetals have a high rate of flow and rapid mold-filling features, heated molds are advisable (to 250°F) to minimize stress and control shrinkage.

Good temperature control in the 400 to 440°F range is imperative for all processing equipment. Material held too long in heating cylinders decomposes and gives off formaldehyde vapors. Even small amounts of gas released in the molding plant can cause severe eye and respiratory irritation, and greater amounts may require evacuating the premises.

Anticipated delays in operation of equipment call for immediate purging of

---

[1] Delrin, an acetal homopolymer, is the registered trade name of E. I. du Pont de Nemours & Company.

[2] Celcon, an acetal copolymer, is the registered trade name of Celanese Plastics Company.

heating units and disposal of purged material in cold water. Heating chambers and operating procedures need to be such that no danger of an exploding cylinder exists. A combination of a heater stressed to near fatigue limits by repeated heating and cooling, plus a blocked nozzle or feed chamber can create an extremely hazardous condition.

Properties

Mechanical   Acetals do not embrittle with long-term exposure at elevated temperature. Tensile strength is substantially the same after 1 year in air at 240°F or 1-year immersion in 180-deg water. Once ignited, acetals are difficult to extinguish. Recommended-use temperature is 220°F. These values are important in applications where molded or machined dimensions must be maintained in severe operating conditions. Impact strength compares well with nylon; retained shock resistance at −40°F is 85 percent of room-temperature value. Aerosol packaging molded in two parts and spin-welded can contain pressure without creep because of impact resistance and tensile strength.

Chemical resistance   Acetals cannot be dissolved by organic solvents, which precludes solvent bonding. They withstand staining by most common household foods and resist discoloration by industrial oils or grease. Moisture absorption is extremely low, leading to usage in shower heads, sprinklers, and pump assemblies. Acetals have limited resistance to strong acids and oxidizing chemicals, but both types resist weak alkalies.

Weather resistance   Acetals show a decrease in mechanical strength after exposure to prolonged periods of ultraviolet light, with chalking of exterior surfaces occurring with shorter exposure, even when compounded with carbon black.

Thermal   Continuous operating temperature is at 185°F for Delrin and at 220°F for Celcon. Both are slow burning, emitting a sharp, irritating odor.

Electrical   Both types compare favorably with each other in dielectric strength (500 V/mil), but Celcon resists a high-voltage surface arc for a longer period.

Optical   Resins in the natural state range from translucent white to opaque. A wide range of colors is possible.

Machining   Acetals machine similar to free-cutting brass. They can be sawed, drilled, tapped, and turned without the need for cutting fluids. They are tough enough to withstand riveting and staking of metal components.

Cementability   Structural adhesives based on modified epoxy or urethane are adequate to bond acetal to itself. Nonstructural adhesives based on neoprene cements are used for parts not subject to high temperatures or when little or no stress is involved.

Cost   See Table 19-2.

TABLE 19-2   Cost comparison

	Specific gravity	Cost in $/lb, base resin	Cost/cu in.
Homopolymer	1.425	0.65	0.0375
Copolymer	1.41	0.65	0.0375
Nylon type:			
6	1.13	0.90	0.0367
6,6	1.14	0.90	0.0372
6,10	1.08	1.26	0.0492
11	1.04	1.54	0.058
type 6, 40% glass	1.60	. . .	0.089

GLOSSARY

Amide   A chemical in which a —NH$_2$ group is substituted for an —OH (hydroxyl) group of an organic acid.

Purging   Nonproductive operation of a machine until correct conditions for molding are established.

REVIEW QUESTIONS

1   What is meant by the expression "a miracle material from coal, air, and water"?
2   What do the numbers 6,6 ; 6,10 mean when referring to nylons?
3   Which type of nylon is used as the standard of comparison between different nylon resins?
4   Which type of nylon is used for brush bristles? For coatings?
5   Are nylons self-extinguishing, or will they support flame?
6   What characteristic odor serves to identify nylon when burned?
7   What is the effect of moisture absorption on nylon?
8   What effect does crystallinity of nylon have on melting characteristics? On processing?
9   Which nylon has the highest melt point?
10   Explain cold drawing. What products utilize the benefits derived from this technique?
11   What properties of nylons lead to their selection for items such as gears and rollers?
12   List the outstanding properties of polycarbonate.
13   What solvents are used to join polycarbonate? Are these solvents applicable to nylon also?
14   If you were to select a material for automotive tail-light lenses, which choice would you make between nylon and polycarbonate? Why?

15   What effect does moisture have on the processing characteristics of polycarbonate? What procedures are followed in drying this material?

16   What applications utilize the impact resistance of polycarbonate?

17   What other applications now served by glass, other plastics, or metal might, in your opinion, be better served by polycarbonate?

18   Which type of material is used to bond acetals: solvents or adhesives?

19   What hazards exist in overheating acetal?

20   How is hot, purged material disposed of?

ACRYLIC PLASTICS

20

Few materials can compete with the jewel-like brilliance and weather resistance of acrylic. These qualities have brought about new concepts in outdoor advertising and architectural innovations. Acrylics are among the easiest of sheet stocks to fabricate with machinery common to wood or metal working, with the added advantage of being readily softened for forming to complex shapes. The largest single outlet for acrylic sheets is signs; business firms, banks, service stations, markets, and office buildings have accepted acrylic signs as superior to neon, wood, or metal. The animation possible in neon is more than compensated for by acrylic's neater appearance and legibility during daylight hours. Acrylic signs are less susceptible to damage from storms or vandalism, and they require less electrical maintenance.

Acrylic monomers have been known for 100 years. In 1901, Dr. Otto Rohm began working with acrylic materials, and in 1927 the first resin, acryloid, was announced by the Rohm and Haas Company. Cast sheets became available in 1927.

PROCESSING

The term acrylic is descriptive of a large class of resins, the most important of which is *polymethyl methacrylate*. At least 30 acrylate and methacrylate monomers of varying importance can be synthesized. A few find commercial outlets as ingredients in floor waxes, car wax, water-emulsion type paint, and adhesives. Other resins, notably polyester and styrene, are modified with methyl methacrylate monomer to improve stability in lighting or weathering applications.

20-1 Extensive use of acrylic sheets is shown in skylighting and store signs. (*Reprinted by permission of American Cyanamid Co.*)

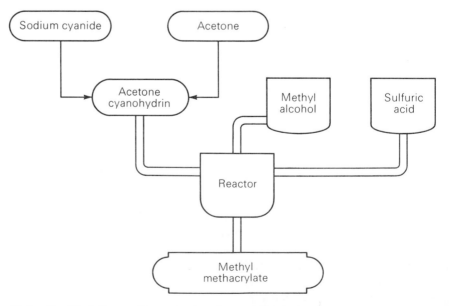

20-2 Simplified diagram for preparation of an acrylic monomer.

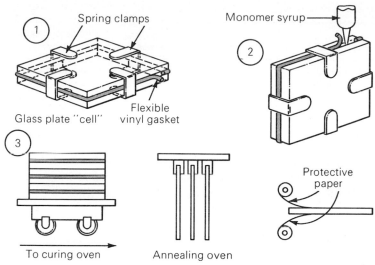

20-3   Cell casting of acrylic sheet.

The monomer is a crystal-clear liquid useful for embedding biological specimens, sales display items, and novelties, and for manufacturing dentures—made by
dissolving finely pulverized polymer in the monomer to form a workable consistency.
The uncured monomeric portion polymerizes when a peroxide catalyst and a promoter
are added. The monomer is also used as a solvent-cement[1] for fabricated or molded
parts when catalyzed with benzoyl peroxide in conjunction with a promoter.

Polymethyl methacrylate is obtained as cast or extruded sheet rod and tube or
as molding pellets. Cast sheets are almost universally used for fabrication of signs,
store display fixtures, and thermoformed products, although extruded forms are also
used in these applications.

Rods and tubes are cast in glass or metal cylinders. Sheets are cast in widths
to 102 in., lengths to 144 in., and in thicknesses ranging from 0.0265 to 4 in. Rods
are produced from 0.75 to 6 in. in diameter. A continuous method of casting acrylic
sheet has recently been developed.[2] The advantage of continuous lengths is in the
manufacture of larger one-piece signs and larger thermoformed parts. Cast sheets are
readily worked by conventional machining methods, although drills, circular saw
blades, and lathe tools must be selected to prevent chipping or overheating.

FABRICATION METHODS

Sawing

Circular   Large sheets are best cut with a traveling blade to avoid uneven kerf (see
Table 20-1). Smaller sheets are readily handled on the conventional table saw.

[1] Cement PS-18 is manufactured by Cadillac Plastics.
[2] Swedcast is the name of the method developed by Swedlow Plastics.

TABLE 20-1 Circular sawing dimensions

Sheet thickness	Type blade	Teeth/in.	Blade thickness, in.	Peripheral speed (ft/min)
0.040–0.080	Hollow ground	8–14	0.0625–0.0937	8,000–12,000
0.100–0.160	Hollow ground	6–8	0.0937–0.125	8,000–12,000
0.187–0.375	Spring set	5–6	0.0937–0.125	8,000–12,000
0.437–0.750	Spring set	3–4	0.125	8,000–12,000
1.000+	Carbide tipped	3–3.50	0.125–0.1506	8,000–12,000

Cast sheets are shipped with protective masking paper on both sides to prevent scratching during shipping and handling. Paper should not be removed when cutting with a table saw. Scratching of unmasked sheets is minimized when sheets are clamped in a traveling saw.

Band and contour   Thick, flat sheets are cut on a band saw to avoid generating too much frictional heat. When metals are machined, frictional heat is dissipated by the cutting tool and the metal. Because plastics are poor heat conductors, cutting tools heat more rapidly and dull sooner. Temperatures are sufficiently high to melt thermoplastics, creating a dangerous situation when sheets more than 0.50 in. thick are cut by circular saws. Melted material tends to expand and fuse together as the cut emerges at the rear of a circular blade, causing blade warpage and binding.

In a band saw only a portion of the blade cuts and there is ample opportunity to cool. Most band saws have provision for variable speed from 2,000 to 5,000 ft/min (see Table 20-2). In general, thick stock is cut at lower speed.

Contours are cut readily by selecting the appropriate blade to suit the minimum radius. Thin sheets are stacked and held together with double-backed tape for simultaneous contour cutting of duplicate parts. Stacks to 4 in. thick are cut routinely in the manufacture of sign letters and shapes.

Contours are also cut with sabre, jig, or coping saws, utilizing blades designed for soft metals. Blades used for wood cutting are not recommended for acrylic.

TABLE 20-2 Standard band-saw dimensions

Radius of cut, min.	Saw width, in.	Teeth/in.	Blade thickness, in.
0.50	0.1875	7	0.028
0.75	0.25	7	0.028
1.50	0.375	6	0.028
2.25	0.50	5	0.032
3	0.625	5	0.032
4.50	0.75	4	0.032
8	1	4	0.035
12	1.25	3	0.035
20	1.50	3	0.035

Drilling

High-speed twist drills commonly used for drilling metal are good for drilling acrylic. Mass-production drilling is more economical and forms smoother holes if drills are modified. Speed, feed, and drill shape are varied according to the type of hole.

Drilling speed is established by trial. Generally, optimum speed is based on the ability of the drill to dissipate heat. Unless coolants are used, the drill must be backed out frequently to produce deep holes. For large-diameter holes, pilot holes can be drilled to 90 percent of desired depth and filled with water-soluble oil to lubricate the cut and flush chips. (See Table 20-3.)

Sanding, buffing, and polishing

Cleanly cut edges that are to be solvent bonded require no further treatment such as sanding. Minor saw marks can be scraped with a sharp tool if necessary. Rough edges can be restored to high polish by using 60- to 80-grit dry sandpaper followed by wet sanding with medium (150) to fine (400) grits. Scratches are removed by buffing on an *unstitched* muslin wheel wiped with a bar of tripoli-wax composition. Unstitched wheels are preferrred to avoid excessive heat build-up on the plastic surface, which causes waviness. Final polish is done on a second wheel treated with wax or left untreated.

Parts that are to be solvent bonded should not be buffed with wax compounds since deposits may interfere with bonding. An acceptable polished edge is obtained with flame polishing by passing a small jet-type flame (as from a butane torch) rapidly along the edge.

Thermoforming

Acrylic sheets are easily manipulated manually or with simple fixtures. Heating the material until limp in an oven at 325 to 360°F enables formation of many interesting shapes. Zone heating is accomplished by passing the section to be formed over an inverted funnel placed over a gas flame. The low heat conductance of plastics keeps remaining portions rigid. Sheets may be bent to any angle by heating the section that forms the angle with a device designed to heat linear sections.

Hemispherical and modified dome shapes may be made by free blowing a softened sheet upward through a clamping ring. Use of a split ring produces helmet or fish-bowl shapes. Restricting the stretch produces flat-bottom objects, and changing

TABLE 20-3 Drilling recommendations

Depth-diam. ratio	Hole depth	Included tip angle (point)	Lip clearance (relief)	Speed	Coolant
1.50:1 max.	Shallow	55–60	15–20	Medium	None
3:1 max.	Medium	60–140	12–15	Slow	Air
Above 3:1	Deep	140	12–15	Slowest (2–3 in./min)	Soluble oil

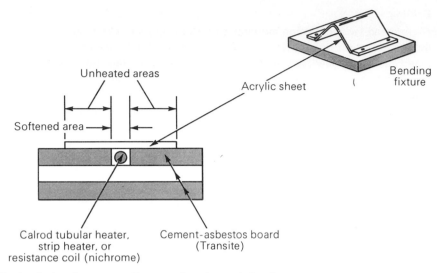

20-4 Device for heating narrow, linear sections for angle bends.

the periphery of the hold-down ring provides many interesting variations. Aircraft canopies, helicopter blisters, and skylights are produced by drawing heated sheets into a female mold with vacuum.

Cast acrylic sheets, free of internal strains because of the annealing process, exhibit an elastic memory. When heated sheets are formed and cooled, stresses are locked in by the stretching. When reheated to the initial softening temperature, sheets or rods revert to their original shape as stresses are relieved and molecules realign to original positions. Pieces incorrectly formed can be heated repeatedly until the desired shape is produced. Elastic memory is not permanent, however, since stresses relieve naturally with time until original molecular orientation no longer exists.

Free-form shapes are produced by bending strips or rods around wood dowels located at predetermined positions.

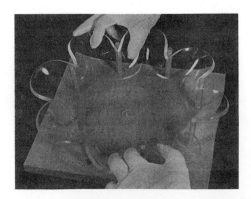

20-5 Forming a dish by hand. (*Reprinted by permission of Los Angeles Trade-Technical College.*)

Cast sheets are normally smooth and high gloss. Textured surfaces are furnished from monomer cast into glass cells having a textured pattern. Extruded sheets are given a post-extrusion embossing with press plates or engraved rollers. Light diffusers, shower doors, room dividers, and building panels are made from extruded sheets.

Light bending

The ability of cast or extruded acrylic to pipe light[1] throughout its internal structure has long been recognized by makers of medical and automotive equipment. Fiber optics, the process whereby small-diameter filaments convey light from source to destination, is utilized by automotive designers. A filament pipes light from a directional signal to the instrument panel to indicate signal flashing. Decorative items also make use of light-piping principles.

Automobile and aircraft instrument dials, clock and radio faces, and interior signs are edge-lighted so that light conveyed from an external source will transmit through the sheet or molded shape, becoming visible only where surfaces are etched or raised. Optical clarity, freedom from distortion, high-shatter resistance, and ease of forming make acrylic a good choice for boat and snowmobile windshields. Acrylic's use in automobile or motorcycle windshields is precluded by its tendency to become frosted by dust or grit or to become scratched from improper cleaning.

Bonding

A number of solvents and techniques are used to bond acrylics. Bond strength increases only as solvent evaporates. Depending on the solvent used, a joined structure

[1] This phenomena is discussed in detail in Chap. 35.

20-6 Automotive dial cluster of chrome-plated acrylic is illustrative of light piping. (*Reprinted by permission of Rohm and Haas Co.*)

can be handled within 30 min; however, full strength may not be realized for several days, depending upon the depth of penetration. Chlorinated solvents based on methylene chloride or ethylene dichloride are widely used, but extreme caution must be taken to avoid inhalation.

Injection molding

Injection-molded acrylic articles are numerous. Volume items include automotive tail and backup lenses, horn buttons, and medallions, and magnifying lenses, bathroom accessories, faucet knobs, brush backs, and emblems for home appliances.

Injection molding of acrylic is accomplished by conventional methods, with thick sections (to 1 in.) being readily produced without undue shrinkage or bubble formation. This is partly because of the high-melt viscosity of the molten polymer. The high-melt stiffness of acrylic is also an advantage in purging operations to thoroughly clean the machine injection cylinder. Cast sheet scrap is ground to make a purging compound, and special purging compounds based on acrylic are available.

PROPERTIES OF POLYMETHYL METHACRYLATE

Cast sheets and molding resins are among the most expensive of thermoplastic materials, costing approximately five times more than general-purpose polystyrene.

Specific gravity, at 1.18, is intermediate among thermoplastics.

Color

Transparent grades are crystal clear without blue or green tinge. They are also produced in a limitless range of translucent and opaque colors, including fluorescent.

20-7 Automotive escutcheon, injection molded in "reverse relief," exhibits three-dimensional effect when painted or plated on reverse or underside. (*Reprinted by permission of Rohm and Haas Co.*)

Mechanical properties

Mechanical values of polymethyl methacrylate rank with those of the average thermoplastic, having impact strength somewhat lower than most transparent materials, except general-purpose polystyrene. (See Table 20-4.)

Dimensional stability

Molded and fabricated products exhibit a greater degree of permanence in severe outdoor weathering than any other material. Corrosive environments or salt spray has no effect. Acrylic can be exposed for years to strong ultraviolet light without yellowing or discoloration.

Thermal properties

Service temperatures range from 160 to 250°F; flammability is rated as slow burning or self-extinguishing, depending on type. When burned, acrylic has a characteristic fruitlike odor.

Chemical properties

Weak acids and alkalies have no effect; concentrated forms attack most grades. Acrylics dissolve readily in chlorinated hydrocarbons (methylene chloride, trichlorethylene, and ethylene dichloride; ketones and proprietary solvents[1]). Acetone, glacial acetic acid, and chloroform are not recommended for solvent cements because they create residual stress at the bond line.

Electrical properties

Insulation values are adequate for any application that places molded or fabricated parts in contact with normal voltages. Because of comparatively higher material costs, acrylics are normally not selected for electrical components on the basis of insulation alone.

TABLE 20-4 Comparison of major mechanical properties

Property	Acrylic	Cellulose acetate	Polystyrene, general-purpose	Polysulfone	Polycarbonate
Tensile strength, psi	7–11,000	2–9,000	5–12,000	10,000	8–9,000
Flexural strength, psi	13–17,000	2–16,000	8–14,000	15,400	13,500
Compressive strength, psi	12–18,000	2–36,000	11–16,000	13,900	12,500
Impact strength, ft-lb/in.	0.3–0.5	5.0	0.25–0.40	1.3	12–17.5
Elongation, %	2–10	6–70	1–2.5	50–100	100–300

[1] Cement 1-A is a solvent of Rohm and Haas Company. Weld-On, types 3, 4, 11, and 12, are solvents of Industrial Polychemicals.

MODIFIED ACRYLICS

Methyl methacrylate monomers are copolymerized with styrene monomers to reduce cost, increase heat resistance of styrene, and provide light stability without sacrificing transparency. Special impact grades for molding and extrusion are produced in opaque colors. A combination, or alloy, of acrylic and polyvinyl chloride is available in sheet stock for thermoforming products requiring toughness in thin section and a high degree of stretch during forming. Acrylic film extruded to 6 mil thick can be laminated to many substrates, including other plastics, to provide a durable facing for building siding, outdoor signs, and doors.

GLOSSARY

Casting   The process of pouring liquids into a mold to harden without benefit of pressure.

Kerf   Cut made by a saw blade.

Traveling saw   Circular saw mounted on parallel rails manually or mechanically drawn across the work.

REVIEW QUESTIONS

1   List several reasons why acrylic sheets are widely used for signs.
2   Why are saw blades for wood not recommended for acrylic?
3   What is the advantage of using a band saw for cutting thick sections, as opposed to using a table saw?
4   Why is the width of a band-saw blade reduced as the radius of cut is decreased?
5   What governs the speed and feed when drilling holes?
6   How can drilling large, deep holes best be accomplished?
7   In buffing or polishing, what causes waviness of the plastic surface?
8   How can elastic memory of cast acrylic be used to advantage?
9   What solvents are used to bond acrylic?
10   How does the abrasion or scratch resistance of acrylic compare with that of polycarbonate? With that of melamine?

20-8   Motor cover of Kydex acrylic PVC alloy. (*Reprinted by permission of Rohm and Haas Co.*)

POLYPHENYLENE OXIDE
POLYSULFONE
POLYURETHANE

21

POLYPHENYLENE OXIDE

General Electric Company, pioneers in the development and application of phenolic, alkyd, and silicone resins, produced a series of materials in 1965 based on new methods of combining long molecular chains. One method is termed *oxidative coupling*.

We have seen how thermosetting phenolics are polymerized by the formation of *methylene* linkages to provide heat-stable resins. Oxidative bridging is a result of reacting oxygen with molecules containing active hydrogen atoms in a condensation-type polymerization. Linear molecules are produced as water is eliminated.

Polyphenylene oxide (PPO) polymer is opaque beige and capable of being colored, although high-processing temperatures (to 650°F) limit choice of pigments. The resin is characterized by an extremely high heat-distortion temperature (375°F), plus dependability in environments as low as −275°F. This outstanding thermal stability, coupled with low water absorption and good resistance to detergents, dilute acids, and alkalies, makes PPO valuable for surgical equipment and as a replacement for more expensive stainless-steel hospital items that must withstand repeated sterilizing.

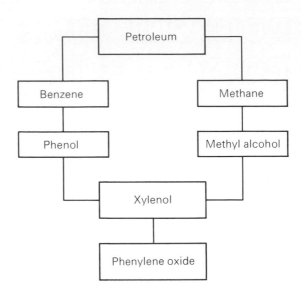

21-1   Flow diagram for preparation of phenylene oxide.

Polyphenylene oxide is self-extinguishing, has high-dielectric strength, and is approved by Underwriters Laboratory for critical applications. Its specific gravity (1.06 g/cc) ranks PPO among the lightest of all engineering thermoplastics.

Thermal expansion of PPO is extremely low, which suggests its use as a replacement for thermoset laminates and molding compounds in electrical components where dimensional stability is critical. PPO does not withstand strong oxidizing acids and is attacked by ketones, aliphatic hydrocarbons, and strong alkalies. Solvents useful for bonding parts are toluene and ethylene dichloride. Molded or extruded products can also be joined with epoxy or urethane adhesives, or they can be ultrasonically welded.

A modified version of polyphenylene oxide[1] somewhat easier to process in conventional molding and extrusion equipment is available in unlimited, opaque colors. Extremely tough and resistant to creep at elevated temperature, it is used for business machines, power tools, and carrying cases.

Injection molding of PPO and modified types involves relatively high softening temperatures and heated molds (see Tables 21-1 and 21-2). Molding temperatures are critical and confined to narrow ranges.

POLYSULFONE

Introduced by Union Carbide in 1965, polysulfone represents an important step forward in the development of heat-resistant, transparent thermoplastics. The stability of its chemical structure is attributed to the linkages that combine to furnish resistance to oxidation and provide rigidity at high temperatures.

---

[1] Noryl is its registered trade name and it is a product of General Electric Company.

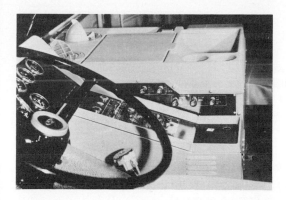

21-2 Centrally located console of tilt-cab truck is injection molded of Noryl ®. Part measures 2 × 4 ft and weighs 9 lb. (*Reprinted by permission of General Electric Co.*)

Molded and extruded products are self-extinguishing. Resistance to oil and gasoline enables usage in automotive "under-the-hood" applications, as well as gasoline pump and valve parts. Heat resistance to 300°F and ability to withstand deformation after thousands of hours under stress conditions suggest uses in home appliances (diswashers, dryers), kitchen-range hardware, and wire insulation. Extruded sheets are used in thermoformed aircraft parts where heat resistance, self-extinguishing properties, and low-smoke generation are of paramount importance.

Polysulfone is resistant to dilute or concentrated acids and alkalies, but is attacked by ketones and chlorinated and aromatic hydrocarbons. A special surface treatment to prepare polysulfone for electroplating requires good molding technique. Molded products must be protected from contamination of any type, including residual release agents. Parts to be plated require annealing, either in glycerine or mineral

TABLE 21-1 Typical injection-molding conditions

Condition	PPO	Noryl
Cylinder temperature, °F	550–650	460–600
Injection pressure, psi	15–20,000	15–20,000
Mold temperature, °F	230–300	160–220
Predry material	Only for improved surface appearance	
Shrinkage/in.	0.007–0.009	0.005–0.007

TABLE 21-2 Typical extrusion-molding conditions

Condition	PPO	Noryl
Cylinder temperature, °F :		
Zone 1 (feed)	580	520
Zones 2 and 3	590	520
Zone 4 (exit)	600	520
Die temperature, °F	500–650	540
Takeoff : cool in air, °F	300–450	200–300

21-3 Astronaut's pressure helmet, thermoformed from extruded polysulfone sheet, protects against solar glare and micrometeroids; it also reduces heat gain with gold coating on inner surface. (*Reprinted by permission of NASA and Union Carbide Corp.*)

oil at 335°F for several minutes, or in air at 335°F for 2 to 4 h. Annealed parts are checked for stress relief by immersion in ethyl acetate for 1 min at room temperature. Incorrect annealing shows up in surface crazing or cracks.

TABLE 21-3   Typical injection-molding conditions for polysulfone

Condition	Range
Cylinder temperature, °F	625–750
Injection pressure, psi	15–20,000
Mold temperature, °F :	
Thick sections	140–160
Av. section, 75–100 mil	200–210
Thin sections or long flow	300–320
Predry	Oven, 250–275°F, 4 h, plus hopper dryer
Shrinkage/in.	0.007
Purging material	Polypropylene

POLYURETHANE

No single group of synthetic resins has a greater potential for changing our way of life than polyurethanes. In less than 15 years these materials, now simply called urethanes, have revolutionized housing, transportation, furniture, and apparel; and wholly new concepts in manufacturing and use have been developed.

Urethanes produced as flexible foams are used for mattresses and cushions, as automotive dashboards, armrest and visor safety features, and as clothing interliner. Rigid and semirigid versions are used to seal mine shafts and to insulate houses, storage tanks, rail cars, refrigerated trucks, warehouses, and pipe lines. Complete lines of furniture, molded to any design, are replacing traditional wooden items.

It is not outside the realm of possibility to envision completely self-contained cities founded on a space platform, totally enclosed in a urethane foam shell. Such shell structures are currently in use by earthbound people in temporary and permanent installations at military posts, radar sites, and commercial establishments.

Urethanes owe their popularity more to ease of production and application than to low cost. Large-volume uses such as covering a huge storage tank involve expensive equipment that proportions, pumps, mixes, and dispenses reactive chemical components at the correct rate for each, and at precise temperatures. Producing flexible

21-4 Rigid urethane foam provides frame for upholstered pieces, as well as credenza and tables. Flexible urethane foam is used for cushioning. (*Reprinted by permission of Mobay Chemical Co.*)

slab stock for mattresses and comfort cushioning, or thin sheets for packaging and clothing, further increases equipment cost. Other units operated on an on-off principle are used to fill multiple molds for the production of furniture parts, picture frames, novelties, wall plaques, and protective packaging.

For evaluation purposes or extremely small production, ingredients can be mixed in a paper cup, using a wood stirrer. Better mixing of larger batches (to several pounds) is accomplished with portable power mixers or "drill press" at 1,000 rpm.

Foaming methods

Methods of producing foams vary, depending on the scope of the operation and the item to be made.

Foam in place   Early applications in aircraft led to pouring urethane between skins of reinforced plastics to provide rigid, lightweight structures. This technique is currently used in on-site pouring, as between parallel walls in building construction, and in hard-to-reach areas of prefabricated structures. Delicate objects are packaged by surrounding them with foam, using the shipping carton as the mold.

Spray and foam   Specially designed spray guns are used to deposit a urethane coating on any surface that requires insulation or buoyancy. Coatings foam almost as soon as applied, and the natural adhesive quality of the resin prevents sagging from vertical or overhead surfaces.

Roofing, either in new housing or as replacement for old, is an entirely new area of application for urethane. The weathering characteristics, color range, and toughness of semirigid foams, plus the improved insulation they offer, make urethane foam a formidable competitor to asphalt or other roofing materials.

Frothed foams   With this method, ingredients are metered and combined with the blowing agent but held under pressure until released to the atmosphere where instantaneous foaming occurs. Incorporating two blowing agents, one to cause initial frothing, the other to continue the foaming action after placement, controls the rise and final foam density.

The advantages of frothed foams, which react much like shaving cream from an aerosol can, are the lowering of pressure as foam expands and the ability to make large pours on a continuous basis. As normal pour-in-place foams rise, pressures can be sufficiently high to bulge thin retaining walls. With preexpanded froth, pressures are held as low as 0.5 psi. Frothing also permits successive build-ups, particularly on large-surface areas; build-ups are easier to produce since each successive "pass" can be supported by the previously laid section. Heat generation is more uniform, leading to cell uniformity and lower overall density.

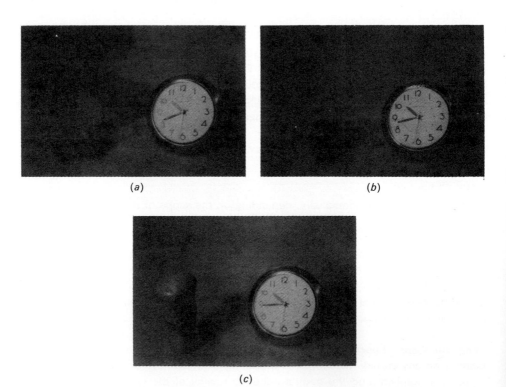

(a)   (b)

(c)

21-5   Mixing, pouring, and foaming. Time from adding component *A* to component *B* until maximum foam rise: 4 min. (*Reprinted by permission of Los Angeles Trade-Technical College.*)

Preparation of urethane foams

Urethane polymers originally were developed by Dr. Otto Bayer in Germany in the 1930s. Today several methods can be used to develop the foamed polymers, which can be rigid, semirigid, or flexible. Chemical formulas vary greatly, and methods of combining ingredients are altered according to process conditions. Regardless of formulation or method, a foam is formed in three distinct phases, which occur in rapid succession:

1   Liquid phase through "cream" phase.

2   Development of gaseous components and foam structure. Gas generation can be induced chemically by evolution of carbon dioxide, or by physical introduction of a volatile blowing agent.

3   Polymerization and stiffening of cell walls, preceded by short gel period.

Foamed urethanes are produced by reacting an isocyanate with a polyol. The most common isocyanate is actually a mixture of isomers of toluene diisocyanate (TDI).

A polyol is an alcohol containing more than two hydroxyl (OH) groups, as we have seen previously in the chapter on formation of polyester resins. Polyols with

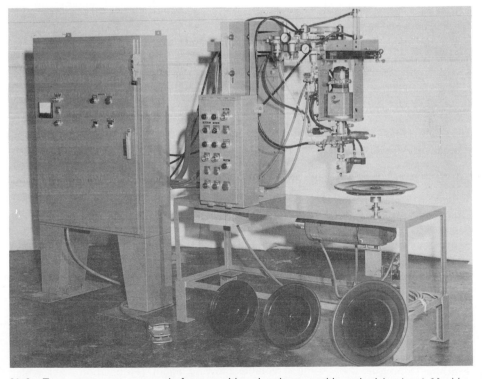

21-6   Two component, automatic foam machine showing turntable and mixing head. Machine shown can dispense 2 to 10 lb/min. Molds shown are for gaskets, which can be made at a rate to 30 per minute. (*Reprinted by permission of Martin Sweets Co., Inc.*)

fewer hydroxyl groups provide fewer reactive sites for a given quantity of isocyanate, resulting in a flexible foam. Typical polyols are polyether glycols, castor oil, and polyester resins.

Rigid urethanes produced in densities to 60 pounds per cubic foot (pcf) can be made by reacting the mixed isocyanate isomer with a polyol such as polyoxy-propylene derivative of sorbital or sucrose, and from other polyethers.

A foaming agent, either in the form of a small amount of water (which reacts with the isocyanate to produce carbon dioxide), or a fluorocarbon[1] (which is inert in the reaction) is used to develop cell structure. Fluorocarbons are used for rigid foam of lower density; CO_2 is used for foams above 5 pcf.

To promote uniform cell formation and prevent collapse, a surface active agent (surfactant) is added to the mix. Based on silicone oils, surfactants are generally used in amounts of less than 1 percent.

Catalysts in amounts to 1.50 percent are added to speed curing. Catalysts used for rigid and semirigid foams are dibutyl tin dilaurate, stannous octoate, and N-methyl morpholine.

Flexible foams are prepared in much the same way, reacting diols and triols, or polyester resins, with isocyanates. Carbon dioxide obtained during the reaction is the primary blowing agent. The amount of water and TDI used determines density. An excess of water leads to an initial foam rise, followed by collapse.

Methods of producing foams

The methods by which the recipe is delivered to processing equipment vary with the extent to which controls can be effected from beginning to end, or methods are dictated by the type of product being made.

One-shot process   In this technique, all ingredients are mixed simultaneously and immediately dispensed as foam. The foam may be metered over a continuous conveyor designed as a trough for production of slabs, or it may be pumped into conveyorized molds, as in the manufacture of seat cushions or arm rests.

Prepolymer technique   The catalyst and other ingredients such as surfactant or color are premixed and then added to the blend of TDI and polyol for either a batch or continuous operation. This system is widely used in the carbon-dioxide-liberation technique.

Semiprepolymer technique   Also referred to as quasi-prepolymer method, this system is a three-part operation: (1) all isocyanate is combined with one-half polyol; (2) remaining polyol is combined with catalyst, surfactant, blowing agent, and water; and (3) first part (A) is mixed with second part (B), usually in equal amounts.

Frothing technique   By using a volatile blowing agent (Fluocarbon 12[2]) which

---

[1] Freon 11 is the registered trade name of E. I. du Pont de Nemours & Company.
[2] Dichloro difluoro methane is produced by E. I. du Pont de Nemours & Company.

volatizes at −21°F, foam is produced as the mixture exits from the metering nozzle. Depending on formulation, froth resins can emerge partially foamed (5 to 12 pcf) as compared with other methods, which are viscous.

One-shot technique generates more exothermic heat, which in turn produces foam more rapidly. Prepolymer methods are least sensitive to conditions and can easily be formulated to provide a wide range of foam times.

Producing the foamed article

Virtually any material that will retain its shape under internal pressures developed by the rising foam can be used to form a foamed article. Furniture, plaques, picture frames, encapsulations, and novelties are mass produced in silicone or urethane elastomer molds to obtain fine detail and handle undercuts. Original wood carvings are sandblasted to raise the grain prior to taking rubber-mold impressions. After pouring, open (five-sided) molds can be capped with a removable plate of plywood to contain the rise and control density. Newsprint, waxed paper, or cellophane is used to prevent urethane sticking to the cap plate.

Aluminum, beryllium copper, and other heat-conducting metals are used in automated (or with limited) production runs when control of exotherm, cell formation, and density is required, as in seat cushions, chair frames, and other mass-produced items. For low-density foams, molds may be internally cored for warm (180°F) water circulation to aid in foam rise, create a more rapid exotherm, and develop a "skin" as

21-7 Portable unit with 2- to 10-lb/min throughput capacity. (*Reprinted by permission of Martin Sweets Co., Inc.*)

exterior cells collapse. High-density foam may require mold chilling to create an opposite effect.

At room temperature, cycle time (from pour to demold) may range from 30 min to more than 3 h. Raising mold temperature reduces cycles to as low as 5 min.

Urethanes do not release well from mold surfaces. They require wax, silicone, or fluorocarbon parting agents for many jobs. Molds produced by casting silicone rubber or thermoformed from polyolefin sheet do not require parting agents.

Factors affecting foaming

The problems of foaming and their possible causes are:

Incomplete rise or fill: Insufficient material, mold too cold, component *A* of pre-polymer too cold, incomplete mixing, and loss of blowing agent through incorrect storage.

Internal voids: Incorrect pour procedure (depends on cavity configuration), large amounts of trapped air, usually caused by poor mixing, and overcatalyzed.

External blisters: Water present on mold because of condensation, incomplete drying of water in dispersed-type parting agent such as PVA solution.

Dark central area: Exotherm too high for rapid dissipation, overcatalyzed, and incorrect proportioning.

Cell collapse: Poor mixing, contamination, and excess water in formula.

Irregular cell formation: Inadvertent blending of two untested surfactants.

Poor release from mold: Undercure, incorrect parting agent, and insufficient parting agent.

Hazards in working with urethane

Isocyanate   The Manufacturing Chemists Association lists TDI as an irritating substance in either liquid or vapor form. Any noticeable concentration will irritate the eyes and upper respiratory tract. A totally enclosed, leakproof system should be maintained for handling isocyanates during storage and transport to the equipment. Well-ventilated areas with downdraft exhaust systems are mandatory. Personnel working in confined areas such as experimental areas or involved in spray operations in enclosed spaces must wear adequate face masks. Isopropyl alcohol removes resins that contact skin; cool water is used in case of eye splash.

Polyisocyanates are extremely moisture sensitive. Small amounts may be spoiled by normal storage procedures. When a container has been opened, it should not be recapped tightly until it has been purged first with dry nitrogen. If small traces of moisture remain, they will create carbon dioxide, which may burst the container.

Catalysts   Many catalysts, particularly organic tin compounds and amines, are skin irritants. Adequate clothing must be worn, and isopropanol or soapy water used to remove these chemicals on contact.

Urethane elastomers

Casting resins Resins for casting uses are available as prepolymers requiring the addition of a catalyst to effect the cure. Proportions of catalyst vary with formulation, requiring strict adherence to manufacturers' recommendations. Cast products range from pump liners and solid tires to shoe parts.

Fibers Extruded monofilaments overwrapped with cotton or synthetic yarns[1] are used for foundation garments, elastic supports, and bandages.

Molding resins Extruded, calendered, and injection-molded products are processed

CHEMICAL REACTIONS INVOLVED IN PREPARATION OF POLYPHENYLENE OXIDE

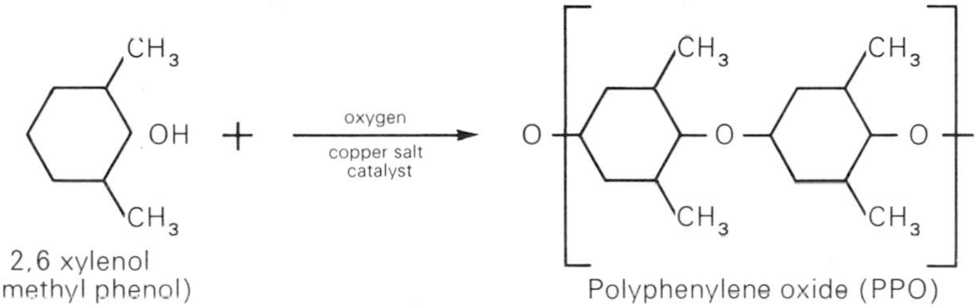

2,6 xylenol
(dimethyl phenol)

Polyphenylene oxide (PPO)

CHEMICAL REACTIONS INVOLVED IN PREPARATION OF POLYSULFONE

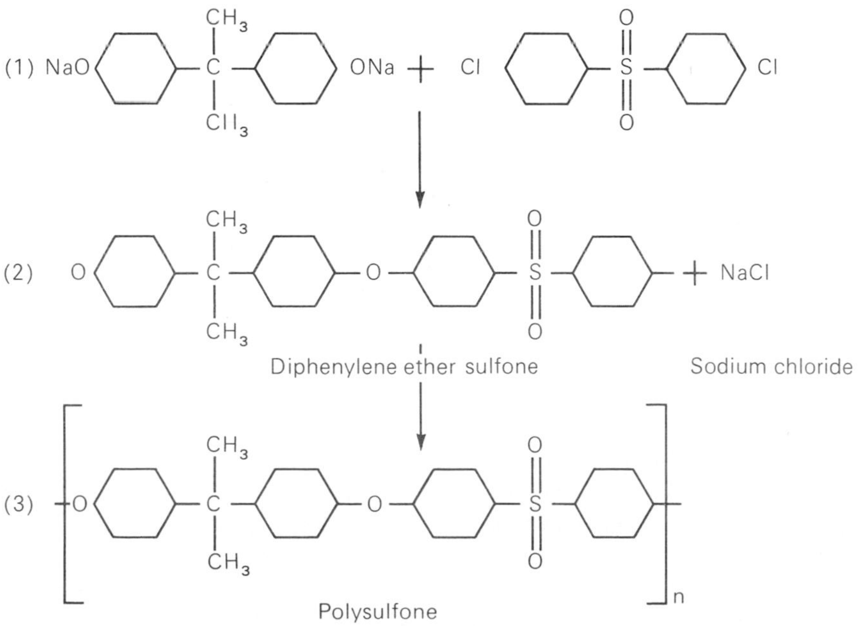

Diphenylene ether sulfone

Sodium chloride

Polysulfone

[1]Spandex is the registered trade name of Uniroyal, Inc.

on conventional equipment. Applications requiring impact strength, abrasion and wear resistance, and oil resistance include friction wheels, belting, wire covering, and gears.

GLOSSARY

Barrier property   Resistance to permeation.

Ductile   The property that describes the extent to which a material can be drawn or extended without fracture.

Isomer   A compound that has the same number and types of atoms as another, but which has a different molecular arrangement, for example, n-(normal) butane, with carbon atoms in a straight chain, $CH_3CH_2CH_2CH_3$, and its isomer (isobutane)

$$CH_3$$
$$|$$
$$CHCH_3$$
$$|$$
$$CH_3$$

Ultrasonic welding   Creating a fusion between mating surfaces by means of high-frequency (20,000 Hz) sound waves.

REVIEW QUESTIONS

1   What applications are suggested for PPO because of its unique properties?
2   Which solvents are used to bond PPO?
3   Why would it not be advisable to mix PPO and Noryl together for processing purposes?
4   What applications are suggested for polysulfone? Why?
5   How is polysulfone annealed? How does this compare with the technique for annealing polystyrene?
6   What applications utilize flexible polyurethane foam?
7   What advantage does a frothing foam exhibit over pour-in-place types?
8   What is a surfactant? What purpose does it serve in the polyurethane formula?
9   Compare the one-shot process with the prepolymer method of producing foam.
10   What effect does excessive exotherm have on foamed articles? What other resins studied thus far exhibit high exotherm?
11   What causes cell collapse?
12   What causes irregular cell formation?
13   How does the Manufacturing Chemists Association classify TDI?
14   What danger exists in storing isocyanates?
15   What precautions must be observed in handling catalysts for urethanes?
16   What is an isomer?

CELLULOSICS

22

Whether cellulose plastics truly qualify as man-made material or should be classified as pseudosynthetic is debatable. There can be no doubt, however, that cellulosics not only occupy an important place in the history of plastics, but that they are a most versatile and high-volume material.

It has already been shown that polymers are derived mainly from natural sources (petroleum, coal, natural gas, and air). Cellulose, although it is a major ingredient for cellulose plastics, can neither be broken down, nor converted, into a resin in and of itself. Instead, it must be modified as a unit or regenerated to form a tractable, processable material having thermoplastic properties.

NATURE OF CELLULOSE

Cellulose is a large, complex molecule ($C_6H_{10}O_5$) that forms the major part of living plants, grass, and trees. It is probably the world's most abundant carbohydrate.

Commercially, purified cellulose is obtained from cotton-gin operations or is extracted from wood pulp. Long cotton fibers called *staple*, ranging from less than one to several inches long, are first removed from cotton seeds for textile yarns and cordage. Shorter lengths called *linters*, which are too short for spinning, are used for plastics. Removed from the seed, they are cleaned, filtered, and chemically purified to form the basis for cellulose plastics.

In the wood method, soft woods (pine, spruce) are reduced to chips, chemically treated, rinsed, and collected on a screen as a felt or pulp. The purified paper is furnished to the resin manufacturer in rolls or bales.

Purified cellulose called *alpha-cellulose* is also furnished to manufacturers of phenolic, urea, and melamine molding compounds as a filler and for production of translucent panels.

Cellulose is a highly crystalline polymer having strong intermolecular bonds. The presence of hydroxyl (OH) groups permits direct substitution of ester or ether groups to form a number of cellulose plastics. Since there are three reactive (OH) sites per molecule, the degree of esterification or etherification yields thermoplastics

22-1   Chemistry of cellulose nitrate and cellulose acetate.

with a wide range of properties. The type of cellulosic that results in each case is therefore a function of the degree and type of substitution.

MANUFACTURE OF CELLULOSE NITRATE

In 1862, an English chemist, Alexander Parkes, exhibited a nitrated cellulose material. He suggested the product as a replacement for tortoise shell, ivory, and similar materials in items of personal adornment and, dissolved in alcohol, as a covering for minor cuts and skin abrasions. The material, Parkesine, and the company of that name that produced it failed to realize any lasting commercial success.

In 1869, the Americans John and Isaiah Hyatt obtained the first American patent for a synthetic resin, nitrated cellulose, which incorporated camphor as a plasticizer. The Hyatts established the Celluloid Manufacturing Company to produce items from sheet stock for fabrication into brush backs, mirror frames, cutlery handles, combs, toys, shirt collars, and side curtains for early automobiles.

Cellulose nitrate, as the material came to be known chemically, appeared briefly as movie film and as safety-glass interlayer, but was later replaced by other cellulosic resins. Although the popularity of cellulose nitrate has waned as other products have made inroads into its markets since early in the twentieth century, it is still produced in moderate quantities for fast-drying lacquers, adhesives, and fingernail polish, and as sheets and rods.

The degree of nitration determines the end use for cellulose nitrate:

% Nitrogen	End use
13–13.5	Smokeless powder, explosives (nitrocellulose)
11.3–12.0	Air-dry lacquers, adhesives
10.7–11.2	Thermoplastic resin

The plastic is highly flammable, burning rapidly with a bright flame and emitting an odor of camphor. Because of this hazard, cellulose nitrate is not processed by conventional methods such as injection molding or dry extrusion. Temperatures required for softening are high enough to cause fire or explosion. This material also tends to embrittle as plasticizers migrate from thin sheets and to yellow with age.

Cellulose nitrate has advantages, however. Extremely tough, water repellent, with high gloss, and depth of color, it is highly valuable in opthalmic frames, purse frames, umbrella handles, table tennis balls, tool handles, and dice. Sheets of pearlescent and solid color are used in quantity to cover wood toilet seats, laundry hamper lids, and shoe heels. Thin sheets softened in an acetone-water solution are stretched around wood cores to shrink tightly when dry. The largest single outlet for cellulose nitrate is for fast drying lacquers. Automotive uses account for much of the volume, particularly repair work, but high-gloss coatings for aircraft, knit or woven textiles (so called artificial leather), window shades, and as a binder for artificial wood putty provide additional volume.

MANUFACTURE OF CELLULOSE ACETATE

The development of cellulose acetate as a commercially feasible material can be traced to the work of G. W. Miles, an American, in 1905. Previously, acetate was synthesized (in Germany, 1865; in France, 1879; and in Great Britain, 1894), but the resulting resin was not soluble in inexpensive solvents. Miles showed that cellulose acetate could be dissolved in acetone and other low-cost materials by adding dilute acetic acid after the acetylation step. Unlike cellulose nitrate, cellulose acetate is widely used for injection molding, extrusion, and film manufacturing. Molding pellets formulated with plasticizers (dimethyl phthalate), flame retardants (triphenyl phosphate), and colorants are obtained in a variety of flow grades to suit molding requirements (see Table 22-1).

Cellulose acetate is tough, scratch resistant, exceptionally clear with high gloss, and is grease and oil resistant. Molded and extruded products are dimensionally stable and stress-free. Cellulose acetate is easily bonded with acetone; it can be plated by vacuum-deposition methods; and it has a slight affinity for moisture. Material must be predried before molding or extruding.

Typical of numerous examples of molded acetate are microphone housings; oil cans; vacuum-sweeper parts; electric shavers, hair clippers, and massager housings; lenses and watch crystals; safety glasses; and sun glasses (frames).

Cellulose acetate film and sheet are made by several methods. Conventional dry extrusion through a slit die with cooling over chill rolls produces an exceptionally fine sheet to 0.1875 in. thick. Casting from solvent solutions over polished cooling drums results in a tough, high-clarity film as thin as 0.0008 in. Table 22-2 shows the uses for cellulose acetate film and sheet.

Films are slow burning and emit an odor of acetic acid. More highly acetylated (triacetate) special formulations are self-extinguishing.

MANUFACTURE OF CELLULOSE ACETATE BUTYRATE

Cellulose acetate butyrate, the third cellulose ester to be developed (1935), is produced by a process similar to that used for cellulose acetate. Butyric acid, butyric anhydride, and acetic anhydride are introduced into the heated reaction kettle along with alpha cellulose. The resin is referred to as a mixed ester, usually containing a

TABLE 22-1   Major grades of molding compounds

Property	Soft-1	Med. soft-1	Med. soft-2	Med. hard-1	Med. hard-2	Hard-2	Hard-4	Hard-6
Flow temperature, °F, ±40 F	266	284	284	302	302	320	338	356
Tensile strength, min, psi	2,000	3,500	4,000	4,200	4,600	5,500	6,000	6,500
Impact strength, min, ft-lb	4.0	3.5	2.8	2.5	2.5	1.2	0.8	0.6

higher percentage of acetyl groups. Like cellulose acetate, cellulose acetate butyrate has excellent clarity, is tough, flexible, and easily molded or thermoformed. Resins are formulated to have a wide range of flow characteristics, mechanical properties, and color (see Table 22-3).

The fact that cellulose acetate butyrate is compatible with plasticizers that are higher boiling and less migratory leads to uses where dimensional stability, improved water resistance, and outdoor weathering are required.

Butyrate is easily identified when heated, giving off an odor of rancid butter. It is not recommended for food containers or food packaging. Its resistance to strong acids and bases is poor. However, dilute solutions do not affect it. The mixed ester is

22-2 Toughness, high gloss, and clarity are reasons for choice of cellulose acetate in this application. (*Reprinted by permission of Eastman Chemical Products, Inc.*)

soluble in a wide variety of solvents, among which are ketones, ethyl acetate, ethylene dichloride, and diacetone alcohol. Like cellulose acetate, butyrate is slightly hygroscopic, requiring predrying of material to preclude blisters. These two resins are not compatible; therefore, care should be exercised to prevent contamination of molding resins.

Applications of cellulose acetate butyrate

Injection molded   Automotive armrests, steering wheels; ball-point and fountain pens; hose nozzles; portable fan housings; tackle boxes; telephone bases; traffic lane markers; truck, trailer, and boat light lenses; and typewriter keys.

Thermoformed   Blister packaging; toys (slot cars, games); utility boxes and covers; and ventilator covers.

Extruded   Pipe for natural-gas transmission; ball-point pen barrels; table trim; and sign-letter trim (metallized).

Based on special formulations of cellulose acetate butyrate with plasticizers,

TABLE 22-2   Uses for cellulose acetate film and sheet

Sheet	Film
Thermoformed articles:	Envelopes
Cake covers	X-ray film
Signs	Recording tape
Letters	Book covers
Blister packs	Page protectors
Boxes	Capacitor insulation
Windshields (motorcycle)	Motor-slot insulators
Face shields	
Machine guards	
Braille pages	
Die-cut products:	
Playing cards	
Index tabs	
Lamp shades	
Laminated products:	
Lamp shades (decorative fabric interleaf)	
Glazing (wire cloth interleaf)	

TABLE 22-3   Major grades of molding compounds

Property	Flow grade				
	Soft	Med. soft	Med. hard	Hard-2	Hard-4
Flow temperature, °F, ±40°F	266	284	302	320	338
Tensile strength, psi	2,300	3,200	3,900	5,000	5,900
Impact strength, ft-lb	7.5	6.0	4.0	2.8	1.0

22-3 Injection-molded lens for blinker light utilizes transparent amber CAB. (*Reprinted by permission of Los Angeles Trade-Technical College.*)

oils, and solvents, strippable coatings are used to protect machined parts (particularly threads), drill bits, milling cutters, and saw blades from damage or corrosion. The material in cube or slab form is heated from 300 to 375°F, and the objects are dipped and cooled. Coatings can be peeled or threads unscrewed.

MANUFACTURE OF CELLULOSE PROPIONATE

Cellulose propionate (sometimes referred to as cellulose acetate propionate) attained commercial value in 1945. Currently it is used for injection and extrusion molding. It is produced by reacting acetic acid and anhydride, plus propionic anhydride with cellulose. The resin is furnished in transparent, translucent, and opaque grades in a range of flow and mechanical properties. Cellulose propionate is often selected by injection molders because of its excellent mold-filling properties.

In general, propionate has similar electrical, impact, and chemical resistance to

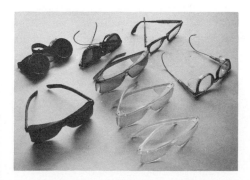

22-4 Safety glass molded in cellulose propionate, which has toughness, dimensional stability, and good mold-filling properties. (*Reprinted by permission of Eastman Chemical Products, Inc.*)

butyrate and is available in weather-resisting grades. It is soluble in ketones, ethyl acetate, and other esters, and is slightly soluble in alcohols. (See Table 22-4.)

Heat resistance of propionate is similar to acetate. It is classified as slow burning, giving off a sweet grapelike odor.

Applications of cellulose propionate

Injection molded   Business-machine housings and keys; tooth brushes; and vacuum sweepers.

Extruded   Sheets, tube, and profile shapes.

MANUFACTURE OF ETHYL CELLULOSE

When alpha cellulose is reacted with caustic soda, alkali cellulose is formed. Further reaction with ethyl chloride produces the ether, ethyl cellulose. As might be expected, ethyl cellulose exhibits properties quite different from the cellulose esters we have been comparing. It is not crystal clear; it is transparent amber in the natural state. A wide choice of colors is possible. The resin is important for uses above a moderate temperature range, excelling the esters in low-temperature ($-40°F$) impact strength. Chemically, it is more resistant to strong alkalies, but it is soluble in a mixture of benzene and alcohol, cyclohexene, and ethylene dichloride. Water-absorption quality is the lowest of the cellulosics.

Ethyl cellulose, like butyrate, is compatible with waxes or oils and is used as a strippable coating. Its toughness as a coating also makes it useful as protection for bowling pins and wood tool handles. (See Table 22-5.)

Applications of ethyl cellulose

Injection molded   Flashlights; luggage parts; refrigerator crisper trays and light housings; rollers, wheels, and casters; soap dispensers; and military uses.

Extruded   Refrigerator bezel and trim.

As seen in Table 22-5, a product that must withstand shock at normal temperature and retain impact strength at subzero, such as a portable lighting unit for military use, might be molded of type 2, grade 9. It must also exhibit good thermal properties, since the same unit might later serve in a tropical climate. Extruded-edge trim or handles for carrying cases must have good tensile strength, heat resistance, and moderate impact strength, as exemplified by type 1, grade 3.

TABLE 22-4 Major grades of cellulose propionate

Property	Grade					
	1	2	3	4	5	6
Tensile strength, psi	5,000	4,500	4,000	3,000	2,500	1,500
Impact strength, ft-lb	1.0	1.5	2.3	6.0	7.0	9.0

In molding or extruding ethyl cellulose, the resin must not be overheated, or an autocatalytic effect will result and cause rapid decomposition. Heating chambers may need dismantling to clear them of charred particles.

Ethyl cellulose can be used for special molded and extruded applications. It is more important as a base for lacquer formulations, as an ingredient in adhesives for frozen food cartons, rendering wallpaper oil- and grease-proof, and as a hair-spray ingredient.

FIBERS AND FILMS FROM CELLULOSE

More pounds of cellulose plastics are converted annually into fibers and thin-gauge films than are used for molding. Wearing apparel consumes the bulk of the output, but considerable quantities are used as yarn for auto and truck tire carcasses.

Acetate fibers

Dissolving cellulose acetate flake in acetone and extruding monofilaments into a current of warm air produce a continuous filament as the solvent evaporates. The acetate fibers contain approximately 50-percent acetyl groups. By comparison, if acetyl content is held at 60 percent, the degree of substitution is nearly 2.9 of a theoretical 3.0 percent. The resultant filament (a triacetate) is used for wearing apparel, wire covering, and heat-resistant fabrics.

TABLE 22-5 Major grades of ethyl cellulose molding compounds

				Grade							
			Type 1						Type 2		
	1	2	3	4	5	6	7	8	9	10	11
Property											
Tensile strength, psi	6,500	6,000	5,400	4,800	4,500	3,800	4,000	3,500	3,000	2,900	2,300
Impact strength, ft-lb:											
At 23°C (75°F)	1.7	1.7	2.0	2.0	2.2	2.8	3.5	4.0	5.5	5.5	6.0
At −72°C (−40°F)	0.5	0.5	0.5	0.5	0.5	0.5	1.0	1.4	1.5	1.0	1.5
Heat-deflection temp. at 264-psi fiber stress, °F	180	170	160	150	140	120	150	170	160	140	120

TABLE 22-6 Cost comparison of cellulosics

Type	Specific gravity	Cost/lb	Cost/cu in.
Cellulose acetate	1.23–1.34	$0.40–0.52	$0.018–0.024
Cellulose acetate butyrate	1.15–1.22	0.60–0.62	0.026–0.029
Cellulose propionate	1.16–1.24	0.62	0.027
Ethyl cellulose	1.09–1.12	0.67–1.00	0.026–0.041

Acetate–rayon fiber

If cellulose acetate is spun into a fiber and drawn (stretched) to orient the molecular structure in the linear direction, a high degree of tensile strength results. Further treatment with saponifiers reduces elongation. Because of greater strength, yarns can be lighter weight and fabrics made thinner. Tents, tarpaulins, parachute cloth, packaging tape, and industrial belting are made from acetate-rayon when weight and heat resistance are paramount. Acetate-rayon fabrics coated with rubber or vinyl are used for inflatables such as rafts, balloons, and life-saving gear.

Rayon

The term *rayon* is generic, describing any fiber made by regenerating cellulose. In a regenerated cellulose, the physical form is changed without altering the basic chemical structure.

Rayon is produced by two different methods:

Viscose rayon (see Fig. 22–5)

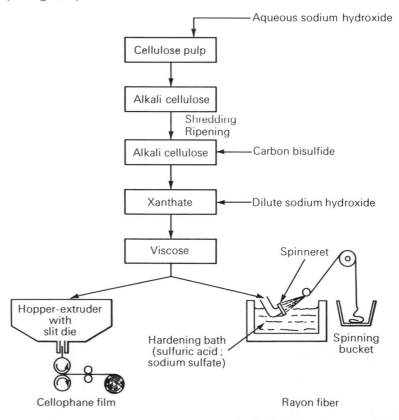

22–5   Viscose process for producing rayon and cellophane.

Cuprammonium rayon   Dissolving cellulose in a solution of ammonia and copper and forcing the liquid through spinneret holes into an acid coagulating bath produce rayon filaments commercially known as Bemberg rayon.

In addition to widespread use of viscose and cuprammonium rayons for wearing apparel, bedding, curtains, and cable wrap, these fibers are blended with cotton, wool, polyester, and acrylic fibers to provide a wide variety of textures for the highly competitive garment industry. Rayon yarn also competes with nylon, glass, and polyester fibers for tire cords and is used as a reinforcement in impact grades of phenolic molding compounds.

Cellophane

If xanthate solution is extruded through a narrow slit die into an acid coagulating bath, film is produced. After rinsing, a plasticizer such as glycerin is added to provide flexibility.

Packaging represents the largest single outlet for cellophane. Of the one-half billion pounds of cellulosics used for film annually, more than 90 percent is used for packaging. Of this quantity, 75 to 85 percent is cellophane wrapping film.

The tremendous demand for cellophane of various grades (moisture-resistant, vapor-permeable, heat-sealable types) also provides an outlet for synthetic resins such as acetate, polyolefins, and vinyls, which are used as lacquer-type coatings or laminated directly to the film.

NONPLASTIC CELLULOSICS

The text is geared mainly to the technician who will be involved with machinery that produces plastics articles. The information on nonplastics is geared more to those whose prime interest is in the chemistry of plastics, and they will be in the minority.

Methyl cellulose and carboxy methyl cellulose

These cellulose ethers are prepared in flake form by reacting methyl chloride or chloro-acetic acid with alkali cellulose. Both products are approved by FDA regulations as being harmless when applied externally or ingested. The following table shows the many uses for these cellulose ethers [a major market (15 million lb in 1968) exists] :

Product	Application	Major characteristic
Methyl cellulose	Cosmetics, hand cream, baby lotion, depilatory cream, paint emulsifier	Improves spreadability
	Dietary supplement (reducing aid)	Provides noncaloric bulk

Product	Application	Major characteristic
Methyl cellulose (*Cont'd.*)	Bulk laxative	Aids water retention
	Ice cream, ice milk, salad dressings, cake icings, mayonnaise	Improves consistency (texture) or retards spoilage
	Icings, pie filling, and meringue	Improves consistency
	Laundry detergents	Acts as sizing and aids in suspending dirt
Carboxy methyl cellulose	Paint (water emulsions)	Stabilizes viscosity and controls pigment dispersion
	Shampoo, liquid soap	Controls viscosity and texture

Although the label may not state specifically that a product contains such an ingredient, it may be mentioned as an emulsifier, stabilizer, or thickener.

GLOSSARY

Emulsion   Suspension of an immiscible liquid in water. Temporary emulsions can be made permanent through addition of emulsifying agents.

Ester   A combination of an alcohol and an acid, with water eliminated. The combination of cellulose with nitric or acetic acid results in the formation of the cellulose ester. Sulfuric acid is used as the dehydrating agent.

Ether   A class of organic compounds having the general formula: R—O—R'. An example is ethyl ether:

$$C_2H_5—O—C_2H_5$$

Ketone   One of a group of chemicals made by oxidizing a secondary alcohol. The general formula is:

$$\begin{matrix} R \\ | \\ C=O \\ | \\ R \end{matrix}$$

Common ketones are acetone, methyl ethyl ketone, and methyl isobutyl ketones, which are used as solvents for some thermoplastics.

Saponification   Reaction of an ester with sodium hydroxide.

REVIEW QUESTIONS

1 What are the sources of cellulose used to produce plastics?
2 List the early uses for cellulose nitrate sheets.
3 What factor led to the successful development of cellulose nitrate plastic in the United States?
4 What are the inherent limitations preventing more widespread use of C/N?
5 What solvent is used to join cellulose acetate?
6 Which cellulose plastic is recommended for applications involving outdoor weathering?
7 Why is C/A/B not recommended for food-contact applications?
8 What advantage does cellulose propionate offer to molders?
9 What are the characteristic odors emitted by C/N, C/A, C/A/B, and C/P?
10 Describe the differences in the methods used to produce a cellulose acetate fiber as compared with the methods used to produce acetate-rayon fibers.
11 Which type of regenerated cellulose is used for tire cords? Which thermosetting molding compound also utilizes this fiber as a reinforcement?
12 Compare the manufacture of cellophane with the manufacture of rayon.
13 What are some of the applications for nonplastic cellulose derivatives?
14 Do you think that cellulose plastics are truly synthetic, or are they pseudosynthetic? Explain your view.

INJECTION MOLDING

23

Injection molding is the single most-used method of producing plastic articles. Beginning in 1927 with the introduction of cellulose acetate molding resin, the injection-molding process has continually occupied a leading position as a mass-production technique. Injection molders as a group annually convert more pounds of resin into useful products and account for greater expenditures for machinery, molds, and auxiliary equipment than any other segment of the industry.

The injection-molding process consists of conveying a powder or pelletized quantity of material from a hopper into a heating (plasticating) chamber where it is melted then injected through channels into a closed mold. The injection process utilizes a comparatively cool or even refrigerated mold to chill the thermoplastic rapidly so that it can be ejected without distortion. No chemical change takes place. This description makes the process appear relatively simple, but complexities arise because of the variety of thermoplastics used and the diversity of product designs that might be chosen.

In order to achieve rapid cycles (as low as 4 sec), material is plasticated in advance of actual injection and held in the heating cylinder while the previously molded part becomes rigid. Until recently, only thermoplastics were used because, as fusible polymers, they can be kept heated to a viscous melt stage for reasonable periods without degradation. Thermosetting resins, by comparison, cannot be held at high molding temperature without premature gelling and stiffening, causing disastrous results in the plasticating cylinder. Recent development in machine design (between 1965 and 1970), aided by modifications in thermosetting molding

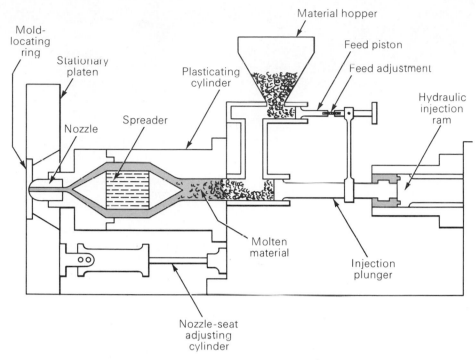

Material hopper

Feed piston

Feed adjustment

Mold-locating ring

Stationary platen

Plasticating cylinder

Hydraulic injection ram

Nozzle

Spreader

Molten material

Injection plunger

Nozzle-seat adjusting cylinder

23-1 Single-stage, in-line, plunger-injection machine.

compounds, makes it possible to mold either class of material by the injection process. Although this chapter and Chap. 24 deal primarily with injection molding of thermoplastics, some attention is directed to thermoset molding.

TYPES OF MACHINES

Since the end of World War II, injection machines have undergone many changes in size and concept, with some designed to suit particular molding requirements.

The in-line, reciprocating, single-stage plunger machine shown in Fig. 23-1 represents one of the earliest designs. It is still in use in many factories throughout the world.

Deficiencies in melting and injecting material were improved to some extent through the design of a two-stage injection system.

More recent designs incorporate a single screw for melting the material. The screw also acts as a plunger to force the melt into the mold. This type of in-line, reciprocating screw press is replacing most of the previous types. Even though it is more expensive, a superior molded part is obtained, usually in less time.

Regardless of the make, model, or size of an injection machine, all have many common engineering features. After a basic knowledge of the sequence of a molding

23-2 Hydraulic clamp, 2500 ton, 350-oz injection-capacity machine represents trend to large presses. (*Reprinted by permission of Farrel Co., Div. of USM Corp.*)

cycle is acquired, any machine can be operated after a short period of indoctrination concerning its special features. As in all molding and laminating processes, the importance of the proper relations among temperature, pressure, and time cannot be overemphasized (see Table 23-2 at end of this chapter). In injection molding, these controls must be more closely balanced and maintained than in compression or transfer molding. Slight fluctuations in a single control factor have a direct influence on the others; therefore, all equipment is designed with as many automatic features as present-day electronic and hydraulic technology permits. However, much of the

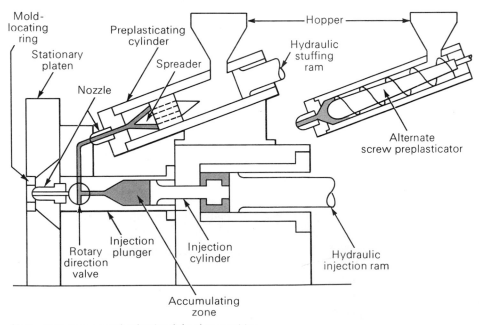

23-3 Two-stage preplasticating injection machine.

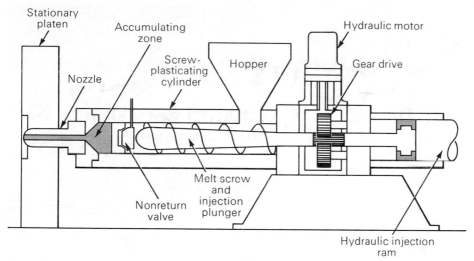

23-4   Single-stage, in-line, reciprocating-screw-injection machine.

cycling control still depends on the skill of the machine setup man and the press operator.

MOLDING CORRELATIONS THAT MUST BE BALANCED

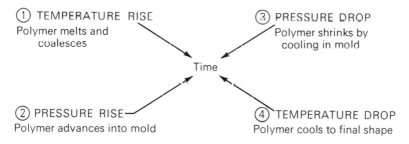

① TEMPERATURE RISE
Polymer melts and
coalesces

③ PRESSURE DROP
Polymer shrinks by
cooling in mold

Time

② PRESSURE RISE
Polymer advances into mold

④ TEMPERATURE DROP
Polymer cools to final shape

TYPICAL MOLDING SEQUENCE (SINGLE CYCLE)

Through a series of electrically connected switches, relays, timers, and valves, the machine can repeat a preset cycle indefinitely or until malfunction. In a typical semi-automatic sequence:

1   The operator manually closes the safety gate activating the primary electrical circuit with switches energized by the gate.

2   The mold closes and is clamped by means of a toggle mechanism or hydraulic ram.

23-5   Removing molded parts from ejector pins. (*Reprinted by permission of Revell, Inc.*)

3   The injection plunger moves forward into the plasticating chamber, at the same time picking up a charge of material to replenish material pushed through the nozzle into the mold.

4   The plunger retracts, then moves slightly forward to push a new charge into the plasticating chamber.

5   The mold opens and the cooled article is ejected by knock-out mechanism.

6   The press operator opens the safety gate for access to parts in the mold.

23-6   Two-plate mold showing considerable length of travel for melted plastic. (*Reprinted by permission of Revell, Inc.*)

THE INJECTION MOLD

Molds are constructed of tool steel to withstand high clamping forces and injection pressures. For limited production, aluminum alloys prove satisfactory, and beryllium copper is often used for cores and cavity inserts when special cooling is required. Construction features are similar to compression and transfer molds except that a more intricate runner and gate system for filling molds can be employed because of the ability of molten polymers to flow long distances.

RATING THE PRESS

Reciprocating plunger machines are rated according to the maximum number of ounces of polystyrene that can be displaced by one forward stroke of the injection plunger. This is called the shot size. Machine sizes range in capacity from a few grams to 300 oz, with the majority of presses from 6 to 28 oz being able to satisfy most molding requirements (see Table 23-1). Larger presses belong in the "special" category for molding such large items as chairs, air-conditioner housings, trash containers, auto dash panels, and pipe fittings. In all machines, plasticating cylinders are designed to hold several times the rated shot size to provide time for complete melting as the pellets progress from hopper to nozzle.

An additional rating called *clamping pressure* is the rated maximum tonnage available to keep the mold closed during the injection stroke and while the shot cools. Injection pressures are high; clamping pressure (mold lockup) must provide a counter-force to prevent resin from escaping at the mold parting line, causing flash and danger to personnel.

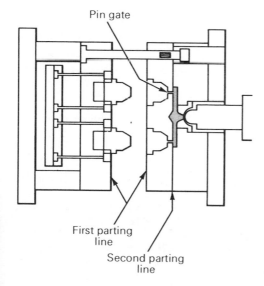

Pin gate

First parting line

Second parting line

23-7 Three-plate pin-gate mold.

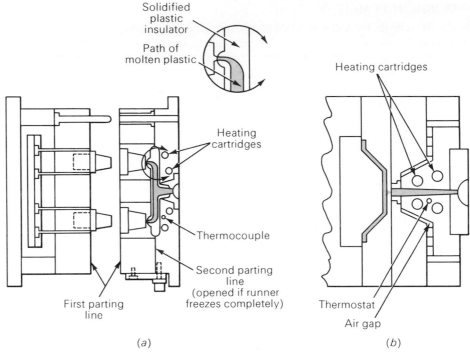

23-8   (*a*) Insulated runner mold; (*b*) hot runner mold.

Clamping pressure is distributed over the parting line of the mold and influences the allowable projected area of molding surface that is permissible. In this respect, the consideration of projected area is similar to transfer-molding thermosets. For example, material flows into a mold cavity at 5,000 psi. If clamping pressure is 350,000 lb (175 tons), the permissible cavity area would be:

$$\frac{350,000 \text{ lb}}{5,000 \text{ psi}} = 70 \text{ sq in.}$$

The total projected area includes the surface area of all runners and the sprue inlet.

TABLE 23-1   Typical plunger press ratings (one- or two-stage)

Injection capacity, oz	Clamping pressure, tons	Platen size, in.
1	10	5 × 4.25
4	125	9 × 10
12	250	17.7 × 17.7
28	375	24 × 20
50	650	30 × 30
300	2,500	48 × 48

FEEDING THE MACHINE

Depending largely on the demands of a machine, materials may be dumped into the hopper from bags or conveyed by suction tubes from 1,000-lb containers or from large outdoor silos. From the hopper, pellets are gravity-fed through a metering unit working in connection with the injection plunger or dropped directly into the feed throat of a reciprocating screw machine. Plunger machines can be fitted with a weigh-feed unit that accurately proportions the amount required for each cycle. Whether the machine operates by gravity feed or weigh feed, the amount of material dropped into the plasticating chamber each cycle should be equal to (or preferably be slightly more than) the amount being injected. Excessive feed creates pressure and mold-filling variations, resulting in product variation.

Injection-plunger travel should be observed for many consecutive cycles. If adjusted correctly, it should stop at the same point on each forward stroke. Feeds are set to prevent the ram from "bottoming out" by adjusting intake to provide a "cushion" of material at the feed end. This is generally between 0.25 and 0.50 in.

THE PLASTICATING CHAMBER

Plasticating chambers are steel cylinders externally heated to melt the polymer. For resins that are corrosive or abrasive, special nickel-chromium alloy metals[1] are used. The forward end of the cylinder is fitted with a nozzle that is mated to the sprue bushing of a mold. Nozzles are interchangeable to suit molding requirements. As a general practice, nozzles should have a smaller spherical radius than the concave

[1] Xaloy is the registered trade name of Industrial Research Laboratories.

23-9 Feed pickup mechanism, above hydraulic injection cylinder, showing adjustment to govern amount of feed. (*Reprinted by permission of Los Angeles Trade-Technical College.*)

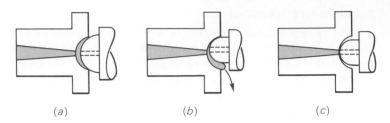

      (a)   (b)   (c)

23-10   (a) and (b) Incorrect, (c) correct nozzle alignment.

spherical radius of the sprue bushing in order to avoid overheating the sprue bushing and chilling the nozzle. The nozzle orifice should be smaller than the orifice of the sprue bushing to ensure a perfect seal in case of minor misalignment. More surface contact at this juncture means greater heat losses from the nozzle to the mold. This influences cooling of the sprue, resulting in longer molding cycles.

Heating the material

Resistance-type heating bands provide heat necessary to melt the plastic. For plunger presses, heater bands should provide 45 W/in.² of surface area. Screw plasticators generate heat through mechanical working of the resin, so that less wattage is needed,

23-11   Variable transformer (rear) controls nozzle temperature on 0.75-oz machine. (*Reprinted by permission of Los Angeles Trade-Technical College.*)

generally 25 to 30. Long cylinders are provided with more than one heating band connected in a series of two or three, with each set monitored by pyrometers that receive signals from thermocouples. For accurate control of the plastic melt, the cylinder is considered as having heating zones, each separately controlled.

Nozzles may have independently controlled bands or may be governed by the adjacent zone control. On most machines, nozzle heat is governed by a variable transformer that supplies current at the desired rate.

Function of the spreader

A streamlined, torpedo-shaped spreader device is positioned inside the plasticating cylinder at the forward end just behind the nozzle. Its purpose is to divide the mass of polymer into a smaller cross section by directing it through small slots parallel to flow. An analogy would be to take a multiconductor cable and separate it into individual wires. The effect is to expose the plastic to more metal-heating surfaces and thus speed the plasticating process. Plastics are a poor heat-transfer medium. Without a spreader, barrel temperatures would be inadequate to raise the center of the mass to a flow condition before decomposing outer sections. The spreader also effects a back pressure to cause entrapped air to work its way back through the material, rather than going forward to the mold. As the pellets move through the plasticating chamber, several changes take place that influence the rate of mold filling and the quality of the molded piece:

1 Material is compressed by the pushing action of the injection plunger.

2 Air is eliminated from the mass to some extent.

3 Pellets soften and become somewhat elastic.

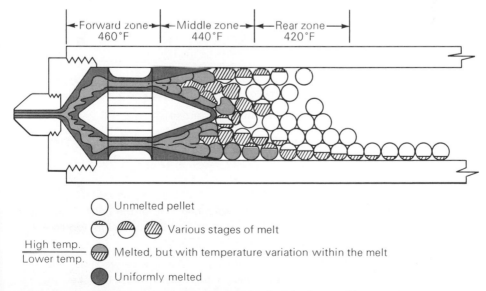

23-12   Melt-temperature variation in plunger-spreader injection machine.

Further heating reduces intramolecular forces and increases mobility.

5 Tendency to adhere to hot inner surfaces of the chamber is increased.

6 Movement of the plastic is laminar in the chamber and in the mold, the extent of which is dependent on the magnitude of temperature differences.

As a result of these temperature differences within the melt, the molded piece may be highly stressed because of uneven heat loss as the part cools. This problem is largely overcome by the use of the screw plasticating press, which melts the resin by turbulent action, as will be seen later.

Mottled colors and marble effects are molded on a plunger machine to avoid the too thorough color mixing of the screw type. By selecting slightly different flow grades (resins having different melt points), distinct mottles and striated effects can be more effectively produced on plunger machines. A mixture of red and white, for example, molded with a screw plasticator, would result in pink moldings.

INSTALLING THE MOLD

Injection presses that have toggle clamps operate on a fixed stroke. The hydraulic ram type of clamp operates like a compression press, with no adjustment required to develop full clamping pressure. In this respect, hydraulic clamps are preferred in situations where many mold changes are required, since less time is used to set a mold.

Each press has a maximum- and minimum-size mold that can be installed. The majority of presses are in-line, horizontal types, which means that molds are hung vertically on the platens. The first important dimension is the mold height (thickness) of the mold in closed position, for example:

Maximum press daylight: 22 in.

Minimum press daylight: 14 in.

Press stroke: 8 in.

Maximum mold height = maximum daylight minus stroke: $22 - 8 = 14$

Minimum mold height = minimum daylight minus stroke: $14 - 8 = 6$

Molds are hoisted by mechanical or electric hoists or carried over the press with an A frame.

Checklist

1 Check the weight of a mold against the safe load limit of the hoist. Figure steel at 0.28 lb/cu in. *Example:* A mold 14 in. square and 6 in. in height would weigh approximately 330 lb ($14 \times 14 \times 6 \times 0.28 = 329.28$ lb).

2 Measure the mold. If it's smaller than horizontal distance between tie rods, then it can be lowered from above the press. Otherwise, it must be taken through either the gate or rear opening and raised into position.

3   Check the eye bolt for cracks. Insert it into the mold to a depth at least equal to the bolt diameter. Use a nut on the bolt for a visual check on depth.

4   Strap or wire mold halves together for safety if it does not have provisions to prevent mold separation. Most molds are unbalanced. If guide pins are short or greased, the danger of mold separation in midair is extremely great.

5   Have all power OFF.

6   Place the mold in the press, using the mold locating ring to seat the mold in the hot half (nozzle side). Locator rings ensure alignment with the nozzle.

7   Clamp the entire mold, by means of the mold-adapter plate, to the stationary platen, removing straps as required. Use socket head or hex head bolts with horseshoe, knee, or slotted clamps. Spacer bars and clamping methods are the same as for compression molds.

8   Before turning power ON to close press:
   (*a*) Set control for MANUAL.
   (*b*) Set injection switch on OUT.
   (*c*) Set mold[1] close switch to OPEN.
   (*d*) Set low-pressure switch to OFF.

9   With power (pump) ON, open mold speed control valve slightly for slow close. Close safety gate and change mold switch to CLOSE.

10   Close off mold speed valve; turn pump OFF.

11   Turn studs that control *press* knock-out plate movement all the way IN.

12   Check press knock-out plate for free movement and examine push-back springs.

13   Measure distance from the front of the movable platen to the front of the plate.

14   Measure distance from the back of the mold-adaptor plate to the back of the mold-ejection plate, with the mold-ejection plate back against any stops.

[1]The words *mold* and *die* are interchangeable. However, mold is preferred.

23-13   Adjustable studs for controlling ejection-plate travel. Newer models utilize hydraulic knock-out bars. (*Reprinted by permission of Los Angeles Trade-Technical College.*)

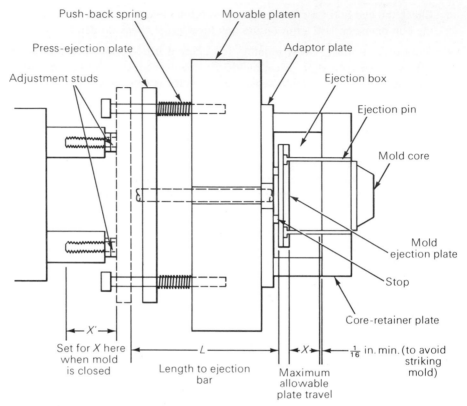

23-14   Setting the knock-out bar.

15 Add measurements for steps 13 and 14 and insert a knock-out bar this length in the center knock-out hole of the movable platen. Some molds are designed for more than one bar. If so, bars should be of the exact same length.

16 The press can now be closed to clamp the other half to the movable platen. (*Check all control switches again*):
   (*a*) MANUAL operation of cycle switch
   (*b*) Mold switch: OPEN
   (*c*) Mold speed control valve: CLOSED
   (*d*) Injection-plunger switch: OUT
   (*e*) Low-pressure mold switch: OFF

17 Turn pump ON.

18 Close safety gate.

19 Change mold switch to CLOSE.

20 Open mold speed control valve slowly until platen closes, then shut completely OFF before turning pump OFF.

23-15 Tie-rod adjustment collar and toe clamp. (*Reprinted by permission of Los Angeles Trade-Technical College.*)

21 Clamp the ejection side of mold to platen. If mold height is too shallow, it will be necessary to make a mold space adjustment by moving the entire clamping end of the press forward.

(*a*) Loosen toe clamps from each of the four tie-rod adjusting collars and hand tighten them out of the way.

(*b*) With pump operating, move mold space adjusting lever to forward position.

(*c*) Open mold space shut-off valve until platen moves slowly. This valve activates a small piston at the rear of the machine.

23-16 Typical machine control panel. Mold space-adjustment lever is shown left of pressure gauge. (*Reprinted by permission of Los Angeles Trade-Technical College.*)

(*d*) When contact is made, close valve, shut pump OFF, and clamp mold.

22 If mold height is greater than will permit toggles to extend fully, loosen toe clamps and spin the adjusting collars back several inches, after first opening the press to take pressure off the rear support.

23 Close the press and spin each collar hand tight against press support. Secure temporarily with one toe clamp on each collar.

24 Open the press at slow speed by keeping the mold switch on CLOSE, but by moving the safety gate slightly off its switch. This allows a single person to observe mold action and at the same time enables him to immediately close the mold if trouble develops.

25 Observe mold opening. There should be no action of ejection plates.

26 Close and open the mold several times, increasing speed by using the mold speed control valve. Observe guide-pin engagement. A bumping noise indicates that the mold is not set correctly and that guide-pin bushings were permitted to rest on the guide pins. This is dangerous, since guide pins can shatter or mold clamps can become loosened.

At this point, the mold is installed, but precise clamping pressure is not adjusted. At high-speed closing with low-pressure switch OFF, correctly set toggles should extend rapidly, but hesitate momentarily at the end of closing action. This adjustment is on a "sight" basis. Toggles that snap into position quickly are set too loose and the mold will flash. Toggles set slightly tight will close with effort, but as much as 1 sec will be unnecessarily added to the cycle. Toggles set altogether too close will prevent activating a limit switch, which in turn prevents energizing the injection plunger. As an added precaution, so that tie rods are not overly stressed during molding, the amount of expansion of each tie rod can be measured when full clamp is achieved. This method is preferred on presses that develop high tonnage.

27 With mold open, loosen the four toe clamps and again tighten them out of action to avoid smashing them.

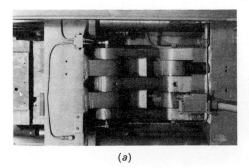 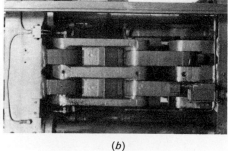

(*a*)   (*b*)

23-17 (*a*) Mold-open position showing retracted toggles and deactivated switch; (*b*) mold-clamped position showing extended toggles and activated switch, which energizes injection sequence. (*Reprinted by permission of Los Angeles Trade-Technical College.*)

28   Using mold space adjusting lever, move toggle end forward about 0.25 in., and switch pump OFF.

29   Each collar has a series of graduations or increment marks similar to a scale, but unnumbered. Major and minor divisions are for greater accuracy in applying an even pressure through the toggle to the mold parting line. Adjusting a toggle to the correct pressure is done on a trial basis for each mold change (see Fig. 23-15). Turn each collar toward the press support 10 major divisions, using a mark or starting point for each. If a permanent mark is scribed into the support at each tie rod, it may mean starting on a minor division line and ending on an equivalent line of the tenth major increment. Hasty workmanship or counting error can result in broken molds, broken toggle link pins, broken tie rods, or mold flashing.

30   Move the assembly back against the collar and lock temporarily with one toe clamp each.

31   Close the mold to observe toggle action. If still loose, repeat steps 27 through 30. If too tight, back collars away three major divisions, or as needed.

As a mold reaches operating temperature, mold expansion may require slight clamp adjustment to prevent straining the toggles and slowing down the closing speed. When toggle is satisfactory, clamp all toe clamps into position.

Adjusting the knock-out plate

Ejection plates must travel smoothly and on a level plane. Failure to set properly results in broken ejector pins or stripping a mold from the platen with possible injury to an operator.

1   Measure maximum permissible distance for ejection plate travel. Some molds are marked with correct travel.

2   Close mold and switch pump OFF.

3   Unscrew top and bottom studs one-half to three-fourths of distance X in Fig. 23-14.

4   Open press slowly, using mold speed control and observe ejection plate and pin travel. If the mold has a sprue pulling pin, check exposure to see that the undercut portion is at least flush with the mold parting line. Ejection plate travel can be increased at this point or molding can be started before making any further adjustment. In any case, the ejector plate must never strike the core retainer plate.

SETTING HEATING-CYLINDER TEMPERATURES

No one set of temperature conditions applies to any material or mold. Temperatures for the various zones are set and adjusted according to the softening temperature of the plastic, its decomposition point, and the length of time required in the chamber.

This is established through trial with each mold and even for different machines with the same mold.

In general, temperatures are lowest at the feed end and increase gradually toward the nozzle. The important factor is the temperature of the polymer, which seldom reaches temperatures indicated by temperature recorders when fast cycles are run. Conversely, long cycles require that resins be prevented from overheating by lowering all temperature zones. Thus, the same resin may be molded over an average temperature range of 350 to 550°F, depending on the overall cycle and mold temperature.

Factors that influence temperature settings

Shot size vs. chamber capacity: For small shots in a large press, use lower temperature.

Mold-filling rate :   Nozzle, sprue, runner, and gate-size control rate of flow.

Part setup or chill rate :   Thick sections (over 0.125 in.) require longer cooling time and are molded at a lower temperature.

Ease of ejection and part removal :   Time may be lost in removing cores or placing inserts, which keep plastic in the heating chamber longer.

It is always good practice to start a cycle at relatively low temperatures and increase each zone gradually until the correct balance of temperatures, pressure, and cooling rate is established. This prevents flashing a mold or degradation of material. Purging dangers are also minimized. It is good practice, too, to set nozzle and forward zones to the desired temperature, while leaving the middle and feed zones either OFF or several hundred degrees lower. This allows material to expand forward and emerge from the nozzle. Otherwise, thermal expansion of polymer in the wrong direction can cause backflow to fill the feed area with molten plastic. This results in a cumbersome cleanup job and, if not detected, creates a danger of hot spatter from injection ram travel. When front zones reach operating heat, set the rear zones.

Under normal conditions, between 45 and 60 min are required to bring a plasticating chamber up to operating temperature. Heats are normally turned on before beginning mold setup so that the machine is ready to cycle without delay. An uncomplicated mold can be removed and another installed in approximately one hour.

Materials suppliers should be consulted when in doubt as to complications arising from overheated resins. Most thermoplastics decompose by charring. This may result in black streaks serious enough to require dismantling the plasticating chamber and cleaning it. Materials containing chlorine cause corrosion and pitting of internal sections, necessitating expensive replacement; acetals emit an extremely irritating odor.

An electric hot plate and thermometer can be used to observe the initial softening point of a polymer, as well as its decomposition point. Usually heating zones should be set between these two positions.

COOLING THE MACHINE

Most production presses provide cooling for three important sections:

Hydraulic system: Hydraulic oil is cooled by passing water through a heat exchanger to maintain the system below 120°F.

Feed throat: Water is circulated through the feed block to prevent pellets from softening and fusing together.

Injection plunger: The plunger travels into and remains in the heating cylinder for a significant portion of a cycle. Partially melted material will stick to the contact end and be dragged back, clogging the next feed.

Water should be directed to the machine hydraulic system constantly when in operation and to the feed block and ram during heating and while cycling.

PURGING THE MACHINE

Purging is performed for the following reasons:

1   To remove overheated plastic before cycling the machine. This prevents flashing a mold. Flashing a mold may be minor, with flash only at the periphery of the part. In some cases, flash can seep along ejector pins or between mold sections, necessitating mold removal and complete dismantling for cleaning.

2   Changing from one material, color, or flow grade to another.

3   To remove contaminated material.

Certain dangers are prevalent:

1   Hot plastic can spatter and ricochet to burn personnel.

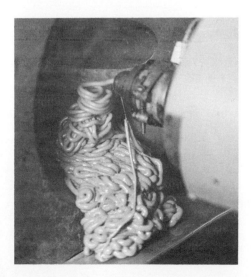

23-18   Purging a heating cylinder. (*Reprinted by permission of Los Angeles Trade-Technical College*.)

2 Material can inject through the sprue and coat the opposite mold face if the mold is open.

Purging is accomplished by MANUAL operation of the injection plunger switch while the injection carriage is in the retracted position. Purging should be done by:

1 Using low-injection pressure.

2 Bringing the plunger forward in several "steps" rather than one continuous stroke. This is done by manipulating the control switch from forward to neutral in quick succession.

3 Protecting one's upper body with face shield, gloves, and adequate clothing.

4 Guarding the sprue bushing and platen from spatter by placing cardboard in the recess.

ESTABLISHING THE MOLDING CYCLE

Through a series of relays, sequence timers control each phase of the molding operation. As one portion of the cycle completes, the next timer is activated. An entire cycle from start to finish may require only a few seconds, while another may involve more than 1 min. Average cycles last from 15 to 45 sec.

Time-controlled sequences actually begin when the mold is first closed and under full lockup. This activates an interlock switch to energize the injection timer. An examination of the sequences of a cycle will show the importance of understanding these interrelations:

1 Mold is closed by closing the safety gate.

2 Injection plunger moves forward and contacts the material.

3 Material is forced from the nozzle into the mold.

4 Pressure exerted by the injection plunger is greatly reduced as material cools and shrinks in the mold.

5 Shrinkage allows more material to follow and pack the cavity.

6 Gate freezes, locking material into the cavity. If the plunger is retracted too quickly, a reverse flow situation is created because of the comparatively soft gate, runner, and sprue. This would create sink marks, internal voids, or poor dimensional control.

7 Ram retracts to pick up new charge of material at feed zone.

8 Mold remains closed for final cooling.

9 Mold opens for part ejection and removal.

A molding cycle is a compromise between the fastest possible production rate and one that is repetitively dependable. The most rapid cycle, with very high heat settings and extremely cold molds to chill parts quickly, can result in an overall lower production rate. In many cases, inferior products result from stresses set up during forced cooling. It must be remembered that operating a press at the upper end of the melt-temperature range results in more serious problems when anything goes amiss to interrupt machine cycling. When minor troubles develop, such as a sprue sticking in the bushing or a part becoming difficult to eject, just a few minutes' delay in correcting the trouble can cause serious overmelting of the plastic. Time is lost in readjusting pressure or in purging the overheated polymer, and many cycles are forfeited while getting pressure and temperatures (mold and machine) balanced.

Setting the timers
Never adjust a timer when it is operating during a cycle.

Plunger-forward timer Mold-filling time is dependent on many factors: (1) melt viscosity of the polymer; (2) shot size; (3) nozzle, sprue runner, and gate size; (4) runner shape and layout (straight, branched, restricted); (5) complexity of cavity (design obstructions such as slots, core pins, or inserts); (6) distance of furthermost cavity from the sprue; (7) mold temperature; and (8) injection pressure and speed.

From these variables, it should be recognized that every mold presents its own set of problems, not only at the outset of molding, but during the entire production run.

An injection ram can complete its stroke in a few seconds, but it is necessary with most molds to hold the ram forward long enough to freeze the gate, preventing backflow of material from the cavities. The length of time the ram remains forward after filling the mold is called dwell time. It is expedient to have a minimum dwell time and retract the ram so that the next charge of material can be brought into the heating chamber. In general, the smaller the gate (thickness), the shorter the dwell time.

Delayed–unload timer Some machines are equipped with a dual control of injection pressure. This is accomplished by incorporating first-stage and second-stage pressure-relief valves in the injection system. By setting each stage individually, it is possible to vary the pressure (and thus the fill rate) during the injection stroke. (On some machines, first-stage pressure is referred to as booster-time.)

If we assume a maximum line pressure of 1,000 psi (20,000-psi injection pressure) and a minimum of 400 psi (8,000-psi injection pressure), then by adjusting each stage separately the injection ram can exert full pressure for the entire stroke or operate at reduced pressure for any portion, depending on the number of seconds allotted to the delayed unload timer or booster. For example:

| | | Injection pressure | |
Condition	Time, sec	First stage	second stage
Plunger forward (total time)	15	1,000	600
Delayed-unload time (booster)	12		
Mold fill (actual time)	5		
Dwell time	10		

All the injection of material takes place at 1,000 psi since the pressure could not drop until 7 sec after the mold filling, as might be the case with thick moldings (0.125 to 0.150). Three seconds before the ram returns, high pressure is reduced to the lower 600 psi. As another example:

| | | Injection pressure | |
Condition	Time, sec	First stage	Second stage
Plunger forward (total time)	7	1,000	600
Delayed-unload time (booster)	4		
Mold fill (actual time)	5		
Dwell time	2		

In the second table, injection is started at 1,000 psi and is reduced to 600 psi for the final second of mold filling, which is maintained during the additional 2-sec dwell. Molds that flash easily or entrap air that must be vented out, or cause difficult ejection because of mold packing can be made to operate more effectively through proper balance of these two pressure controls.

Never set the delayed-unload timer for longer than plunger forward time in order to retract the plunger at normal low pressure. This prevents unnecessary use of high pressure, contributing to longer valve life.

Mold–closed timer   Cooling time is controlled by the mold-closed timer. Some machine controls are designed so that the mold-closed timer is activated at the end of the injection time. Other machines are designed so that the injection timer and the mold-closed timer are both activated at the beginning of a cycle. In this latter instance, the mold-closed time is increased to include the time set for injection.

Mold–open timer   Molds designed to eject parts automatically are operated on a fully automatic cycle by switching to AUTOMATIC at the press control panel. The AUTOMATIC switch brings the mold-open timer into the control cycle when toggles are fully retracted at the end of a cycle. The time the mold remains open is governed by the time needed to eject parts clear of the mold, plus time for mold-stripping mechanisms to return to starting position, usually 2 or 3 sec. The gate remains closed at all times. Parts drop into containers or on a conveyor.

Stuffing timer   Some presses are equipped with a feed-device control to increase capacity of the plasticating cylinder. Figure 23-12 shows that there is vacancy between the plastic and the ram after feeding and before injection. A stuffing timer operates on a numerical rather than a time sequence. Setting this timer for 3, for example, means that the ram will jog forward and back twice from its retracted position, forcing additional pellets into the chamber.

As much as 50-percent more shot size is realized with this device. However, compensation must be made to provide more heat input at the feed end and possibly more injection pressure.

With stuffing, a 4-oz press can plasticate up to 6 oz/cycle. As a common practice, injection molds are designed for use at 75 to 80 percent of press-shot size in order to avoid excessively high chamber temperature and pressure.

Injection pressure   Injection pressures are extremely high in comparison with any other molding process. This leads to dangers from molten material flashing out of inadequately clamped molds or between mold sections, escape at misaligned nozzles, splattering during purging operations, and mold distortion.

Injection plungers operate on a hydraulic pressure of 1,000 to 2,000 psi, depending on the pump used. This translates from 15,000 to 30,000 psi on the material, depending on plunger diameter. Safe press operation demands a full understanding of *effective* pressure rather than pressure indicated by pressure gauges.

Assume that an injection system can deliver 20,000 psi to the injection ram. If this initial force could be transmitted to a liquid resin, the fluid plastic would reach its destination in a mold with only slight pressure loss. Plastics, even when thoroughly

23-19   Machine console with pyrometers and timers. (*Reprinted by permission of Los Angeles Trade-Technical College*.)

melted, are non-newtonian,[1] so that pressure losses as high as 60 percent may be expected as the cavity fills. Several factors affect plastic flow:

1   The pellets are unmelted at the point of ram contact (feed zone). Unequal-feed rates result in more or less "cushion" that the ram must overcome.

2   Hot plastics tend to stick to hot inner surfaces of the chamber, creating a drag condition.

3   Restrictions such as spreader grooves and dispersion disks reduce pressure in proportion to nonsoftened polymer.

4   The melt loses some mobility the moment it leaves the nozzle and enters the sprue because of temperature decrease.

5   Material further solidifies as it contacts runner and cavity walls.

Any change influencing the rate of heating the polymer or its cooling rate changes the *effective pressure* on the injection stroke:

1   Overfeeding at zone 1 decreases effective pressure at zone 3.

2   Raising mold temperature at as low as 10°F increases effective pressure by maintaining polymer viscosity for a longer portion of mold-fill time.

3   Loss of several seconds (or longer) in starting the next cycle increases effective pressure because of additional heating of material.

4   Changes in room temperature affect mold and chamber temperatures, thus influencing pressure.

Other factors to be recognized are less prominent, such as the number of times a material has been processed through a machine, particle size and proportion of reground material used with virgin material, and the rhythm an operator develops in tending a machine.

In general, a balance among cycle times, temperature of mold and chamber, and the pressure sequence must be established. On a new start-up, it is advisable to use a low-temperature profile, low pressure, and short feed, with a lengthy cycle. Observation of material expelled at the nozzle during a purging step is a reliable indication of these relations. If material emerges as a smooth melt without violence, temperature and/or pressure is adequate for trial shots.

Seating the nozzle   The injection assembly can be moved hydraulically into position to align the nozzle with the sprue bushing. During production, the nozzle is kept in contact with the mold.[2] When shutting down, or during temporary delays, back the nozzle away from the mold to prevent chilling the nozzle or overheating the bushing.

To seat a nozzle, close the mold (on MANUAL) before moving the assembly.

---

[1] A newtonian liquid is one that flows in direct proportion to applied force.

[2] Some machines incorporate a "sprue break" in which the injection unit separates slightly at the end of the injection stroke.

This is a precaution against stretching clamp bolts. Move the nozzle forward slowly to avoid damaging it or the mold.

A final checklist should be reviewed to set the machine for a cycle:

1   Material in hopper

2   Toggles set for correct action

3   Ejection mechanism adjusted

4   Mold cavities, sprue, and runners empty

5   Plasticating temperatures set and up to heat

6   Injection, mold close, and delayed unload timers (booster) set

7   Low-pressure close switch ON

8   Injection plunger switch OUT

9   Flow control valve OPEN

10   Cycle switch on SINGLE CYCLE

11   Hydraulic pump ON

12   Injection pressure, 700 to 800 psi, gauge reading

13   Cooling water ON
     (a) heat exchanger
     (b) injection plunger
     (c) feed throat

14   Nozzle seat adjusting lever FORWARD

Closing the safety gate starts the mold closing and activates all subsequent phases of the injection cycle. At completion, and upon removal of molded parts, the next cycle begins when the safety gate is reclosed.

Observe the toggle action and injection rate. If the ram appears to be slowed when contact is made, increase pressure in increments until maximum is reached. If the mold does not fill, increase feed-zone temperature by 10 to 20 deg, and later, progress with temperature increase through the various zones, including nozzle.

AUXILIARY EQUIPMENT

Press operators and mold set-up mechanics may perform other duties in addition to producing parts. The average molding plant requires additional equipment correlated with actual molding.

Mold chillers and mold heaters

The majority of thermoplastics are injected into molds maintained at 70 to 180°F. Unless shot size is very small relative to the mass of the mold, heat cannot dissipate fast enough to assure a rapid cycle. Molds are provided with channels to carry a coolant. A water supply can be connected to each mold half from a manifold. When starting a cycle, water is kept turned off until required mold temperature stabilizes. At that point, sufficient coolant is admitted to maintain the best molding condition. Mold heat is governed by the volume of water valved through channels. When practical, molds are brought to molding temperature prior to molding by connecting a mold heating or chilling unit.

Large shots injected at high temperature or small shots run at very fast cycles require water or antifreeze solutions dispensed through a mold-chilling unit. Heat-resistant thermoplastics require use of mold heaters to pump hot water or oil through the mold. Units are designed to circulate heated liquid to each mold half independently.

Material driers

Moisture is a problem with many resins. Among the more hygroscopic are the cellulosics, acrylics, ABS, polycarbonate, and nylons. Drying temperature is governed by the temperature at which pellets soften enough to melt together or stick to the drying apparatus.

The method of drying material varies from plant to plant. Hopper driers are installed on individual presses; larger units can be part of a central supply system. Cabinet driers and trays placed in circulating air ovens require that pellets not be over 2 in. deep for adequate drying. Several hours are required for moisture removal

Granulators

Scrap made up of sprues, runners, and defective parts is reground for reuse, unless contaminated. Small grinders are placed at the press so the operator can grind scrap as produced. Older scrap grinders are quite noisy, causing large units to be located away from the production area. Newer models are soundproofed to a great extent.

Granulators must be disassembled and cleaned thoroughly for each change of material. When production runs of colors are scheduled, lighter colors are produced first, progressing through the darkest shades. Generally, machines do not require more than vacuuming for adequate cleaning. Plastic dust, held on by static charge, may require wiping with a damp cloth. The most troublesome situation of the molding plant is material contamination. Extra time spent in cleaning away all traces of one plastic to avoid contaminating another pays high dividends in trouble-free production. One gram of one plastic can render 20 lb of another material useless.

Cleaning the granulator   Turn off the machine switch *first*, then disconnect power, preferably from the source. (This rule should also be followed if scrap jams the cutting knives during grinding.) Remove the bin, feed chute, and sizing screen for cleaning.

Before reassembly, check clearance between bed knives and rotary blades. Clearance should be between 0.005 and 0.015 in. to obtain good shearing action. Check for blade looseness, breaks, and other dangers. After reassembly, rotate motor shaft by hand or by jogging the switches to check the work.

23-20 Scrap-grinder, showing rotary and bed knives. (*Reprinted by permission of Los Angeles Trade-Technical College.*)

Hopper magnets are installed in the feed section to trap stray metal. Very soft materials can be chilled with dry ice to become brittle enough for granulating.

COLORING MATERIAL

In-plant coloring of transparent thermoplastics is common practice for several reasons: (1) lower raw-material cost; (2) smaller inventory of colored resins; and (3) reground material can be colored a different color.

23-21 Drum tumbler. (*Reprinted by permission of Beloit Corp.*)

Dry coloring

Pigments and dyes in powder form are added to granules in amounts ranging from 1 to 5 percent, depending on intensity of color desired and the type of molding machine used. Mixing is accomplished in a drum tumbler rotating between 30 and 45 rpm for 10 to 30 min. Drums are slack filled (two-thirds to three-fourths) to facilitate tumbling.

Dry coloring requires individual drums for each color. Undetected caking of pigment or poor dispersion leads to streaking of molded products. Moisture absorbed by the powder can be a cause for rejects.

Color concentrates

To overcome inherent disadvantages of dry-powder pigments, color concentrates are used. Major resin suppliers and some independent producers furnish molding pellets of the required resin type containing a high percentage of color. These are charged into a drum tumbler in proportions of up to 5 lb per 100 lb of plastic. Although slightly more expensive than dry-powder pigments, they are preferred because of their cleanliness.

SAFETY FACTORS

Because of the potential dangers inherent in the operation of injection-molding equipment, the technician responsible for production must be aware of all sources of personal injury as well as potential damage to molds and machinery.

Benefits of a safe operation

1   Prevention of personal injury

2   Less absenteeism

3   Better production (quality and uniformity)

4   Increased production (fewer rejects and less downtime)

5   Less tension on the press operator

6   Lowered insurance premiums

7   Better company reputation

Hoisting and mounting molds safely

1   Check weight of mold against safe load limit of hoist.

2   Check eye bolt for cracks.

3   Insert threads of eye bolt at least at a distance equal to thread diameter.

4   Tighten a nut against the unthreaded portion of the eye bolt.

5   Clamp molds together with removable straps when installing.

6   Check for greased or short guide pins.

7   Check eye bolt hole in the mold for off center, and thus unbalance.

8   Make personnel wear safety shoes.

9   Check mounting bolts for heads in good condition.

10   Insert bolts a minimum of four threads into the platens.

11   Use "knee-type" clamps under heavy molds for additional support when they are clamped at the sides.

12   Double-check tightness of all clamps before opening the press.

Safe operations

1   Pliers, wrenches, cutters, and sprue knock-out rods should be insulated.

2   Remove rings, wrist watches, and other jewelry when working near electrical terminals. Do not allow steel rulers, pens, and other metal to be carried in shirt pockets to avoid the possibility of their dropping into the hopper or granulator.

3   Have operators stand on a wooden platform or rubber mat.

4   Use guards on top of machines when operator can reach over the gate.

5   Use extendable rear guards when molds extend beyond the platen.

6   Never operate without the safety gate in foolproof operation.

7   Never operate without a safety gate.

8   Clean spilled material from work areas immediately.

9   Have grinder operators wear goggles.

Safe purging operations

1   Wear face shield when purging the cylinder.

2   Clear all people from the area.

3   Protect the sprue bushing from spatter.

4   Purge in partial stages :
    (a) Do not completely purge the cylinder with one forward travel of the ram. Use a start-stop technique until the material emerges and can be observed.
    (b) Purge at reduced pressure and ram speed.
    (c) Do not throw hot purgings into a container with combustibles.

5   Never clean a hopper with the machine in operation.

6   Never tamp material or try to dislodge packed pellets in a hopper using metal tools.

Safe start-up and shut-down procedures

Start-up

1 Have nozzle disengaged from sprue bushing.

2 Turn on nozzle and front heats several minutes before turning on central and rear zones.

3 Check to see that the nozzle is heating.

4 Turn on center and rear zones and adjust temperatures to a minimum of 20°F lower than front heat.

5 When molten material drools from the nozzle, reset all zones to desired operating temperatures.

6 If material does not emerge from the nozzle, touch it with a piece of plastic of the same composition to be molded. If the temperature controller shows the nozzle to be hot, check for burned-out heater bands in other zones. Thermal expansion of polymer in the cylinder should cause material to drool out of the nozzle. Frothing, sputtering, or excessive vapors is an indication of moisture or excessive heating.

7 Avoid overheating the center and rear zones to prevent blowout of material from feed zone.

Shut-down

1 Turn off rear heat and continue to mold on cycle until short shots appear. Delay the gate closing for 30 to 60 sec and continue molding until shots again appear.

2 Disengage the nozzle from the sprue bushing.

3 Turn off all remaining heat controls. This procedure preserves the granular separation to the pellets at the rear zone and prevents formation of a solid block of material that would be dangerous on the next start-up.

4 Spray the mold with rust preventive.

Installation of electrical equipment

Drop cords   Drop cords should be:

1 Long enough to prevent tension on the cord.

2 Protected from water.

3 Protected from abrasion by moving mold parts.

4 Electrically grounded.

Electrical connections

1 Conduit and junction boxes should be checked for abraded insulation.

2 Heater bands should have their metal shells grounded.

3   Terminals should have ceramic caps or woven fiberglass covering.

4   Plug-in receptacles should be either vertical or positioned so that inlets face the floor to prevent metal or plastic from getting in.

Safety switches and interlocks

GATE SWITCH

1   Check for defect by pressing switch with the gate open. If mold closes, the switch is defective, since there are usually two switches to be activated.

2   Check all switches near the gate to determine if the operator may accidentally lean on them or bump them when removing parts.

3   Install guards over switches near the operator.

4   Never operate a machine on AUTOMATIC if parts are removed by an operator.

TABLE 23-2   Suggested machine settings for various plastics

Condition	ABS			Acetal		
	Less than 0.125 in.	0.125 in.	Over 0.125 in.	0.0625 in.	0.125 in.	0.25 in.
Max. part thickness						
Cylinder temp., °F:						
Zone 1 (feed)	400–475	390–460	390–470	420–440	Varies slightly, de-	
Zone 2 (middle)	410–480	400–470	400–470	420–440	pending on machine	
Zone 3 (forward)	410–480	410–475	400–470	420–440	and shot size, runner	
Nozzle temp., °F	410–480	410–475	400–470	390	size, type of gate, etc.	
Melt temp. (exit), °F	400–475	375–450	375–470	420	Max. temp., 450°F	
Mold temp., °F	100–200	100–200	100–200	150–250		
Injection time, sec	10–20	10–30	20–60	15	25	30
Total cycle time, sec	30–45	45–60	60–120	30	45	60
Injection pressure, psi		15–30,000			15–20,000	

	Acrylic			Acrylic (high impact)		
	Less than 0.125 in.	0.125 to 0.25 in.	Over 0.25 in.	Less than 0.125 in.	0.125 to 0.25 in.	Over 0.25 in.
Max. part thickness						
Cylinder temp., °F:						
Zone 1	480–500	480–500	480–500	440–500	440–500	400–500
Zone 2	440–450	440–450	440–450	460–510	450–500	430–480
Zone 3	440–450	440–450	440–450	490–540	450–520	440–500
Nozzle temp., °F	375–390	375–390	375–390	480–510	480–510	460–490
Melt temp., °F	400–420	400–420	380–400	480–520	450–500	440–490
Mold temp., °F	175–190	175–190	180–190	150–190	150–180	130–120
Injection time, sec	10–15	15–25	30+	6–15	15–60	50–120
Total cycle time, sec	35–40	40–60	60+	20–45	40–90	60–360
Injection pressure, psi		15–30,000			15–20,000	

TABLE 23-2 Suggested machine settings for various plastics (*Continued*)

Condition	Cellulose plastics		
Max. part thickness	Less than 0.125 in.	0.125 to 0.25 in.	Over 0.25 in.
Cylinder temp., °F :			
Zone 1	340–500	340–480	340–480
Zone 2	350–510	340–500	350–510
Zone 3	350–510	340–510	350–510
Nozzle temp., °F	310–460	310–430	300–420
Melt temp., °F	320–450	330–430	300–400
Mold temp., °F	70–175	70–180	80–150
Injection time, sec	5–30	10–45	15–60
Total cycle time, sec	12–60	20–90	40–180
Injection pressure, psi		8–20,000	

Condition	Nylon 6			Nylon 6/6		
Max. part thickness	Less than 0.125 in.	0.125 to 0.25 in.	Over 0.25 in.	Less than 0.125 in.	0.125 to 0.25 in.	Over 0.25 in.
Cylinder temp., °F :						
Zone 1	475–580	460–580	470–580	600–700	600–660	600–630
Zone 2	450–560	440–560	440–560	550–650	525–600	525–560
Zone 3	450–560	440–560	440–550	525–600	510–530	500–520
Nozzle temp., °F	430–520	430–500	430–500	500–550	500–550	500–550
Melt temp., °F	450–530	450–520	440–510	520–575	510–520	500–510
Mold temp., °F	70–200	70–200	60–190	90–200	130–190	180–200
Injection time, sec	5–20	10–40	20–60	8–20	20–50	40+
Total cycle time, sec	15–40	20–60	40–120	20–30	30–75	60+
Injection pressure, psi		12–25,000			15–30,000	

Condition	Polyethylene (low and med. density)			Polyethylene (linear)		
Max. part thickness	Less than 0.125 in.	0.125 to 0.25 in.	Over 0.25 in.	Less than 0.125 in.	0.125 to 0.25 in.	Over 0.25 in.
Cylinder temp., °F :						
Zone 1	300–450	350–450	300–400	300–600	350–600	350–600
Zone 2	350–450	380–500	320–400	350–600	350–600	350–600
Zone 3	400–500	450–550	350–450	350–600	350–600	350–600
Nozzle temp., °F	400–500	350–500	320–450	400–550	400–550	400–500
Melt temp., °F	450–500	400–500	300–450	400–500	400–500	400–450
Mold temp., °F	40–180	60–200	100–200	40–200	50–210	50–210
Injection time, sec	5–20	10–30	15–40	5–20	10–30	15–30
Total cycle time, sec	10–40	25–80	40–120	10–30	20–90	40–180
Injection pressure, psi		8–25,000			10–30,000	

TABLE 23-2 Suggested machine settings for various plastics (*Continued*)

Condition	Polyphenylene oxide			Polypropylene		
	Less than 0.125 in.	0.125 to 0.25 in.	Over 0.25 in.	Less than 0.125 in.	0.125 to 0.25 in.	Over 0.25 in.
Max. part thickness						
Cylinder temp., °F:						
Zone 1	480–550	480–550	490–550	450–600	450–575	450–575
Zone 2	550–600	550–600	550–600	475–600	475–570	475–570
Zone 3	550–650	550–600	500–600	430–550	410–500	410–500
Nozzle temp., °F	480–510	480–510	470–500	400–500	400–500	400–500
Melt temp., °F	450–500	450–500	450–500	450–525	450 525	450–525
Mold temp., °F	70 200	70–200	70–180	60–180	60–220	100–220
Injection time, sec	5–20	10–30	30–40	3–15	5–30	10–40
Total cycle time, sec	20–40	30–60	60–180	6–45	25–90	60–120
Injection pressure, psi		15–20,000			8–30,000	

Condition	Polystyrene (general purpose and high impact)			Styrene-acrylonitrile		
	Less than 0.125 in.	0.125 to 0.25 in.	Over 0.25 in.	Less than 0.125 in.	0.125 to 0.25 in.	Over 0.25 in.
Max. part thickness						
Cylinder temp., °F:						
Zone 1	350–550	325–500	300–475	400–550	400–500	400–500
Zone 2	375–575	350–550	325–510	425–550	425–550	400–550
Zone 3	400–600	375–550	350–525	425–525	425–550	425–450
Nozzle temp., °F	375–575	325–450	325–500	400–530	400–500	375–475
Melt temp., °F	350–500	325–450	310–420	380–500	380–475	380–475
Mold temp., °F	90–150	90–150	90–150	100–180	100–180	100–180
Injection time, sec	2 30	8–40	15–60	2–20	5–40	10–60
Total cycle time, sec	5–50	20–90	40–160	10–60	20–80	40–180
Injection pressure, psi		10–20,000			10–20,000	

Condition	Polyvinyl chloride (flexible)			Polyvinyl chloride (rigid)		
	Less than 0.125 in.	0.125 to 0.25 in.	Over 0.25 in.	Less than 0.125 in.	0.125 to 0.25 in.	Over 0.25 in.
Max. part thickness						
Cylinder temp., °F:						
Zone 1	290–330	290–330	300–330	310–330	320–340	320–340
Zone 2	300–340	300–340	300–340	320–340	320–350	320–340
Zone 3	310–350	310–350	310–350	330–360	340–360	330–360
Nozzle temp., °F	320–350	320–350	340–350	340–360	340–360	330–350
Melt temp., °F	320–350	320–350	340–350	350–360	330–360	330–350
Mold temp., °F	110–140	100–140	80–140	140–150	120–150	120–150
Injection time, sec	3–10	4–20	30–60	5–10	10–30	30–60
Total cycle time, sec	20–45	30–120	60–200	20–40	30–60	60–240
Injection pressure, psi		10–20,000			10–20,000	

GLOSSARY

Plasticate   To soften a thermoplastic by means of heat and mechanical working.

Reciprocate   Travel in two directions on the same plane, such as back and forth.

Striated   Having a series of threadlike forms in the direction of flow.

Toggle   A mechanical device that exerts pressure by applying force on a knee joint. It is used for opening and closing presses and for applying pressure.

REVIEW QUESTIONS

1   Trace the development of injection-molding machines from the original concept to present-day types.
2   What two methods are used to clamp injection molds?
3   What two factors are used to rate an injection machine?
4   Why should a nozzle have only minimal contact with a sprue bushing?
5   What is the purpose of a spreader?
6   What safety precautions should be observed when installing a mold?
7   Why does safe practice dictate that cylinder temperatures be set low on start-up, then increased as required?
8   Name three areas of a molding machine that require cooling.
9   When is a cylinder purged?
10   What are the dangers to be recognized in a purging operation?
11   What is meant by dwell time?
12   What governs the length of time that dwell time is applied?
13   What is the importance of first-stage and second-stage injection pressures?
14   Why are injection machines operated at 75 to 80 percent of rated capacity?
15   What is meant by effective pressure?
16   What factors contribute to a significant reduction in the initial pressure exerted by the injection ram?
17   Develop a checklist for:
   (*a*)   Installing a mold
   (*b*)   Purging a heating cylinder
   (*c*)   Establishing the optimum cycle
18   Which thermoplastics require drying before using?
19   Why should scrap grinders be cleaned thoroughly for each material change?
20   Explain the difference between the use of dry colorants and color concentrates.
21   What benefits are derived from a safe manufacturing operation?
22   What safety factors should be recognized in:
   (*a*)   Operating the machine
   (*b*)   The electrical areas of the machine
   (*c*)   Shutting down the machine

INJECTION MOLDING (CONTINUED)

24

The injection-molding process is constantly undergoing changes, either in the design of the machine or in new mold-design innovations. The most important departure from established technique has been the shift from the plunger-spreader method of plasticating material to the more efficient screw plasticator. Because of the characteristic melting action of a screw, several benefits are derived to improve the cycle and quality of product: (1) more uniform melting of polymer; (2) better color dispersion; (3) lower injection pressure and larger projected area permissible; (4) more uniform cooling of molded parts; (5) fewer stresses in molded parts; and (6) faster overall cycle in most cases.

Operation of screw-plasticating injection presses is essentially the same as for the reciprocating plunger press described in Chap. 23, although an understanding of the melting principle is necessary to avoid molding problems.

FACTORS AFFECTING SCREW PERFORMANCE

A screw performs the same function as a pump. In the plasticating process, the screw reduces solid pellets (or powder) to a molten, flowable condition while conveying it to the discharge end. In rotating, the screw performs a great amount of work, resulting

24-1   Turbulent mixing action of a screw.

in a turbulent shearing action and a complex material-flow pattern within the cylinder. Mechanical energy of the screw is converted to heat in the plastic.

Compression zone

The section where material is picked up from the feed throat and undergoes a gradual transition from large to small volume is called the compression zone. Forcing the solid polymer through the constantly reducing channel results in a high rate of shearing, and so most melting occurs here.

Compression ratio

Compression ratio means the percentage increase in root diameter between the flights of the compression section. A general-purpose screw may have a ratio of 2:1; a special-purpose screw may be rated at 3:1. General-purpose screws are used for polystyrene and styrene copolymers, polyolefins, ABS, and cellulosics. Special screws are needed for vinyls, nylons, and many glass-filled thermoplastics.

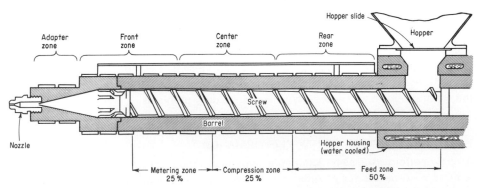

24-2   A typical in-line, reciprocating screw in cross section showing the three primary barrel zones. (*Reprinted by permission of Modern Plastics.*)

Metering zone

This section controls screw output in terms of plasticating capacity in pounds of melt per hour. Its function is to convey molten plastic forward, without surge, and at a controlled rate.

Length–diameter (L/D) ratio

The inside diameter of the plasticating cylinder corresponds closely to the diameter of the screw flights. Therefore, L/D ratio can be taken to mean the length of the screw (from the end of the feed section to the last flight of the metering zone) compared with the diameter of the cylinder. Screws are rated as having ratios of 14:1, 16:1, 18:1, 20:1, etc., with the smaller presses having 16:1 ratios. Larger capacity units have screws to 4 in. in diameter.

Screw speed

Screw speed influences shear rate and thus the melt viscosity of the polymer. Speeds range from 20 to 150 rpm. Figure 24-3 shows two controls that, in combination, make possible 150 different speed settings. The torque selector is a three-step torque control that can be compared with a gear-shift lever of a car:

Low range: Maximum power, limited speed

Medium range: Medium power, higher speed

High range: Least power, highest speed

24-3 Torque-selector control lever for low, medium, and high ranges. Flow-control valve adjusts for 50 different rpm settings in each torque range. (*Reprinted by permission of Los Angeles Trade-Technical College.*)

Polystyrene and polyolefins are usually run in the high range; cellulosics, nylons, soft grades of acrylic, polysulfone, and acetal are usually melted in medium range. Low range is needed for rigid vinyl, polycarbonate, hard-flow acrylic, and some glass-filled resins that prove difficult to melt quickly.

Back pressure

As the screw rotates according to rpm and torque settings, it generates pressure in the melt. When we consider back pressure as a resistance to being moved forward, rather than an actual surging backward, we can visualize that it is affected by: (1) viscosity of the polymer; (2) restrictions imposed by screw design; and (3) weight of the screw-drive assembly. It is back pressure that forces the screw backward even as it is conveying polymer forward and may be thought of as a pressure differential. The screw actually unscrews itself from the melt. The greater the pressure differential, the faster the screw retracts at a given rpm. By throttling or choking the oil being discharged from the injection cylinder, back pressure is increased. This applies a counter pressure at the drive end of the screw, in effect, making the screw work harder. Plastic is retained within screw flights for more revolutions, so it is being exposed to more mechanical shearing that increases its temperature.

It is important to start a cycle with low back pressure, increasing only to get the quality of melt desired. Back pressures greater than 200 psi indicate that torque and rpm settings are incorrect for the cycle. It should be regarded as a fine tuning device rather than as the primary melt control. Excessive back pressure can degrade the polymer through overheating or create excessive mold flashing or mold distortion. Slow mold filling, poor color dispersion, and weak weld lines indicate that back pressure should be increased.

Screw rotation and retraction should begin when the injected polymer has initially set up and end just prior to mold opening. This is achieved by balancing mold temperature, plasticating temperature, screw speed, back pressure, and injection speed. It is poor practice to retract the screw, then have it idle while the molded part cools. This results in changes in the uniformity of polymer temperature.

SETTING PLASTICATOR TEMPERATURES

Temperature settings for screw-plasticating presses (see Table 24-1) are quite different from temperature settings for plunger machines. Although the temperature profile for the latter rises from the feed zone (zone 1) to the nozzle, heat generated by the screw (according to its design) will create major variables in one machine as compared with another. For example, a short L/D screw (14:1) may require an inclined heat profile (lower at feed and central zones). This is similar to plunger-machine profiles. Medium L/D screws (16 or 18:1) may use either an inclined or a hump profile (feed and forward zones low, middle zones high). A long L/D screw (20 to 24:1) may require a nearly flat or even a declining profile.

In general, cylinder temperatures are lower for a mold used on a screw-plasticator press than for the same mold and material in a plunger press of equivalent capacity. Because of the danger of cracking or distorting a screw, the press operator must be absolutely certain that the polymer is molten before rotating the screw.

INJECTION PRESSURE

On machines equipped with first- and second-stage pressure-regulating valves, the injection-pressure sequence is exactly opposite that for reciprocating plunger machines. The effective injection pressure is higher because all the shot to be injected is at the forward end of the cylinder and is unrestricted by any spreader device. Injection speeds are up to three times faster, even though hydraulic line pressure is the same as on a plunger. By using lower pressure on the first stage until cavities fill, flashing the mold is minimized. Second-stage pressure is increased above the first stage in order to pack the cavity and aid in final flow, while keeping dimensions more uniform.

SAFE SHUT-DOWN PROCEDURE

To shut down a screw plasticator safely, leave cylinder temperatures at set points. Close off feed intake plate. Retract the nozzle and rotate the screw (manual operation) until the cylinder empties. Leave the screw in a forward position. This prevents damage to the screw on the next start. Shut heat controls off. Spray the mold with rust preventive.

In summary, the screw plasticator, although more expensive than an equivalent plunger press, has these advantages:

More uniform melting of polymer: The mass temperature of the melt can be appreciably lower, resulting in reduced cooling time; faster injection, often at lower pressure, improves on laminar flow; and more uniform cooling means fewer stresses in the product, which, in turn, means a more serviceable item.

More effective color dispersion: Mixing action of the screw eliminates streaking.

TABLE 24-1 Typical temperature settings, 10 oz, 16:1 L/D

Material	Rear, °F	Center, °F	Front, °F	Nozzle, °F	Polymer, °F
Polyethylene:					
Low density	350	385	395	400	405
Polystyrene:					
General purpose	380	410	395	420	420
Acrylic	380	420	420	420	425
Nylon 6/6	555	590	545	535	540

Faster purging: Nonproductive time is greatly reduced because there is no place for the polymer to hang up, as there is no spreader device. Less material is wasted.

MOLDING DEFECTS: CAUSES AND REMEDIES

Injection-molding cycles are very rapid, and rejectable parts are produced at an alarming rate when conditions are not correct. Some corrections can be made instantly to take immediate effect, but some corrections do not become apparent until later. For example, raising injection pressure may eliminate "short shots" on the next cycle. Yet increasing or decreasing mold or cylinder temperatures may not stabilize conditions for 20 min. In making a decision about the correct step to take, it is usually best to make only one change at a time and to observe the effect over several cycles. When possible, keep the machine on cycle since even a few seconds' delay increases melt temperature and decreases mold and nozzle heat. For instance, changing the cooling time, barrel temperatures, and pressure at the same time generally results in confusion since one cannot determine which change should or should not have been made if the analysis proves wrong.

The following pages can serve as a guide to help pinpoint molding problems and suggest possible remedies.

Parts sticking to the nozzle (hot) half

Molds are generally designed with the male section (force) on the clamp side. Parts are ejected by using pins or stripper plates. As the product cools, it shrinks around the force and travels back with the movable platen as the mold opens. Failure to release from the cavity half is caused by:

Injection rate too fast: Reduce pressure or speed of injection, or slow the flow by reducing mold or cylinder temperatures, or back pressure if a screw plasticator. High-injection speed results in packing the cavity, which forces the polymer tightly against the cavity walls. A high vacuum is created as the part cools.

Cavity surfaces too highly polished: Some plastics, notably polyolefins, release better when the mold has a satin or matte finish.

Sprue sticking

Two-plate, multicavity molds contain a cold slug well and a sprue pulling pin. The cold slug well is designed to trap the first portion of the polymer as it enters from the sprue. This slug is several degrees cooler than the polymer that follows because of nozzle contact with the mold. If permitted to enter a cavity, the cooler slug results in poor weld lines, flow lines, and stresses.

If the sprue does not release from the bushing, check to determine that the cold slug well is clean. Sprue bushings should be highly polished and be free of nicks, burrs, or gouges, which result from using sharp steel objects for sprue removal. Good nozzle alignment to prevent polymer escape at the sprue entrance must be maintained.

    Sufficient cooling of the sprue allows it to shrink away from the bushing surfaces. Remember that a sprue is the last section to be molded and therefore has the least time to cool. Sprues have a larger cross section than runners and parts, so cooling sequences are actually determined by the length of time it may take a sprue to cool.

    To remove a stuck sprue:

1   Put the machine on MANUAL. Retract the cylinder. Use a brass rod and tap the sprue out.

2   If a sprue is obstinate, a brass screw can be twisted into the large end and pulled out.

Parts cracking when ejected

Brittle materials such as general-purpose polystyrene may crack as they are ejected. This defect is usually a mechanical problem. Check to see that ejector pins move smoothly and travel the same distance at the same rate. Ejector plates can become bent. If ejector pins are worn, flash may work down along the pins, making release difficult. Flash may also work between close-fitting mold parts, thus causing sticking.

    Low impact strength resins are somewhat stronger at elevated temperature. If a warmer part can withstand ejection without warping, then cracking may be overcome. Slowing the mold-opening speed may also prove helpful.

Contamination

Contamination can be classified in two ways:

1   Foreign matter (grease, oil, dirt, metal, tobacco)

2   Dissimilar plastics

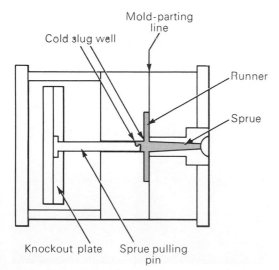

24-4   Cold slug well and sprue pulling pin.

Foreign matter is obvious and the source can be identified and eliminated. Contaminating one polymer with another that is dissimilar is a constant problem in the average molding plant. Contamination is easily detected because molded parts can be peeled apart in layers (delaminated). Moldings have poor strength, and if two clear resins are mixed, they may appear cloudy. Sprue break off and part sticking are common.

The most common sources of contamination are:

1   Placing scrap for regrind into containers previously used for another material and not cleaned

2   Failure to thoroughly clean a granulator

3   Failure to clean hoppers and feed units between material changes

4   Using unclean drums for dry coloring

5   Blowing dust from one machine to another during cleanup

6   Leaving material containers uncovered

7   Spillage from drying operations

8   Broken bags in storage area

Surface blisters

A common defect variously identified as blisters, splay marks, streaks, mica marks, or other terms requires analysis for cause. These surface imperfections are gaseous substances expelled by hot plastic. The gas is caused by (1) thermal degradation of polymer; (2) presence of solvents; (3) moisture; and (4) excessive use of mold-release agents (mold lubricant).

Polymer degradation   No plastic material can be held indefinitely at normal operating temperature without some decomposition. Depending on the material, the decomposition point may vary from several minutes to several hours. Blisters occurring in material known to be dry are good evidence of polymer breakdown.

24–5   Contaminated parts. (*Reprinted by permission of Los Angeles Trade-Technical College.*)

The causes of degradation are:

1 Too large a heating cylinder for low-volume shots. This results in polymer remaining in the cylinder too long (residence time).

2 Running a mold that requires full capacity of the heating cylinder each cycle. This requires maximum heating of the previous feed and extra-high temperature at feed zone.

3 Long cycles because of slow mold filling and/or thick sections; or slow removal of complex parts such as those with complicated inserts; or molding many inserts, creating long open-mold time.

4 Mold filling too fast, creating high frictional heat through small runners and gates.

Generation of gas from solvents

Solvents can prove to be a source for rejectable products because

1 Reground material previously solvent bonded can be a source of defect.

2 Solvents used to compound or dye material may not be completely removed.

3 Plasticizers may be volatized if overheated.

4 Traces of monomer may develop through degradation of polymer.

Generation of gas from water   Materials absorb moisture from various sources:

1 Atmosphere (in storage or open containers).

2 From water-quenched parts, later reground.

3 From extruded pellets cooled in a water bath. This can occur in reprocessing operations where companies reconstitute scrap polymers.

4 In the hopper when humidity is high.

5 From dry colorant or color concentrates that are permitted to absorb moisture in storage.

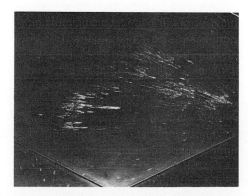

24-6   Surface blemish caused by moisture. (*Reprinted by permission of Los Angeles Trade-Technical College.*)

Violent frothing as melt emerges from a nozzle or a cracking noise within a cylinder is evidence of high moisture content. This poses an extreme danger during purging.

The mold itself can be a contributing factor. Faulty water connections may permit leakage to the mold. Condensate can form on molding surfaces as the mold opens, if the mold is cold and the ambient condition of warm air and high humidity prevails in the molding plant.

Bubbles and voids

Internal bubbles and voids are more common in a thick section (above 0.125 in.), but they can occur in any molding. The causes are: (1) directly caused by gas; (2) shrinkage during cooling; and (3) physical entrapment of air during the fill sequence.

Previously described was the avoidance of overheating the polymer, which causes surface imperfections. Of concern now is the internal formation of blisters, which are obvious in transparent moldings, but are nonetheless present in opaque resins.

Directly caused by gases   Volatiles develop in the plasticating cylinder and result in a porous melt. Volatiles may result from moisture, plasticizer separation, solvent entrapment, or depolymerization of the polymer. They enter the cavity where the initial high-injection pressure is constantly decaying as the part cools and the gates freeze. When this occurs, injection pressure is no longer of value since it cannot overcome the vapor pressure of the gas.

Voids of this type are made up of groupings of small, randomly distributed spheres mainly concentrated at the molded section that retains its heat longest. Corrective remedies are:

1   Use thoroughly dry material.

2   Avoid overheating and degradation. Do not run a small mold in a large press unless extremely low melt temperature is used.

3   Reduce overall cycle if possible, allowing maximum plunger dwell time.

4   Inject at higher pressure and/or speed.

5   Raise mold temperature to delay setting time and to avoid "skin" effect.

6   Increase gate depth and possibly runner, sprue, and nozzle orifice to maintain effective pressure longer. This also reduces frictional heat generation.

Shrinkage during cooling   Material in direct contact with the mold cools first and then stiffens. As the center of the section cools, it shrinks without benefit of pressure since pressure can be transmitted only by molten polymer. The lowering of pressure results in:

1   Material contracting in all directions from the center toward the mold walls, resulting in an internal vacuum bubble

2   Accumulating gases at the hottest point (center of mass), which is the weakest area at that moment

   Corrective measures are :

1   Filling the mold with low-temperature plastic and using higher effective pressure.

2   Reducing front heat zones and increasing middle- and feed-zone temperatures. This results in higher effective pressure on a lower viscosity melt.

3   Supplementing these remedies with the same correction listed under "voids caused directly by gases."

Physical entrapment of air   This type of void can occur when low-viscosity melt is injected at high speed or "jets" into the cavity in a nonuniform wave front because of inadequate gating. Obstructions to flow encountered with core pins, slots, or other design features can create a spot for turbulence to entrap air.

   Corrections are made by :

1   Using any mold temperature or machine control that slows fill rate. Machine control means back pressure, injection speed, pressure, or changing first- and second-stage injection sequence.

2   Increasing or relocating gate in order to decrease jetting.

Flashing of molded parts

Flash of molded parts is caused by many factors. As a result many corrective measures must be considered.

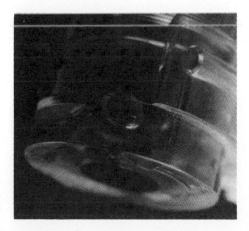

24-7   Internal bubble. (*Reprinted by permission of Los Angeles Trade-Technical College.*)

Part design   Products designed with very thin walls (0.020 to 0.040 in.) are usually molded from melt heated to the upper end of the temperature tolerance to avoid premature chilling. This low-viscosity material can squeeze out at mold parting lines. Low-viscosity material increases the effective pressure developed by the injection plunger, which conversely is the same as decreasing the clamping pressure. Operating a mold with too large a projected area for a particular press also causes flashing.

Fast solidifying of partly filled cavities creates high-cavity back pressure. This can be recognized when a partially filled cavity still flashes. Nonuniform mold cooling may create this situation. Gates may require modification.

Low-volume requirement (small cavities) means more rapid filling; therefore, build-up of cavity back pressure is rapid.

Product designs involving ribs, bosses, holes, corrugations, and grooves will slow the fill rate and develop back pressure. Parts with irregular surfaces tend to flash more.

Mold design   A poorly designed (or built) mold is expensive to operate. The following checklist points out some underlying causes for flash:

1   Lack of sufficient cavity backup plates, support pillars, or metal thickness results in mold distortion.

2   Cavity insert blocks should be raised a few thousandths of an inch above the mold parting line to increase effective clamping pressure.

3   The backs of cavity blocks should be flush with the cavity-retainer plate to prevent movement.

4   Molds originally designed for plunger presses may not be adequate for higher effective pressures generated by reciprocating screws.

5   Venting molds to permit air escape reduces fill time and cavity back pressure.

6   Thin, slotted parts should have cavities built from poorer heat-conducting steel alloys to prevent too rapid cooling that results in a need for higher injection pressure and temperature. Thick parts can be made to chill rapidly to form an outer "skin" by using beryllium copper or aluminum for mold surfaces.

24-8   Flashed parts. (*Reprinted by permission of Los Angeles Trade-Technical College.*)

7   Molds should be designed for independent water cooling of not only each mold half, but often for zone control of various mold sections.

8   Gates can be enlarged to permit lower pressure and melt temperature.

9   Sprue orifice can be enlarged to accommodate a larger nozzle.

MOLDING CONDITIONS

1   Thin sections should be filled rapidly with melt at highest viscosity. Packing of the cavity should be avoided by close control of ram dwell time and cutoff of high pressure (second stage) earlier.

2   Thick sections should fill more slowly to permit a "skin" to form before full pressure is developed.

GENERAL CONSIDERATIONS

Any machine or mold control that affects the flow of polymer can be used to remedy flashing. It should be remembered that one change in conditions will affect some other area of cycle. The entire sequence must be monitored in order to make the necessary compensation. It should also be recognized that it takes time for conditions to stabilize. Among those to be considered are:

1   Improper feed adjustment. Even though plunger machines are self-regulating, feed intake should be adjusted to meter the same amount for each cycle. Otherwise, a mold may flash on the "short" intake cycle because of the nonuniformity of melt conditions. When the cylinder is at its low-volume point, effective pressure is increased because the plunger is working against less material.

2   Pressure too high on the second stage of injection.

3   Second-stage pressure applied too long: reduce delayed-unload time.

4   Mold too hot.

5   Clamping pressure low.

6   Mold faces (parting line) not aligned because of faulty die setting, resulting in uneven clamping.

7   If only one of several cavities flashes, check for distortion caused by carelessness. Closing a mold and injecting material into a cavity already filled with solidified polymer can distort a cavity.

8   Projected area too large for clamp.

9   Cycle too long.

10   Ratio of regrind to virgin too high, necessitating high heat and pressure.

11   Nonuniform granulation of material. Fines will heat faster and more uniformly than large particles, thus affecting pressure.

Short shots

As in other defects, the molder must check mold design and all molding conditions to determine the cause of shorts. Any control affecting flow must be analyzed. Factors to be analyzed are:

1 Improper feed adjustment. Too much feed on plunger machines results in over-working the heater. It also reduces effective injection pressure by forcing the plunger to work on more unmelted polymer than necessary. Observe forward travel of plunger for many consecutive cycles to see that it stops at the same point on each stroke. In some instances pellet shape can cause overfeed, as when polymer is in spherical form. Externally lubricated pellets also create problems with gravity-feed systems.

2 Low-injection pressure. If maximum pressure does not fill cavity:
   (*a*) Increase zone 1 temperature 10 to 20°F.
   (*b*) Increase plunger delayed-unload time.
   (*c*) Increase plunger speed (if independently controlled).
   (*d*) Increase back pressure on screw.

3 Mold temperature too low.

4 Misalignment of nozzle and bushing causing flashback.

5 Overall heat profile too low. Increase all zones and nozzle in increments of 5°F.

6 Nozzle orifice too small.

7 Incorrect flow grade of polymer.

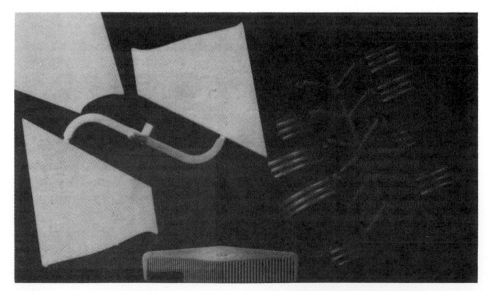

24-9 Short shots. (*Reprinted by permission of Los Angeles Trade-Technical College.*)

8   High ratio of regrind to virgin polymer.

9   Burned-out fuse or heater band.

10   Clogged nozzle.

11   Trapped air in cavity.

12   Small particle broken off in cavity, particularly in slot sections.

13   Cycle too short. Increase mold-open time or mold-closed (cooling) time. Do not increase injection time. It is desirable to retract ram or screw as soon as possible to prepare the next charge for injection.

14   Runners or gates too small; or runners designed with sharp bends, or not polished.

Sink marks

There are two general causes for surface imperfections known as sink marks.

1   *Uneven Cooling.* Parts designed with ribs, bosses, lettering, or other similar features create an uneven cooling effect because section thickness changes. Sink marks appear on the surface opposite the rib or boss. They cannot always be eliminated, but they can be minimized as follows:
(*a*)   Increase injection pressure.
(*b*)   Increase mold temperature.
(*c*)   Increase injection time (total) with corresponding increase in second-stage pressure.
(*d*)   Reduce plasticating temperature, particularly nozzle and forward zone.
(*e*)   Enlarge gates to prevent rapid freeze-off.
(*f*)   Reduce screw back pressure.
(*g*)   Reduce injection speed.

2   *Collapse of internal air bubble.* When a trapped air bubble is dissipated, one or both external surfaces may cave in, thus creating a depression. Corrective measures outlined for trapped gases apply to this condition.

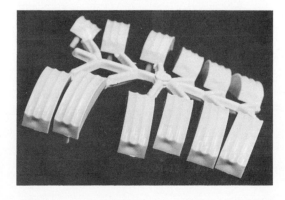

24-10   Sink marks. (*Reprinted by permission of Los Angeles Trade-Technical College.*)

Pressure burns

When any mold closes, the cavity contains air that must be eliminated so molten material can occupy the space. If air is pushed ahead by the flowing polymer, it eventually is compressed into a corner or crevice, or it is surrounded by plastic. Compressing air at high pressure generates localized heat, which is further increased by temperature of the melt. The resultant temperature is high enough to decompose the plastic, resulting in a void surrounded by charred material. Sometimes the defect is merely a small, charred spot. This defect is identified as a pressure burn.

Any corrective measure that reduces flow rate into the mold or allows air to escape can be tried. Usually changes in operation where quick response can be noted are made first, such as injection pressure, speed, screw rpm, and back pressure.

If the burn occurs in a deep cavity well below the mold parting line, slow filling should be done by:

1   Increasing feed slightly on plunger machines.

2   Reducing plasticating temperatures, starting with the nozzle.

3   Decreasing mold temperature.

4   Decreasing second-stage pressure time.

5   Increasing plunger forward time (dwell) and decreasing mold-closed time the same amount. This maintains the same overall time cycle, but it decreases the time the material heats.

6   Grinding a "flat" on ejector pins (0.001 to 0.002 in.) nearest trapped air to allow escape from the back of the mold.

If the burn occurs at the parting line, in addition to remedies listed:

1   Decrease clamping pressure slightly to allow air escape between mold halves.

24-11   Burn marks. (*Reprinted by permission of Los Angeles Trade-Technical College.*)

2   Grind or scribe a vent (0.002 in. deep by 0.010 in. wide) from cavity edge to mold edge.

3   Provide an overflow tab at burn location, to be trimmed later.

Weld lines

A weld line occurs if the path of flow in the cavity is interrupted by some obstruction (such as a core pin), or when material flows around a cavity in opposite directions to unite at some distant point. Weld areas can also be caused by polymer doubling back after contacting a cavity edge at high speed.

On edge-gated parts, welds develop opposite the gate or behind a core pin or other obstruction. On sprue-gated or pin-gated parts, welding takes place on one or more sides, depending on how uniform wall thicknesses are and the concentricity of the cavity and force. Slight shifting of cores reduces wall thickness of the part on one side. The polymer flows through the mold section with greater clearance to unite with the polymer on the opposite side, thus creating an unnecessary weld. Nonuniform cooling of the cavity or core sections can also contribute to variable flow.

Weld areas can be as much as 80 percent weaker than rated tensile strength of the polymer, depending on the stock temperature at the time of welding. Poor welds account for many premature failures of products that are otherwise sound.

Any control that insures faster mold filling improves a weld and strengthens the product proportionately. Thus the following steps should be taken:

1   Increase injection pressure and speed.

2   Increase screw back pressure.

3   Increase cylinder heat, starting with forward zone and nozzle.

4   Increase mold temperature.

5   Use preheated pellets.

6   Increase overall cycle time (as a last resort) to provide more residence time.

7   Check cold slug well. If clogged, a cold slug from the nozzle enters the cavity. Sprue pulling pin can be shortened to provide a larger slug if necessary.

8   Provide an overflow tab so the weld occurs outside the critical area.

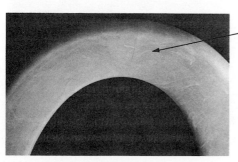

Weld line.

24-12   Weld line. (*Reprinted by permission of Los Angeles Trade-Technical College.*)

Warping of molded parts

Warping is caused by relief of stresses molded into the product. It may occur immediately after manufacture or after a period of time. Highly stressed parts may warp when exposed to moderate temperature or when in contact with solvents that normally would not affect them. Uneven melt temperatures or forcing material not at its optimum melt condition by using unnecessarily high pressure contributes to stress formation. If cooling is uneven because of poor mold control or caused by thick–thin wall sections, then stresses develop. Locked-in stresses tend to relieve themselves once pressure is removed, as when the mold opens.

Specific causes

PRODUCT DESIGN

It is not always possible to design products with the same wall thickness throughout, although this is desirable. Thinner sections cool first and solidify under initial injection pressure. As heavier sections cool, they shrink more (because of gate freeze-off and ram retraction) and warp the part in the direction of the thick section.

MOLD DESIGN

If a cavity is aluminum or beryllium copper, the core should be made of the same metal; otherwise polymer can cool faster from one surface, unless differences can be controlled through mold-temperature variation. Water-cooling channels should be equidistant from molding surfaces to maintain uniform chill rates.

MOLDING CONDITIONS

Molded shape and dimensions can be controlled by differential temperatures on each mold half. By operating one side appreciably colder than the other, parts can be made

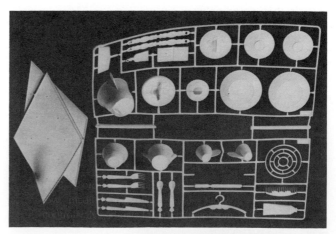

24-13 Warped parts. (*Reprinted by permission of Los Angeles Trade-Technical College.*)

to warp or unwarp in the required direction. Increase of pressure, dwell time, and delayed-unload time can control the extent of packing the cavity to influence warping.

UNEVEN EJECTION

Products with undercuts, raised thin sections, shallow draft angles, or polymers with high shrinkage may cause momentary sticking as the mold opens. Uneven release from mold surfaces may result, causing warpage or distortion. If distortion is caused by sticking in a female section, further cooling is indicated. Molded parts that distort because of difficult ejection from male mold sections require that polymer be warmer when ejected.

POSTMOLDING OPERATIONS

Parts can be ejected directly into a controlled water bath or placed on a shrink fixture to prevent warp. Flat parts can be weighted on a chilled aluminum plate. Extremely thick sections can be placed in sawdust for slow cooling.

Examination of stress

Molded-in stresses are readily detected by polarized light. Specific polymers, namely polystyrene, polyethylene, and cellulosics, are stress-optical and provide the molder with an inexpensive method of inspection. Location of welds and severity of stress concentrations can be analyzed for correction.

Annealing

Stresses can be reduced or eliminated by the process of annealing. Depending on the material, different methods are used. A circulating hot-air oven adjusted for temperatures below the heat-distortion value for the plastic is common. Certain plastics are annealed in oils or wax. Nylons are soaked in wax; acetals are soaked in mineral oil. Plastics unaffected by moisture, such as polystyrene, may be annealed in water. Annealing may require as much as 48 h or as little as 10 min. Molded parts are placed on trays or in wire baskets.

      Generally, annealing benefits products that may warp because of mold design

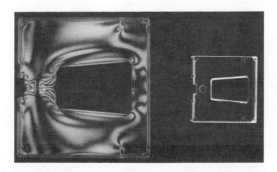

24-14 Stressed areas as seen under polarized light (right-hand photo is the molded part as seen normally). (*Reprinted by permission of Los Angeles Trade-Technical College.*)

or molding conditions, but it aggravates warpage caused by poor part design. Frequently parts that require machining, such as drilled holes, crack unless annealed. Products that craze when exposed to solvents either during bonding or in service can be improved by annealing.

INJECTION-MOLDING THERMOSETS

From 1961 to 1965, major advances in screw designs made it possible to mold thermosetting compounds by the injection process. In theory, the process is not new since machines operating on a fully automatic injection principle were designed and used during World War II. This early version dropped an electronically heated preform into a cylinder, from which it was pushed into the mold by a hydraulic plunger. Now, by incorporating the screw as the means of softening the material, powder can be fed directly from a hopper, eliminating preforming and preheating costs and variables.

Plasticating screws are shorter (17:1 L/D or less) and operate at 45 to 90 rpm to 50-psi back pressure to plasticate the compound. Excessive back pressure results in rapid screw wear because of the abrasive nature of thermosets, particularly mineral-filled types.

To avoid precure of plastic in the barrel or loss of flow from the nozzle, heating cylinders are maintained at 150 to 240°F by hot oil or water circulated through an outer jacket. Barrel temperature, screw rotation, and back pressure must be correlated to bring the plastic up to, but not beyond the optimum flow point (220 to 260°F) in 30 sec or less and avoid precure at the nozzle. If the compound stiffens and cannot be ejected, the nozzle is removed and the shot purged.

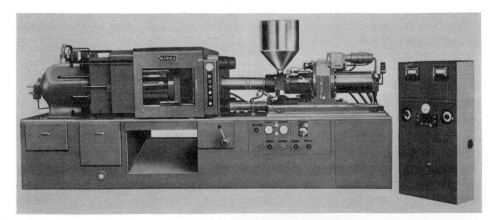

24-15 Screw-plasticating machine for molding thermosets. Hydraulic clamp develops 290 tons; plasticating cylinder is jacketed for hot water. (*Reprinted by permission of Bipel International, Inc.*)

24-16 Phenolic compound purged from injection cylinder. Note granular consistency compared with densified portion. (*Reprinted by permission of The New Britain Machine Co.*)

Molds are heated electrically at 325 to 400°F, depending on the material molded. Venting of molds is more important than for compression or transfer molding because of the speed of injection and pressure (to 30,000 psi).

Injection times are experimentally determined, usually between 5 and 15 sec, depending on mold design. Too rapid injection causes entrapped volatiles or air; too slow injection may result in partial precure, leading to a granular rather than densified part.

Dimensional control presents an area for much experimentation. Fast injection speeds at high pressure with long plunger dwell time, in effect, pack the cavity. After ejection, the part can expand because of an elastic recovery, causing oversize dimensions. Short dwell time adjusted to coincide with gate freeze-off is preferred.

Parts can be molded from conventional transfer grades of compounds, although resin producers offer special formulations having long periods of plasticity with short cure time. As in compression or transfer molding, urea compounds are processed at lower cylinder and mold heats than melamine or phenolic (see Table 24-2).

Rapid cycles, especially for heavy sections, enable the molder to use a thermoset compound in place of a more expensive thermoplastic because the thermoset can be cured to an ejectable stage faster than a thermoplastic of comparable thickness can be cooled. Sections 1 in. thick are molded in less than 1 min; sections of 0.125-in. thickness are molded in as little as 25 sec.

TABLE 24-2   Typical machine settings, °F

Plastic	Zone 1 (feed)	Zone 2	Zone 3	Nozzle	Mold
Urea	165	170	180	200	290–300
Melamine	175	220	240	240	300–320
Phenolic	150	210	220	220	335–350

24-17   Injection-molded circuit-breaker covers with sprue, runners, and gates attached. (*Reprinted by permission of Hull Corp.*)

GLOSSARY

Cold slug   Small cylindrical portion of melt that enters the mold first, but is colder than the melt that follows because of contact of the nozzle and sprue busing.

Draft angle   Taper in a mold wall parallel to the draw (ejection) direction. Shallow draft angles impede withdrawal of a molded part.

Shearing   The process in which one portion of a mass slides (or is torn) away from an adjacent portion.

Torque   Power supplied to rotate the screw.

REVIEW QUESTIONS

1   What advantages can be cited for the screw-plasticator machine as compared with the plunger-spreader type?
2   How does the screw convert mechanical energy into heat within the plastic?
3   What happens to the pellets as they enter the compression zone of the screw?
4   What is meant by the compression ratio?
5   What does the term L/D ratio mean?
6   What is torque?
7   Which plastics require low torque?
8   How does back pressure affect the quality of the melt?
9   How is back pressure controlled?
10   Compare the use of first- and second-stage pressures on a screw plasticator with the plunger type of injection machine.
11   Summarize the safe shut-down procedures for the screw type of machine.
12   Why is it important to make only one machine or cycle correction at a time?
13   Prepare a ready-reference chart listing each type of defect, probable causes, and possible corrections.

14 How are stresses observed in a molded part?

15 What is the purpose for annealing?

16 What is the extreme danger in molding thermosetting compounds in a machine designed for thermoplastics?

17 What are the economic advantages in molding thermosetting compounds by the injection method?

18 What contributes to rapid screw wear in molding thermosetting compounds?

THERMOFORMING

25

The accepted definition of thermoforming is heating a thermoplastic sheet to its softening point and pressing the pliable material against the contours of a mold, where it is held until cool. But a simpler way is to manipulate a strip, rod, or sheet manually for free-form shapes, twisted bars, angle bends, and curves, without using any mold. Thermoforming with vacuum and other aids represents a rapidly expanding process partly because of the variety of sheet materials available and partly because of sophisticated equipment.

The earliest thermoformed products appeared toward the end of the nineteenth century. when sheets of cellulose nitrate were inserted in a metal mold to produce baby rattles, dolls, toy animals, and toilet articles. The process was essentially a blowing process, but interestingly, the method is being revived, using two sheets and forcing air between, aided by vacuum which draws the sheet against mold surfaces.

Thermoforming, as a mass-production process, did not reach fruition until the 1940 to 1960 period, when rigid vinyl, polystyrene, ABS, and polyolefins became available in large quantity. Prior to that time, cast acrylic and extruded cellulosics were used in limited amounts for aircraft enclosures, signs, and advertising displays. Technological improvements, including beta gauge control of sheet thickness, solid-state cycle controls, and fully automatic cycle sequencing, enable manufacturers to thermoform hundreds of thousands of small containers in a single day.

The advantages of thermoforming sheets are numerous, regardless of the type of equipment: larger parts with thin-wall sections are more feasible than injection moldings; lower cost tooling, often single-surface, is common; experimental tools of wood or plaster can be built until the design is set; and sheets can be printed or vacuum metallized prior to forming.

Some disadvantages accrue to the process: scrap losses are high unless extruders are part of the process to recycle material; cost of sheet stock is high initially as compared with pellets for injection molding; it is impossible to have different part configuration on opposite sides; holes or other openings cannot be formed; secondary operations such as blanking, drilling, or sawing are required; thermoforming results in built-in stresses because of the semiplastic state of the sheet during forming; products vary in wall thickness because of stretching inherent to the process; and corners and sharp edges are subject to extreme thinning.

METHODS

Machines for manipulating sheets range from small units for single sheets 18 in. square to giant machines for sheets 10 × 30 ft, and roll-fed formers that form, trim, and stack parts automatically.

All single-station and rotary units have these common features: a clamping frame to hold the sheet in position for heating and forming; mechanical, pneumatic, or manual methods of bringing the sheet in contact with a mold; a vacuum system often in connection with an air supply; and controls for heat, speed of travel, and safety.

More versatile machines designed for several types of forming include upper

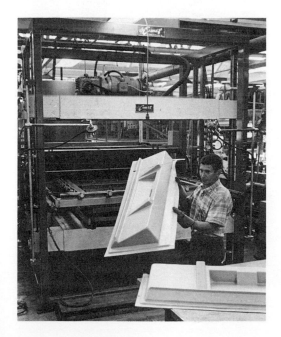

25-1   Single-station thermoformer.
(*Reprinted by permission of Beech Aircraft Corp.*)

and lower reciprocating rams, sheet cutoff devices, and sequential timers to monitor each phase of a cycle.

There are several basic methods by which a sheet may be formed, with many modifications. This chapter outlines some of the major techniques and reviews why each has been developed.

Matched-mold forming

This method is used for large, corrugated sheets, outdoor signs, and other shallow draws.

Straight vacuum forming

This method proves to be inexpensive in that light-duty, single-piece molds are used with atmospheric pressure forming the part. A review of Chap. 12 (vacuum-bag laminating) explains the theory of differential pressure, which is also applied in vacuum forming.

Forming with vacuum alone in a female cavity is limited to a draw ratio (depth to part width) of half the width of the part. This avoids excessive thinning or the

25-2   Three-stage thermoformer. As one sheet is being heated, another is formed and cooled, while a third is removed. (*Reprinted by permission of Brown Machine Co.*)

Clamping frame

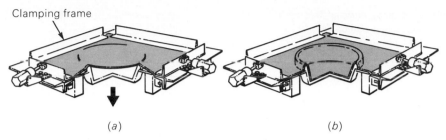

(a)   (b)

25-3 Straight vacuum forming. This technique is used to involve clamping the sheet directly over a female mold, then heating the sheet while in contact with the mold (*a*). When the center of the sheet becomes hot, vacuum is applied, thus sucking the sheet into the cavity (*b*). This method of heating results in long heat cycles and uneven heating. The deeper the cavity, the more the material is thinned out at corners, and thick sheets are generally required. This technique also has severe limitations regarding depth of formed part. (*Based on a figure from "Modern Plastics Encyclopedia." McGraw-Hill, New York.*)

necessity for using extremely thick sheets to maintain minimum-part thickness at corners and bottom. Products made by this method are containers, housings, light diffusers, and products where shrinkage of the cooled part away from the side walls permits less taper in the design. Parts with simple undercuts, such as shallow letters, may be formed from material with sufficient flexibility.

Drape forming

Heated sheets may be mechanically stretched over male plugs or drawn to conformity by suction. Male molds are less expensive to machine and polish, but they must have more taper to permit easier removal. Variations of drape forming include plug and

Clamping frame

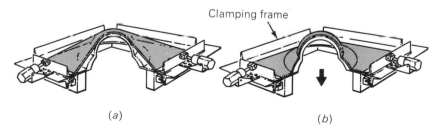

(a)   (b)

25-4 Drape forming; male molds are used. The heated plastic sheet is draped or forced, by mechanical means, over the mold (*a*) and vacuum applied (*b*) to complete the forming. In this method a greater area of plastic is available for distribution on the mold, as compared with vacuum forming. Drape forming therefore provides for more material distribution on parts designed with liberal corners and tapered side walls and it makes for better quality parts. When molds have relatively flat tops and straight side walls, excess material is left on the top or flat of the mold, and the side walls become increasingly thin. (*Based on a figure from "Modern Plastics Encyclopedia." McGraw-Hill, New York.*)

25-5 Inverted drape forming. This process is basically the same as drape forming, except that the mold enters the heated plastic sheet from the top. This technique is ideally suited for working with relatively shallow draws over a large molding area or where sheet sag is excessively great. As the mold enters the clamping frame, vacuum is drawn through the mold, forcing the sheet up into the mold. This eliminates markoff on the corners of the formed part and reduces wrinkling and webbing. (*Based on a figure from "Modern Plastics Encyclopedia," McGraw-Hill, New York.*)

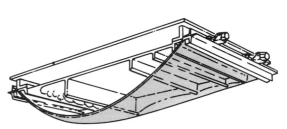

ring, slip ring, and snap-back methods. Drape forming is used for a wide variety of products including signs, advertising displays, contour maps, housings, and other items where draw ratio does not exceed a ratio of 4:1.

Plug-assist vacuum forming

This method utilizes a cored plug, or "helper," with or without air vents. The plug is not heated, so material "locks" to it on contact. Air and vacuum are applied to obtain fine detail and thick bottoms. Forming with plug assist is used for deep tote boxes, housings, and luggage.

Billow forming

This is a method of uniformly prestretching a heated sheet with air pressure and forcing it into a cavity with a plug. Extremely deep draws such as refrigerator interiors are produced on single-station or rotary equipment.

Trapped-sheet forming

One method of producing small parts such as lids for food containers involves using biaxially oriented sheets of polystyrene to control expansion and flow. We have seen how molecular alignment results from cooling a filament or sheet under tension during extrusion. Stretching a warm sheet in the transverse direction by means of clamping chains moving on an angle away from the machine direction aligns the molecules during the two-way stretch. When heated, the sheet expansion and subsequent shrinkage are uniform, which allows flat sheets to be preprinted, then formed so that printing stretches evenly in two directions.

Roll-fed machines

High-production units can produce over one million small parts a day on 200-cavity molds on a fully automatic basis. Small vials or jelly containers for individual use on airlines and in restaurants are examples.

In this operation, the sheet is usually clear or impact polystyrene 0.005 to

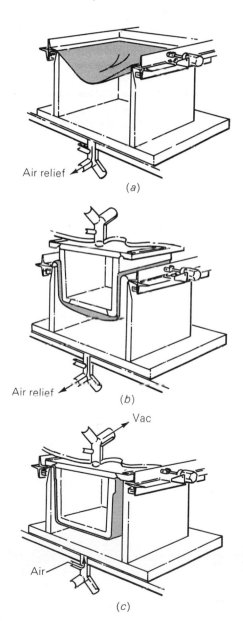

Air relief

(a)

Air relief

(b)

Vac

Air

(c)

25-6 Vacuum snap-back forming. This technique offers complete control and uses sheets of minimum size relative to the formed part size. A predraw box (a) contacts the heated sheet and vacuum is applied through small holes in the box; this draws the sheet into the box for prestretching. In this operation, the plastic is drawn one-half to two-thirds its eventual depth. On certain parts that require odd-shaped predraws, a predraw box may be contoured to fit the particular mold. The distance that the plastic is predrawn into the vacuum box controls material thickness at the base of the mold and also provides for greater thickness of the side wall. When the proper depth of predraw is reached, the mold enters the predraw box (b), where it seals off on the clamp. Vacuum is then automatically reversed from the predraw box to the mold. Air pressure is next introduced into the predraw box, thereby breaking the vacuum in the lower chamber and increasing the detail of the formed part (c). (*Based on a figure from "Modern Plastic Encyclopedia," McGraw-Hill Inc., New York.*)

0.040 in. thick and capable of being rolled into jumbo rolls. An extruder is part of the arrangement, although not necessarily integrated with the forming step since malfunction would create rejects difficult to detect in the formed article.

OTHER APPLICATIONS OF THERMOFORMING

Blister pack

Machines for packaging thousands of articles, from toys and hardware to food, have revolutionized marketing methods. The blister is designed slightly larger than the item and may duplicate the shape or have circular or rectangular designs. Single- or multi-cavity molds are used, and straight vacuum forming permits the greatest number of cavities with the least waste. The blister and item are fixed to a card by several means, depending on the cost allotted to the package, weight, and shape of the product, plus aesthetic considerations. Transparent cellulose acetate, cellulose acetate butyrate, cellulose propionate, and polystyrene are used.

Skin pack

To avoid the expense of a mold or for packaging purposes, more flexible sheets such as polyolefins, polyvinylidene chloride, or vinyl copolymers are used to conform tightly to the enclosed item. In this way, food items, toys, games, and complete pallets of industrial goods are protected from dirt and weather.

25-7   Contour map made by drape forming   (*Reprinted by permission of Union Carbide Corp.*)

MOLDS FOR THERMOFORMING

The least expensive and also the least durable molds are hardwood, pressed wood, and plaster. For short runs or experimental work, they are ideal materials. Cast phenolic, epoxy, or polyester, with or without heat-conductive fillers are also used. Sheet metal, cast and machined aluminum, iron or steel are preferred when durability, rapid heating, and cooling are required. Beryllium copper (hobbed, cast, or machined) provides the ultimate in rapid cooling. Inserts or complete mold sections of room-temperature vulcanizing (RTV) silicone or other elastomers can be incorporated for parts with undercuts or that have minimum draft.

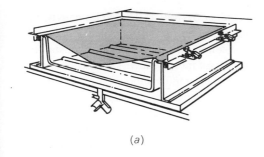

(a)

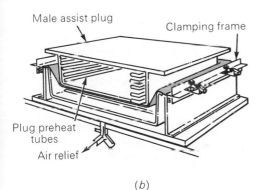

Male assist plug

Clamping frame

Plug preheat tubes

Air relief

(b)

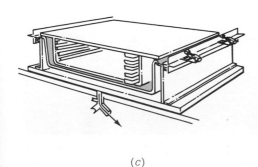

(c)

25-8 Plug-assist forming. This is an improvement over vacuum forming in that it provides for better material control and more rigid corners on a part. The sheet comes into contact with the cavity mold after heating (a). In some instances the plug may be used to prestretch the sheet first. The upper ram or platen on the machine forces a male-assist plug into the sheet (b), thereby pushing it into the cavity. Vacuum then draws the sheet from the plug into the contours of the mold (c). The distance that the plug enters into the cavity controls the bottom and corner thicknesses of the formed part. The result is a piece with a heavy flange, thin side walls, and heavy bottom and corners. (*Based on a figure from "Modern Plastics Encyclopedia," McGraw-Hill, New York.*)

Mold design

As in most product designs where flow, release, and mechanical strength must be considered, taper, radii, and section changes, when properly incorporated, lower reject rate, speed production, and improve product performance. Each material has its own characteristics of hot stretch, tensile strength, and adherence to tool surfaces. Problems associated with these features can sometimes be overcome in the forming process by slower application of vacuum or plug travel. Flexible polyethylene has a tendency to stick to smooth male surfaces, partly because of high shrinkage and partly because of its high conformance. Lightly sandblasted molds alleviate this condition.

Parts made in a female mold can have less taper (as low as 0.5 deg from vertical) since shrinkage is away from the cavity. Male molds should include draft angles of at least 2 deg on all vertical walls, unless air ejection can be used. A radius should be provided at all intersecting planes to assist in forming, reduce stress concentration, and increase impact strength. At least 0.0625-in. radius should be provided for thin sheets and a minimum radius equal to sheet thickness for sheets over 0.125 in. thick.

A series of holes is incorporated in each mold section for air or vacuum. These are strategically located in the deepest recesses. Spacing is optional and may be determined by trial, depending on part detail, sheet thickness, and heat retention.

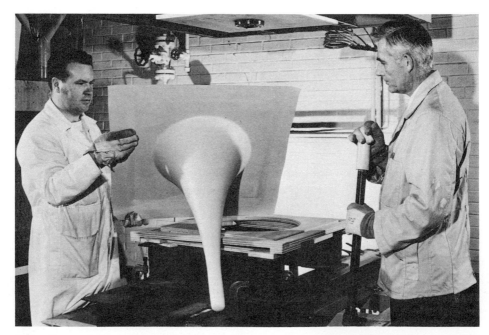

25-9   Extremely deep draw with plug assist, shown as piece of pipe. Material is Kydex (acrylic-PVC). (*Reprinted by permission of Rohm and Haas Co.*)

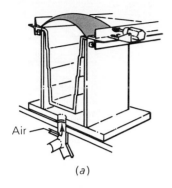

Air

(a)

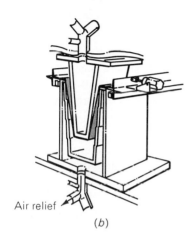

Air relief

(b)

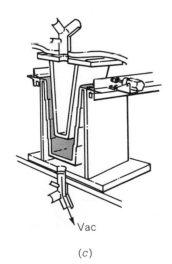

Vac

(c)

25-10 Billow forming. This process is similar to plug-assist forming, but it offers greater draws and more uniform material distribution. The sheet is first sealed off against the mold and a small amount of air pressure is applied into cavity (a). As the plug's impact contacts the slightly preblown bubble, the sudden increase of air pressure causes the sheet to balloon upward. Air pressure in the mold is held in check by an air-relief valve that provides for measured removal of air from the mold as the plug descends (b) at a speed of about 8 in./sec. At the end of the plug stroke (c) full vacuum is introduced, pulling the sheet into the mold. (*Based on a figure from "Modern Plastics Encyclopedia," McGraw-Hill, New York.*)

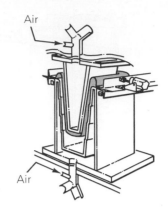

25-11 Billow forming with air cushion. This method is very similar to billow forming except that the sealoff around the mold and clamp frame is not positively made. As the plug enters the mold, air is forced through holes at the base of the plug, thereby providing a minute air space between the plug and the sheet. As the plug enters the cavity, air is also forced into the cavity, providing for better wall thickness in the part. When the plug reaches the end of the stroke, full vacuum is applied. (*Based on a figure from "Modern Plastics Encyclopedia," McGraw-Hill, New York.*)

For soft vinyls and polyolefins, hole sizes from 0.010 to 0.025 in. are used; for thin cellulosics, vinyl copolymer, styrene, and other relatively stiff sheets, hole diameters are increased to 0.040 in. Thick-gauge sheets to 0.50 in. or more in ABS and acrylic can be formed with holes to 0.060 in., or with narrow slots for faster application of vacuum.

Holes in cast plaster or plastic molds can be incorporated during the casting process by using greased wire or monofilament. In all cases, faster evacuation is realized by back-drilling the small-vent holes with a 0.1875-in. or larger drill, depending upon spacing.

For some shallow parts without surface detail, holes are unnecessary if drape forming is used. For example, in making small signs, individual letters are held to a perforated plate or placed on wire mesh. The sign letters are magnetic to prevent movement. Flat molds such as a shallow tray may be placed on a piece of open-weave glass fabric or wire cloth and the sheet drawn around as air is evacuated through the weave to an outlet.

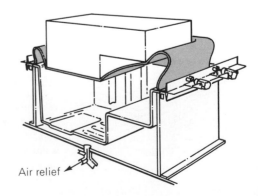

25-12 Billow forming with draw into cavity. In this method the sheet is formed into a cavity mold rather than over a positive mold. The plug used to control billowing of the plastic is normally heated to prevent markoff or chilling of the sheet. (*Based on a figure from "Modern Plastics Encyclopedia," McGraw-Hill, New York.*)

VACUUM SYSTEMS

The system for evacuating air is normally a self-contained unit in all but extremely large machines. Briefly, the system consists of a vacuum port, shut-off valve, surge tank, vacuum pump, and motor. While a sheet is being heated, the port valve is closed, permitting the pump to evacuate the air in the tank. To draw the vacuum on the mold and shape the part quickly before it cools, the port valve is opened. The air sealed between the sheet and mold is evacuated in as little as 0.50 sec because of the empty tank. How quickly the part forms depends on hole size, number of holes, and how

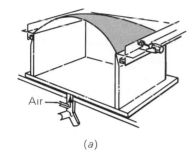

(a)

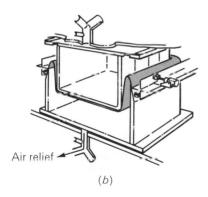

Air relief

(b)

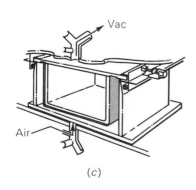

(c)

25-13 Billow forming with reversed draw. This process exaggerates the use of the predraw by blowing a large bubble about half the height of the part (a). Height of the bubble controls material distribution on the lower end of the part being formed. If a part with strong bottom and corners is desired, a bubble of lesser height is blown. If side wall and bottom thickness are to be more uniform, a higher bubble is blown. As the mold contacts the preblown bubble (b), excess air pressure is controlled with the air-relief valve through a fixed outer setting; this controls billowing of the sheet as the mold closes. At the end of plug's stroke (c), full vacuum is applied. (*Based on a figure from "Modern Plastics Encyclopedia," McGraw-Hill, New York.*)

well leaks are prevented, as well as on the slowing of the evacuation rate with the port valve. Larger sheets, for example, 2 × 5 ft, require a surge tank of at least 25-gal capacity and a pump capable of moving air at 150 cubic feet per minute (cfm).

HEATING THE SHEET

Close control of stock temperature is important to avoid tearing, excessive sagging, blisters, loss of plasticizer, and dimensional variation. A starting point for each type of material might be its heat-distortion temperature, although other factors such as color, specific heat, and thickness influence the rate of softening. Most equipment uses radiant heat applied to one or both sides of the sheet. Light colors reflect, while dark colors absorb, so the best forming temperature is arrived at by trial and error. As sheets soften, they sag. Observing the degree of sag or incorporating a photocell or other device to activate the machine elements is general practice.

Sheets may be hung vertically in an oven or laid flat on flannel (to avoid mark-off) and transferred manually to the machine. Thick sheets are effectively heated by convection ovens and still retain heat long enough to form.

SAFETY

The fact that few thermoplastics are noncombustible should make it imperative that sheets do not contact heating elements by sagging or by being blown against them.

The accident rate should be low in the well-equipped shop because of the relatively low pressures used and the safety features built into the machines. Dual switches to activate moving platens, plugs, and frames are built into the system.

CALCULATING AREA RATIO

It is difficult to establish strict rules for redistribution of stock, which is what thermoforming does, because of the many methods of manipulating the sheet. The proper thickness of sheet to start with and its area are based on required minimum thickness at a particular section and part configuration.

Area ratio is associated with depth of draw and is a numerical factor that describes the proportion between the original sheet area and the area of the thermoformed part. For example, if a sheet 15 × 20 in. (300 sq in.) yields a formed part 600 sq in. in total area, the area ratio is 2:1. The thickness of the part averages one-half the original sheet thickness.

Example: Part to be drape formed is 24 × 20 × 2 in. deep. Thickness: 0.090 in.

Area of finished part = 656 sq in.

To determine blank size:

Allow 0.75 in. for clamping and mold clearance.

Blank size = 25.50 × 21.50 in.

To determine sheet thickness:

Usable sheet stock = 24 × 20 in. = 480 sq in.

Area ratio: 656 : 480 ─ 1.37:1

Use 0.125-in. thick sheet to net 0.090 in.

Example: A part 50 × 24 × 20 in. deep × 0.090 in. thick is to be made by billow forming. Prestretching the sheet results in uniform thickness of material at the height of the billow. This thickness conforms closely to the thickness of the bottom of the finished part. Allow 12 percent for clearance between the sheet clamping frame and the mold. Allow 0.75 in. for clamping the sheet (1.50 in.).

Width: 24 in. × 12% = 3 in.

Length: 50 in. × 12% = 6 in.

Blank size = (24 + 3 + 1.50) 28.50 × (50 + 6 + 1.50) 57.50 in.

With plug assist, figure 100-percent use of material that contacts the plug to form the bottom and 50-percent use of the remaining area.

Bottom area = 50 × 24 in. = 1,200 sq in.

Remaining area = 27 × 56 in. = 1,512 sq in.

1,512 − 1,200 ÷ 2 = 156 sq in.

Area equivalent total = 1,200 + 156 = 1,356 in.

The finished part contains 3,160 sq in.

Area ratio = 3,160 ÷ 1,356 = 2.3:1

Sheet thickness = 0.090 × 2.3 = 0.210 in.

GLOSSARY

Heat-distortion temperature   The temperature at which a standard test sample deflects 0.010 in. under controlled thermal conditions.

Specific heat   The quantity of heat required to raise the temperature of 1 g of a substance 1°C.

REVIEW QUESTIONS

1   List several advantages for thermoforming products as compared with injection molding.
2   What is the difference between straight vacuum forming and drape forming?
3   What is the advantage of using a plug assist, as opposed to straight vacuum forming?

4   Describe the vacuum snap-back process. List the advantages of using this method.

5   List the advantages of billow forming as compared with plug-assist vacuum forming.

6   What is meant by the term *biaxially oriented* sheet stock?

7   What is the advantage of using biaxially oriented sheet as compared with conventional extruded sheet?

8   List several common items packaged in blister packs.

9   What is the difference between a skin pack and a blister pack?

10   Which materials are used as tooling for short production runs?

11   What advantage does aluminum tooling offer for thermoforming?

12   Which thermoforming method requires greater taper on sidewalls, drape or vacuum forming? Why?

13   What design features dictate where vacuum holes are located?

14   Which type of product can be drape formed without resorting to vacuum holes? How is this accomplished?

15   What factors influence the heating rate of thermoplastic sheet stock?

16   Why should sheets be prevented from coming in contact with heating elements?

EXTRUSION

26

Anyone who has squeezed toothpaste from a tube has some idea of extruding a product. Many food products, like spaghetti and sausage, are also extruded from a batch.

Extrusion of plastics is a process of continuously forming rod, tubing, sheet, monofilament, and profile shapes in a continuous cross section. Extrusion of cellulosic fibers and ram extrusion of fluorocarbon powder have been previously discussed. The extrusion method adaptable to almost every type of thermoplastic is called dry-melt extrusion. In this process, powders or pellets are fed from a hopper or silo to a conveying screw that melts, meters, and forces the hot melt through an orifice of the required shape.

Extrusion molding ranks as one of the high-volume outlets for thermoplastic resins. If compounding of raw materials and feeding calenders is included, extrusion accounts for processing more pounds of plastics than any other single process. Extrusion processes are used in the following operations: pipe and tubing; wire and cable coating; profile shapes; film and sheet; monofilament; foam; coating substrates; compounding; forming parisons for blow molding; heating material prior to compression or transfer molding; and plasticating material in injection molding.

TERMINOLOGY

The basic functions of a screw are reviewed in the chapter dealing with screw-plasticating injection machines (Chap. 24). Extrusion screws are similar, with variations to suit particular demands. In many cases, screws are longer, with L/D ratios

405

L/D ratio 12 : 1

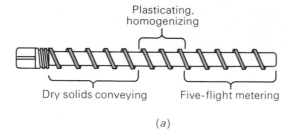

(*a*)

L/D ratio 20 ; 1

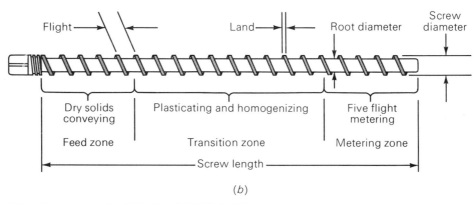

(*b*)

26–1   Comparison of (*a*) 12 : 1 and (*b*) 20 : 1 *L/D* screws.

as high as 26 : 1, and higher for resin compounding. Screw geometry is complex. Its dimensions, based on the rheology of the melt and required output, are calculated by computer. Twin screws, used in Europe, are less widely used in the U.S. Single-screw extruders range from 1-in.-diameter laboratory models to large units capable of processing more than 20,000 lb/h. Power to turn the screw is furnished by gear drives or by direct coupling to heavy-duty motors.

DIES

Extrusion dies

Dies that shape the extrudate may be simple and inexpensive, or extremely complex. They are machined from tool steel for long life, although a die for uncomplicated shapes may be brass or boiler plate for short runs and experimental work. Provisions for electric resistance heat using cartridge, strip, or band heaters are made for large dies. Small, flat dies fixed to the machine head are heated by conduction from an adaptor.

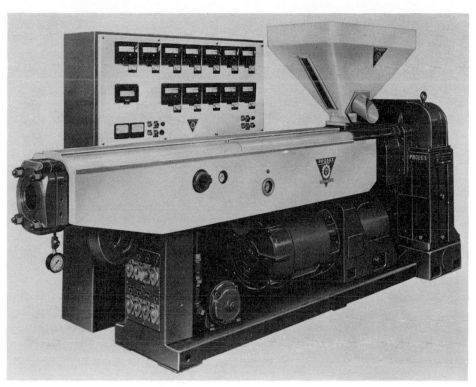

26-2   Extruder and control panel. Gauge inserted into the melt stream near the end of the screw measures melt pressure. (*Reprinted by permission of Prodex Div., Koehring Co.*)

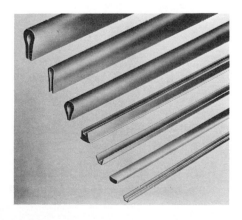

26-3   Examples of extruded profile shapes.
(*Reprinted by permission of Polymer Corp.*)

Profile

Channel, table edging, welting, rod, filament, and other profiles, preferably those not having severe thickness variations, are formed with flat dies machined to the contour, with allowances made for shrinkage and drawdown. It should be noted that polymers are limp when they emerge and are subjected to stretch and orientation by takeoff mechanisms.

Tubing

Flexible tubing, garden hose, rigid pipe, soda straws, and modified hollow shapes are made with an annular die with an internal mandrel to form the inside surface. Air introduced through the mandrel prevents collapse of the extrudate. Flexible materials can be formed to larger diameters than that of the die orifice by regulating air pressure. Closer control of pipe and tubing is maintained by a sizing box, which exerts a vacuum on the extrudate, or by means of sizing mandrels.

Wire and cable

Solid- or multiple-strand conductors are coated with vinyls, polyethylene, fluoro-carbons, and other insulation, and cabled conductors are sheathed by extrusion.

Sheet

Sheet dies capable of producing sheet up to 12 in. wide (for laboratory analysis) and up to 100 in. for production are built by the processor or purchased as part of the extrusion line.

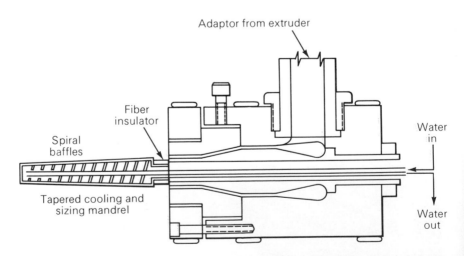

26-4 A tapered cooling and sizing mandrel is used in this type of extrusion die for producing pipe or tubing. (*Based on data from Union Carbide Corp. and a figure from "Modern Plastics Encyclopedia," McGraw-Hill, New York.*)

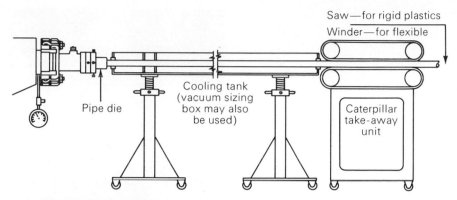

26-5 Method of extruding pipe or other hollow annular sections. (*Based on data supplied by Prodex Div., Koehring Co.*)

BLOWN FILM

Mainly used for polyolefins, but also for vinyls, nylon, and others, blown film is replacing countless cellophane and paper products in the packaging field. Extruded through an annular ring and carried upward to pinch rolls, lay-flat or gusset tubing is produced, which can be slit for sheeting to 140 in. wide. Wide slit sheet is used for lining reservoirs, ditches, and ponds, in construction work, protecting flatcar freight, and other uses where extrawide, continuous lengths are required.

COMPOUNDING EXTRUDERS

Single- or twin-screw extruders are used to compound raw materials for the various plastic processes. When polymers are manufactured, compounding extruders are used to remove final traces of moisture, solvents, or unreacted monomers. Batch or color blending, as well as incorporating fiberglass reinforcements, are also functions of compounding to provide thermoplastics for various processes.

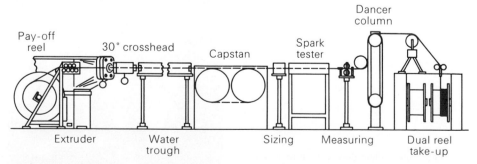

26-6 Typical installation for wire coating or applying sheathing to multiple cable. (*Based on data supplied by Prodex Div., Koehring Co.*)

26-7 Close-up of flexible plastic extruded from film die directly onto paper. Large roll on left provides water cooling. (*Reprinted by permission of Dilts Div., Black Clawson Co.*)

SCREEN PACKS

A group of wire screens backed by a breaker plate make up a screen pack that is often inserted between screw tip and die. Its function is to prevent unmelted particles from extruding, help blend color, prevent foreign matter from contaminating extrudate, and build up compound pressure.

26-8 Blown film dies with air rings. (*Reprinted by permission of Dilts Div., Black Clawson Co.*)

Screens are changed when necessary by unbolting the head or by an automatic screen changer. Dirty breaker plates are cleaned in solvents, or by burning off decomposed resin to prevent contamination and excessive pressure build-up.

TAKEOFF

All manner of devices carry extrudate away from the die. Narrow, flat stock falls on a metal belt or is transferred on rollers while being air cooled. Profiles and monofilament enter a water trough prior to windup or cutting. Other shapes, such as pipe, are pulled with caterpillar tracks.

Speed, cooling temperatures, and distance from die to first contact are adjusted to prevent flat spots on round sections or dimensional changes in irregular shapes. Extrusion is tricky, and good yield depends in good measure on the ingenuity of takeoff controls. Even unusual air currents affect some critical sections.

Extruded wire, tubing, and monofilament are wound on reels, while profiles are cut manually or with a traveling saw activated by an electric eye or limit switches.

HEATING AND COOLING SYSTEMS

The extruder barrel is heated with resistance heating bands, by induction heating, or with thermal fluids. Dry-melt extrusion is adiabatic; that is, the heat that melts the material is provided by the material itself because of friction that develops through shearing. External heat is necessary on start-up and to provide for heat losses during

26-9   Automatic screen changer. (*Reprinted by permission of C. J. Beringer Co., Inc.*)

minor shutdown. In most cases, melt temperature is higher than barrel heat in the plasticating section.

As in all processing, the important temperature is that of the melt. Thermocouples imbedded in the stream are an excellent process control.

Air blowers are used to carry adiabatic heat away when resistance heating is used. Liquids are channeled through induction-heated barrels. Feed throats are cooled, as is the screw. Cooling the screw often results in better heating of plastic since the polymer tends to stick and create more shear. Exit water temperature is an effective clue to machine performance when water cooling the screws.

On start-up, it is good practice to set barrel heat somewhat above that to be used in production and to reduce it later. This is to ensure that unmelted polymer has not locked screw to barrel, which may crack or distort a screw when drive members rotate. Extruders are often equipped with shear pins that connect the screw to drive members. Shear pins are designed to break if the screw cannot rotate, thus preventing screw damage.

Barrels should be allowed to heat soak until material is thoroughly melted. A delay of 15 to 30 min after indicating pyrometers reach set points is good insurance. As with injection molding, heat profiles established by trial and error with new dies or materials are changed to suit extrusion conditions. Profiles may be inclined (lowest at feed throat, highest at head), flat, or downhill.

SCREW ROTATION

A balance of melt and die temperatures, screw cooling, screw rpm, and internal pressure must be established and maintained to avoid overworking the scrap grinder. Slow screw speeds on start-up, between 10 and 30 rpm, are adequate. Screw rotation is recorded by tachometer. Pressure gauges inserted at the die adaptor, or behind and ahead of screen packs, are monitored to provide clues to constant performance. A sharp drop may indicate need for changing screen or a clogged feed.

FEEDING THE HOPPER

Pellets or powders manually or automatically fed to a hopper drop directly between flights of the screw-feed section. Cooling of the feed throat is essential to prevent premature melting, sticking, and bridging. If clogged, a rotating screw must be stopped before dislodging particles. Screws and barrel liners, which are extremely expensive, are easily damaged beyond repair. Avoid using any metal but brass, and be alert to tramp metal getting to the screw. Partially clogged feed sections result in surging of the extrudate or in a lower density product.

CHECKING THE PRODUCT

Profile shapes are checked by caliper, special gauges, and by weight of specific amounts. Sheet stock is monitored constantly by beta gauge, often interconnected to machine control for screw rpm. Sheets that pass through adjustable chill rolls are checked by micrometer. Coated wire passes through a spark tester that signals when the coating is too thin, or when the conductor is off-center.

Visual checks for moisture are made by noting cloudiness or striations in clear and opaque resins, respectively. Fisheyes or other ovoid defects indicate plasticizer exudation usually caused by overheating or clogged extruder vents.

A review of defects and remedies associated with injection molding can assist in trouble shooting some extrusion problems such as contamination, moisture, and bubbles, but corrective measures are not necessarily the same.

SHUT-DOWN PROCEDURES

Operations vary from plant to plant, with a view toward what is to be extruded next. With polyvinyl chlorides, an inexpensive method is to purge completely with low-melt polyethylene, which softens readily on start-up. Vinyls are corrosive, and highly plasticized grades decompose rapidly when overheated. For styrenes, cellulosics, and other less heat-sensitive resins, it may only be necessary to operate until feed empties.

Remove die, breaker plate, and screens (if used), exposing screw tip to atmosphere, before purging with acrylic purging compound.[1] This compound has a high melt viscosity and will build up dangerous pressure if restricted.

GLOSSARY

Adiabatic   Denotes a process or transformation in which no heat is added or permitted to escape from the system. The term is used somewhat incorrectly to describe extrusion and injection molding in which no external heat is applied to keep melt temperature constant. Heat is removed by external cooling to keep melt temperature constant.

Caliper   An instrument having a graduated scale for precise measurement, such as a micrometer or vernier calipers.

Extrudate   Material or product formed by an extrusion die.

Gusset   V tucks in the sides of bags, which permit the forming of a rectangular shape when opened.

Induction heating   Method of heating an electrically conductive material by placing it in a high-frequency magnetic field.

Rheology   The study of flow as related to stress and strain per unit of time.

Tachometer   An instrument employed to measure the velocity of machines and convert it to rpm.

REVIEW QUESTIONS

1   What is meant by the term *dry-extrusion process*?
2   What products are made by dry-melt extrusion?
3   What method is used to prevent tubing from collapsing as it emerges from the tubing die?

[1] Rid is the registered trade name of American Cyanamid Company.

4   What is the purpose of a screen pack?

5   Why must screens be changed periodically?

6   What are several methods used to heat an extruder barrel? How are barrels cooled?

7   What is meant by the term *adiabatic heat*?

8   How is adiabatic heat controlled?

9   What is the purpose of a shear pin?

10   What factors must be balanced to obtain good extruded products?

11   Why must cooling water be circulated through the feed throat of an extruder?

12   What are the operations performed to check size or dimensions of extruded products?

13   What dangers should be recognized in extruding vinyl plastics?

14   What steps should be taken in shutting down an extruder?

BLOW MOLDING

27

Blow molding is a process by which hollow articles are formed by forcing a heat-softened tubular shape to the contours of a mold. Many variations of this concept exist, depending on size or quantity of items or on type of machinery. A blow-molding process is like blowing up a balloon inside a square cardboard box.

As an art, blow molding dates back to the use of cellulose nitrate tubes expanded into a closed mold to produce dolls, baby rattles, and small novelties. But volume applications did not occur until low-cost polyethylene made it possible to compete with glass for the tremendous packaging market. This potential led to many sophisticated and automated machines designed to produce bottles of all sizes and shapes. Some of the techniques used by the plastics industry are adaptations of methods used for blowing glass containers, and, in fact, some glass-bottle producers have established production lines to manufacture plastic containers.

A trend has also been established where companies producing detergents, fruit juice, or milk install an appropriate type of machine to produce their own containers. This eliminates a presterilizing step in the filling process when bottles for consumable liquids are produced, and it minimizes the inventory problem.

Blow-molding processes are divided into two basic types: extrusion blow and injection blow. In each method, the underlying principle is the formation of a tube called a *parison*. As the description implies, an extruder or a screw-plasticator injection machine melts a polymer and forms the required tubular shape. In another method, tubing is extruded continuously, cooled, and stored for later reheating and blowing.

The entire cycle can be fully automatic, including trimming, scrap grinding, and

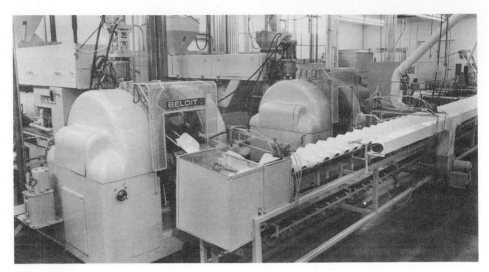

27-1   Gallon-sized containers being conveyed to trim station. (*Reprinted by permission of Beloit Corp.*)

recycling. In some cases, an operator tends one or more machines to remove parts, control cycles, or to place inserts in the mold.

MACHINE TYPES

Blow-molding units vary from small, single-cavity types that produce 250 parts or less per hour to multimold rotary machines that hourly produce as many as 8,000 small containers or other hollow shapes.

Single-station

A single-cavity mold mounted vertically on platens is centrally located below an extrusion head to receive the parison. An extruder forms a vertical tube, which descends between open-mold halves. When the tube is long enough to be pinched top and bottom, the mold closes, and air is injected.

In some processes, screw rotation is stopped when the mold closes and started prior to part ejection. A typical machine that operates on an intermittent screw rotating basis is shown in Fig. 27-2. In a continuous screw operation, melt is extruded as needed by a reciprocating piston.

Another version called the *rising table type* extrudes a parison continually. The mold halves are carried upward by a platen (table), while being closed horizontally by air cylinders. Air is introduced through a mandrel, which also rises with the table, and enters the bottom of the extruded parison.

Multi-station

These units are commonly called T head or H head, which alternately feed two or four molds, respectively.

The screw produces melt continuously, and diversion valves in the manifold direct it to each parison head. These machines are popular because parisons are relatively easy to control, production rates are good, and multiple molds can vary somewhat in size and shape. Rotary machines, with a set of molds mounted on a horizontal table or fastened to a vertical wheel, are used for extremely fast production.

Injection-blow machines are becoming more widely used, mainly because of the success of screw plasticators, which also serve as a ram to form a parison by the injection-molding principle. The higher cost of equipment is offset by features unmatched by extrusion blow. Parisons can be made to unsymmetrical shapes, with

27-2   Single-station blow molder. (*Reprinted by permission of Rocheleau Tool and Die Co., Inc.*)

varying or extremely uniform wall thickness. Molded threads are formed prior to blowing, and flash is minimized. After a parison is formed, it is indexed to the blowing mold either manually or as part of the machine function.

BLOW-MOLDING PROCEDURES

The procedures outlined are based on operating a T head machine, which serves as a compromise between single-station and rotary types. Cycles for all extrusion-blow machines are programmed similarly: (1) parison is extruded; (2) mold closes to pinch top and bottom; (3) air is injected to form the part; (4) parison is sheared by knife or by moving mold; (5) blowing air is cut off; and (6) mold opens and the part ejects.

MOLD DESIGN

Inflating pressures are between 10 and 150 psi, so molds are made of materials less rugged than tool steels since mold-clamping pressure need only be slightly higher to

27-3   T head, two-station blow molder. (*Reprinted by permission of Los Angeles Trade-Technical College.*)

prevent blowout. The most widely used metals are aluminum and kirksite, which are cast or machined. Other mold materials are brass, beryllium copper, cast iron, and steel. Aluminum-filled epoxy is practical for experimental designs.

Molds are usually cored to control cooling or heating with water, as for injection molds. Less widely used is a shell-type cavity, with flood cooling between its outer wall and a mold frame. The necessity for zone cooling to control shrinkage and stress by cooling heavy sections as quickly as thin areas makes channel cooling preferable. Circulating colder water through a mold section where the need for heat transfer is greatest or regulating flow volumetrically is usually sufficient.

Most molds, irrespective of the metal they are made of, are fitted with removable, steel, pinch-off plates, thread-forming sections, and other parts subject to breakdown or damage.

EXTRUSION HEADS

The extrusion head, in a sense, is a modified die for extruding pipe, except that the internal member (mandrel) is especially designed to change melt-flow direction from horizontal to vertical. In addition, the mandrel is grooved to provide a "surge chamber," which promotes uniform parison flow.

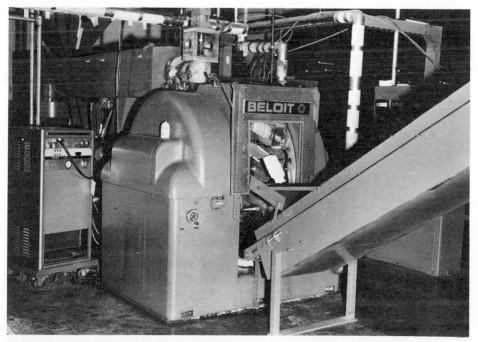

27-4   Six-station, rotary-wheel blow molder. (*Reprinted by permission of Beloit Corp.*)

BLOWING THE PART

There are several methods of injecting air into the sealed parison:

1   Through a channel in the mandrel. Commonly called top (or bottom) blow, this method is used for bottles and other shapes where air is introduced through the neck.

2   By using a blow pin. A small-diameter (0.0625- to 0.25-in.) tube sharpened to a point is fixed in the mold cavity. As the mold closes, the point punctures the parison above the neck of a bottle or in a noncritical area of other items such as a furniture arm or chair seat.

3   By means of a hypodermic needle used in the manner of a blow pin, where small holes must be less obvious or require sealing later.

4   By causing the parison to be pinched off at the bottom on the previous cycle, and introducing air into the parison before the succeeding pinch-off. The trapped air bubble forms a blown product without any opening in it.

INSTALLING A MOLD

Blow molds contain guide pins and bushings for alignment, and adapter plates or grooves for clamping. Heavy-duty clamps are replaced by thin, steel plates unless mold weight requires them. Mold clamping is the same as for compression and injection molds. A knee plate is helpful in supporting blow molds during setup.

27-5   Aluminum blow mold with steel pinch-off plates and thread-forming inserts. Note sharpened blow pin to pierce parison above neck. (*Reprinted by permission of Los Angeles Trade-Technical College.*)

Before setting a mold, adjust the stationary platen so that the mold parting line is at the center of the mandrel, then set that mold half with a 0.375- to 0.50-in. clearance from the extrusion head.

The final clearance between mold and head is determined by trial. Normally, drawdown and shrink-back characteristics are factors to consider. Drawdown is a function of the polymer's hot tensile strength. As a parison falls, its weight may cause elongation as it moves down by gravity, thus reducing diameter and thickness. This is counteracted somewhat by the tendency of a parison to shrink back, increasing its wall thickness as molecules return to an unoriented state accelerated by cooling slightly as it emerges. These characteristics vary with each grade and type of resin and are controlled by manifold pressure, temperature, and speed of extrusion. Parison programmers, as shown in Figs. 27-6 and 27-7 help stabilize parison thickness in relation to its length. More scrap is generated when clearance is more than adequate, and parison cools more.

Next, match the other mold half, strap, or wire halves temporarily, and adjust movable platen to engage, then clamp. Remove straps, and operate the mold slowly several times with manual control for checkout. Connect air line to mold blow pin (if used) and check for bent, dull, or clogged pin. Connect water lines for cooling.

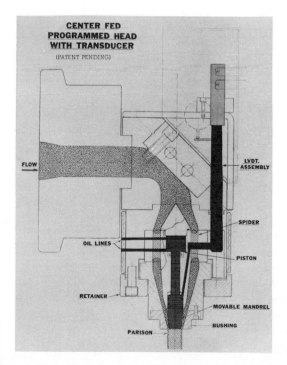

27-6   Diagram of programming unit. (*Reprinted by permission of Strong Plastics, Inc.*)

The programmer permits variation
of the thickness of the extruded
parison to suit individual bottle
configurations, to provide maximum
strength with minimum material
as shown in the illustrations.

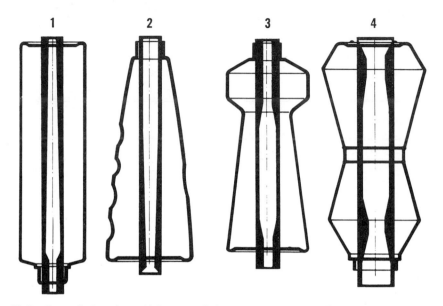

27-7   Controlled parison-thickness variations permit blowing of parts with extreme wall or peripheral changes. (*Reprinted by permission of Strong Plastics, Inc.*)

HEATING THE MACHINE

Review methods for extruder start-up as outlined in Chap. 26. In addition, the T head manifold and extruder head are also heated with resistance heating bands. All units must be brought to operating temperature before extrusion can take place. If the diversion valve (also air-operated) can reverse freely, and the extruder barrel is up to temperature, a parison can be extruded using manual controls for one mold only.

Start the extruder and observe parison formation. A tendency to curl, instead of a straight fall, indicates that the die must be moved toward the mandrel to adjust flow. This is accomplished with four studs, as shown in Fig. 27-8.

When parison fall is satisfactory, and this can be accomplished by adjusting screw speed, temperatures, and manifold heat, the mold can be closed and a part blown, using all manual controls. Air pressure for blowing is correlated to the degree of forming required. On new jobs it is prudent to start at 20 to 40 psi for most plastics.

When the part is produced satisfactorily, the machine can be changed over to automatic.

SETTING A CUT-OFF ATTACHMENT

Parisons are sheared by a movable blade or wire while soft, generally after the mold closes and extrusion stops. The cutoff is automatically controlled by a solenoid-activated movement energized from the control panel.

SAFETY CONSIDERATIONS

Blow-molding machines incorporate safety features such as safety gates which prevent mold closing unless an interlocking switch is activated. Machines should not be operated without this safety precaution in working order. Danger from burns is prevalent, from both the hot parison and the excessively thick flash. The possibility of scratches or punctures from sharp blow pins or needles is also a hazard.

Build-up of polymer, which leaks at any point in the extrusion system, should be avoided. Some materials can ignite, and all can char to the point where a short circuit could occur if in contact with electrical terminals.

27-8 Adjusting studs in extrusion head. (*Reprinted by permission of Los Angeles Trade-Technical College.*)

MOLDING DEFECTS AND REMEDIES

Defects can be classified as obvious, dimensional, or those determined by performance tests.

Obvious defects

The defects usually associated with extrusion and injection molding, such as color degradation, contamination, moisture, and decomposition, also occur in blow molding. Surface roughness and axial streaks can be attributed to unclean diversion valve, die head, or mandrel.

Excessive flash is caused by excessive blowing pressure or incorrect size of mandrel and die head. Inadequate clamping caused by low pressure or mold not closing properly because of flash on a contact surface is another factor.

Products that fail to blow to the required shape should be checked for the following:

1   Parison temperature too low.

2   Extrusion rate too slow.

3   Mold too cold.

4   Blowing air insufficient, applied too late, or not long enough.

5   Blowup ratio too high (ratio between the maximum part periphery and parison periphery or diameter). As in thermoforming, blowmolding is merely a redistribution of material and should be held to a ratio of 3 or 4:1.

6   Mold is positioned incorrectly. Some molds can be inverted or used upright. If the widest section of a part is at the top, parison drawdown may cause starvation at the section that needs the largest parison diameter.

7   Insufficient blow-pin penetration.

Dimensional defects

Parison wall thickness should be commensurate with the minimum part thickness required. This is important not only to conserve material, but to hold bottles between the minimum and maximum weight tolerances demanded by packaging specifications. Trimmed bottles are weight-checked and cut into sections for thickness checks. Molded threads are checked for size by a thread gauge or by using an actual closure.

Wall thickness is controlled by paying strict attention to each mold-zone temperature. If a mold section is too cold, the material will freeze off at that point, resulting in a thicker section at the expense of adjacent sections. Outside dimensions are a function of the inherent shrinkage of a particular resin, plus the time blowing air is kept on the part.

Polyethylenes shrink from 0.020 to 0.050/in., depending on density. High-density resins shrink more than low-density resins. Melt index for these resins is between 0.2 and 2.2 for blow-molding grades. Mixing regrind with virgin material or blending resins made by different companies without regard to differences in flow rate and shrinkage causes a shrinkage-control problem. Applying blowing air longer helps decrease shrinkage. To avoid longer cycles, parison temperature should be reduced to obviate longer blowing time. Vacuum systems can be built into mold cavities to evacuate air quickly between the parison and mold surfaces. This helps to give better detail on the outside of a part and reduces cycle time appreciably.

Performance defects

Functional rejects are determined by laboratory tests standardized by the Professional Activities Group (PAG) on Blow Molding (a subdivision of Society of the Plastics Industry, Inc.). Other standards such as thread, neck, and shoulder sizes for bottles are developed to conform with glass-container specifications and packaging requirements.

Drop or impact tests Toys, flashlight cases, housings, and furniture may be dropped from successive heights or subjected to impact from falling objects. These are practical tests. Bottles are filled with water and dropped from increasing heights until failure of 50 percent of those tested occurs. Bottles are dropped on ends, shoulders, and sides to render complete evaluation.

Column crush test This test is designed to provide information about the crushing properties of containers. Rigid materials may fail because they are brittle; softer plastics may show a high degree of deflection. In the former, values are expressed in pounds or kilograms that cause fracture. Flexible containers are tested for percentage of yield (decrease in length) in inches or centimeters.

Permeability Manufacturers of resins for all thermoplastic processes have performed extensive testing with hundreds of standard reagents to learn their effects. However, cosmetics, tanning lotions, hair preparations, and thousands of other products are made from closely guarded formulas. Each must be tested for its effect on a particular container.

Permeability describes a characteristic by which a molecule in a contained liquid or vapor can escape through the container wall without dissolving it. This results in loss of the product, which may occur rapidly or over a long period. Filled packages are tested, and the results plotted as weight loss and loss rate over a period.

A gain in weight of the net contents of a container indicates leaching out of low-molecular-weight fraction of polyethylene or plasticizers and other additives in other thermoplastics. In some cases, this is accompanied by slight or severe collapse of the container.

Stress crack resistance The polyolefins, in particular, are investigated for environmental stress cracking resistance (ESCR) by immersion in specific test fluids. Details are given in Chap. 34 (chemical properties).

MATERIALS USED FOR BLOW MOLDING

The criteria for selecting a thermoplastic for a blow-molding application are based upon its suitability, cost, and blow-molding ability. Extrusion grades are not necessarily adaptable to forming parisons because of the differences in handling the extrudate, or in the stiffness of a polymer melt. It will be recalled that profile shapes are cooled in water or by forced air; sheet is directly cooled on cooling drums; and pipe is sized. Parisons are usually unsupported and kept hot purposely.

FINISHING AND DECORATING

Finishing blow-molded items is relatively simple and is often automated into the blowing line.

Pinch-offs in the mold reduce thickness to make trimming with one or more shear dies practical. In some cases, the machine tender tears flash manually. Necks of bottles are cut above the threads when the part is blown. Final trimming to provide a flat seal is done with a rotating knife. Inside diameters may require reaming for internal corks.

Reaming and facing operations generate chips that fall into the container. Static developed by cutting causes chips to cling to inner walls. Inserting a deep probe and blowing chips out is one method of handling the problem.

Decorating blown items is the same as for injection-molded or thermoformed parts. Hot stamping, painting, silk-screen printing, and offset printing are common methods of embellishing products or supplying information.

Polyolefins do not accept printing inks or silk screening readily unless surfaces are treated prior to decoration. Passing the blown article through gas flames or through electrical discharge units oxidizes the surface to improve adhesion.

GLOSSARY

Diversion valve   A rotary valve used to direct extrudate from one blowing mold to another.

Drawdown   Decrease in parison thickness and diameter because of stretching created by gravity.

Mandrel   Central core of a parison die, which forms the inside of a hollow shape, prior to extrusion blow.

Parison   Extruded or injection-molded hollow shape that is subsequently blown to final configuration inside a closed mold.

Solenoid   An electromagnetic device that opens or closes a valve.

REVIEW QUESTIONS

1   What are the two major methods for blow molding a product?
2   What is a parison?
3   Why is it feasible to use aluminum for blow molds?

4 List the three methods of injecting air to blow a part and cite the reasons for using each method.

5 What is meant by drawdown? Discuss the factors that affect it.

6 List several reasons for failure to blow a part to specified shape.

7 Name the three classes of defects that are causes for rejection of blown parts.

8 What is the melt-index range for polyethylenes used for blow molding?

9 How are bottle threads checked for size?

10 Describe the column crush test.

11 What is meant by the term *permeability*?

12 How are chips removed from trimmed bottles?

13 What are the major applications for blown articles?

CASTING AND RELATED PROCEDURES

28

The process of casting, in its most elementary form, means pouring a liquid substance into an open mold to harden without benefit of pressure. Heat may be used to hasten hardening. There are variations of casting in which some form of light pressure is applied, or by which a casting is built up to final size in multiple steps. Either thermosets or thermoplastics may be used when available in liquid or meltable form (see Table 28-1). Casting processes are used to produce a wide variety of products ranging in weight from a few grams to several tons.

28-1 Cast-nylon industrial parts. (*Reprinted by permission of The Polymer Corp.*)

Embedding is the process whereby sales display items and mineral, botanical, or biological samples are preserved for ease of handling and observation.

Encapsulation involves surrounding a component (usually electronic) with resin for protection against shock, moisture, or chemicals. Encapsulation is also performed by forcing soft-flow epoxies, heated in powder form, into closed molds via the transfer-molding process.

A variation of encapsulating called *potting* generally involves vacuum impregnation of porous or internal sections of an object in addition to forming the protective outer shell.

Slush casting is similar to straight casting of solid objects, except that excess is dumped from the mold once sufficient wall thickness has formed. Solvent casting is used to produce thermoplastic films by depositing a thin layer of resin solution onto a carrier such as a metal belt or drum, evaporating the solvent and winding the film into rolls.

28-2   Cast epoxy tool for forming aluminum. (*Reprinted by permission of Rezolin Div., Hexcel Corp.*)

28-3 Potting electrical component in clear epoxy permits visual inspection. (*Reprinted by permission of Emerson and Cuming Co., Inc.*)

CASTING WITH POLYESTER RESIN

From a hobby standpoint and in commercial venture, polyesters have gained wide popularity as a casting medium. Crystal clarity of base resins permits a wide range of decorative and color effects.

Polyester casting resins differ from laminating or molding types in that they contain wax or flexibilizers that minimize cracking and reduce air poisoning of ex-

TABLE 28-1 Resins used in casting

Resin type	Application
Thermosetting:	
Polyester	Novelties, art forms, table tops and sinks (referred to as cultured marble), displays
Diallyl phthalate	Electrical encapsulation, table and sink tops, counter tops
Phenolic	Sheet, rod, tube, profile shapes, patterns, and tools
Epoxy	Costume jewelry, tooling, patterns, encapsulation, potting compounds
Thermoplastic:	
Acrylic	Embedments, novelties, models, art forms, dentures
Cellulosic	Tooling
Polyethylene	Tooling, slabs for machining
Polyvinyl chloride:	
Hot melt	Novelties, flexible molds
Plastisol	Fish lures, toys, numerous industrial shapes, seals
Silicone (elastomer)	Molds, gaskets, seals
Urethane (elastomer, rigid)	Molds, gaskets, seals, furniture decoration, and trim

posed surfaces. Casting resins shrink volumetrically between 7 and 12 percent and are not recommended for embedding objects likely to crush or distort. This high shrinkage aids in releasing a casting from a female mold, but it creates a problem when untapered cores are used.

Liquid polyesters are inhibited by the manufacturer with hydroquinone to prevent premature gelling. Storage at 40°F extends shelf life to a year or more. Most manufacturers include a promoter such as cobalt naphthenate to effect a rapid cure at room temperature. Because of resin clarity, the promoter imparts a blue tinge to uncatalyzed resin, which disappears because of the bleaching action of peroxide catalysts.

Processors who prefer using unpromoted resins must exercise extreme caution to avoid the slightest contact between any promoter and any peroxide catalyst because this invariably causes dangerous fire or explosion. Even fingers barely wet with catalysts touching a container with spilled and dried promoter on the exterior can result in severely burned skin.

Artists or people in firms producing high-volume castings often prefer specific proportions of promoter and catalyst to control cure rate. When promoter and catalyst are to be added, always add promoter first (0.02 to 0.5 percent), stir well into the resin, then add catalyst.

Catalysts are the same as for wet lay-up laminating, namely methyl ethyl ketone peroxide, cumene hydroperoxide, or benzoin. The percentage of promoters such as cobalt naphthenate is usually higher in resins used for laminating than for casting grades. For this reason, lay-up resins should not be used for casting because an extreme exotherm will develop, which will damage the product and possibly cause spontaneous combustion.

As In laminating, heat is a by-product of polymerization. For straight casting, exotherm is more troublesome because of the mass involved. Embedding metal is more likely to cause cracking because of heat concentration and differences in expansion between the two materials. Casting resins harden by the same mechanism as in laminating, that is, gellation, exotherm, and final cure, which may require several days (see Table 28-2).

Proportions of catalyst used are dependent on volume of resin poured, ambient temperature, and inclusion or exclusion of heat-conducting fillers.

TABLE 28-2   Typical gel and cure time (10g sample)*

Catalyst, %	Gel time	Cure time
0.50	6 h	2 days
1	27 min	180 min
2	23 min	63 min
3	15 min	28 min
5	8 min	Cracked
10	6 min	Major cracking

*Using 2 percent catalyst for 50 g resulted in cracking, excessive smoking, and near incendiary.

Table 28-2 makes it obvious that the volume cast (i.e., the thickness, since a large volume could be spread thin over a large area) dictates the amount of catalyst. Mass castings are produced with as low as one-tenth of 1-percent catalyst. Exotherm can be reduced by refrigerating large castings or packing in dry ice. Small, thick castings can be placed in a waterproof membrane and immersed in cold water.

Polyesters manufactured for casting contain an incompatible wax that migrates to the surface as the resin cures. Its purpose is to exclude oxygen, which creates a tacky surface. Covering the casting with cellophane, Mylar,[1] waxed paper, or melted paraffin will help avoid this condition. Tacky surfaces can be corrected by heating the removed casting at 160 to 180°F until it withstands fingernail indentation. Uncured surfaces become cloudy if exposed to moisture.

Mold materials consist of metals (preferably aluminum to help dissipate heat), zinc alloys, sprayed metals, glass, ceramic, and other plastics, notably polyethylene, polyvinyl chloride (plasticized), silicone rubber, flexible urethane, epoxy, cast phenolic, polyester, and such materials as plaster, cement, cardboard, and natural rubber latex. Materials containing sulfur or copper should be avoided since they inhibit curing. Some flexible vinyls may cause tacky surfaces because of migration of the plasticizer.

Porous substances require sealing, otherwise good release is difficult. Depending on the extent of porosity, sealants may be shellac, alkyd, or vinyl paints, paste wax, polyvinyl chloride,[2] cellulose lacquers, polyvinyl alcohol solutions, and many proprietary liquids or pastes.

In addition to sealants, a parting agent is usually necessary since polyesters are adhesive. Glass, ceramic molds, and flexible types such as vinyls and silicones may be used without a parting agent. Film-forming types such as polyvinyl alcohol serve a dual purpose of sealant and release if used in sufficient quantity.

Paste wax, silicone or fluorocarbon sprays, cellophane, lecithin, stearic acid, and mineral oil have been used successfully as release agents in various ways.

CONTROLLING THE EXOTHERMIC REACTION

A typical polyester generates as much as 150 cal of heat per gram of resin, resulting in a peak-exotherm temperature to 500°F. Unless controlled, these undesirable results occur: (1) castings are brittle or discolored; (2) castings are cracked or contain voids; (3) castings froth and foam; and (4) castings ignite.

The secret to successful casting or embedding rests more in the recognition of exotherm than in any other factor. The total weight of a casting is not so important as its mass. For example, a gallon of resin weighs approximately 9.4 lb. If 1-percent catalyst is added to the gallon container, one of the above-mentioned, undesirable results almost always follows. If, on the other hand, the same amount of catalyst were added, and the resultant mix poured into a shallow tray to a depth not over 1 in., a trouble-free casting would result.

[1] Mylar is a Du Pont polyester film.
[2] Tygon is the registered trade name of U.S. Stoneware Company.

The answer to this is the ratio of heat-dissipating area to volume. In the gallon can, we have a total surface area of approximately 250 sq in. If we pour the contents into a metal tray measuring 12 × 24 in. to a depth of 1 in., the external surface is increased to almost 650 sq in. of heat-dissipating area.

Exothermic heat is always concentrated in the center of any mass, with the result that the center of the casting is expanding rapidly because of thermal influence, while at the same time it is trying to contract as molecules join under chemical influence. Additional problems may be created if styrene monomer boils and evaporates.

How, then, is one able to control exotherm? First, reduce promoter and catalyst concentrations in line with other conditions. Second, complete the casting in a series of successive pours, rather than as a mass casting. Third, incorporate heat-conductive fillers (see Table 28-3), and use heat-conductive molds.

Larger sections can be poured if opacity can be tolerated by adding small particle fillers. Catalyst should be mixed prior to loading to insure uniform dispersion. Otherwise, catalyst may not be mixed properly as viscosity increases thus leading to hot spots.

It should be remembered that different fillers and some pigments either accelerate or inhibit curing, and they should be investigated on small batches before using in full-scale work.

PROCEDURE FOR MAKING A SMALL EMBEDMENT

Assuming the mold is properly prepared with sealant and parting agent, and exotherm is now fully understood, a successful embedment can result.
Materials consist of:

1 pint polyester resin, casting grade, promoted

TABLE 28-3 Fillers for casting resin

Cost reducers (extenders)	Thixotropic types*	Miscellaneous
Calcium carbonate	Asbestos floats†	Aluminum powders
Clay (kaolin)	Cab-O-Sil‡ (fumed silica)	Microballoons (spheres):
Diatomaceous earth		Phenolic
Marble dust		Glass
Plaster		Saran
Sand		Glass, flake or ground
Slate dust		Glass, milled fibers
Talc		Iron filings
Walnut-shell flour		Thermoplastic pellets
Wood flour or sawdust		Steel wool (clean)
(thoroughly dried)		Vermiculite

*A thixotropic substance is one that is fluid when agitated and immobile when still. Thixos do not drain from vertical surfaces.
†Type 244 is the registered trade name of Union Carbide Corporation.
‡Cab-O-Sil is the registered trade name of Cabot Corporation.

1 vial catalyst, MEK peroxide

Acetone (to clean equipment)

Equipment includes:

Gram scale.

Eyedropper or needleless hypodermic syringe (50 cc). *Note:* in place of eyedropper or syringe, calibrated "kitchen type" measuring spoons can be used. They should be metal or polyethylene. (Polystyrene dissolves in polyester resin because of monomer.) Details for calibrating eyedroppers have been given previously under laminating procedures.

Paper cups.

Stir sticks.

1   Weigh out only enough resin for the first pour (base layer) in a paper cup. This may be for any thickness from 0.0625 to 1 in.

2   Catalyze the resin with 0.50 to 3 percent, depending on thickness. For thin sections, use the higher percentage. Avoid pouring catalyst into paper containers. It may be absorbed, thus reducing calculated percentage. Instead, add catalyst directly and stir thoroughly, but not vigorously, to avoid air occlusion. *Stir thoroughly,* scraping resin away from sides and bottom, for several minutes.

3   Pour into one side of the mold, or allow to move from center to edges to avoid entraining air. Air will generally rise to the top before gellation. Stubborn bubbles can be pin-pricked.

4   Allow to gel, exotherm, and cool. Unless gelled, heavy objects sink. Objects lighter than the resin can be dipped into catalyzed resin and held in position by pinning until curing promotes adhesion to base pour. Irregular objects tend to trap air. Dipping one or more times in precatalyzed resin before embedding prevents bubbles.

5   Catalyze sufficient resin for the second (or final) pour. A safer procedure is to reduce the percentage of catalyst to avoid any undue exotherm that might occur in contact with air-inhibited resin at the interface. Low profile objects can be embedded in two pours. Deep objects or multiple embedments can be made successively in as many pours as needed.

6   Proceed as for the first pour. Vibrating or tapping the mold aids in releasing air.

7   Cover the top with cellophane or other film, if possible.

8   When hard to the touch, tap the mold, or use an air blast to free the casting. Failure to release is attributed to:
    (*a*) undercure
    (*b*) poor mold preparation
    (*c*) rough mold surface or undercut.

Placing the mold in hot water or in an oven at 120 to 180°F further polymerizes the resin for additional shrinkage and also expands most molds in stubborn cases. Tacky surfaces are improved by heat treatment, either by oven, warm air blast, or lamps.

WATER-MISCIBLE POLYESTERS

More recently, water-extended polyesters (WEP) have gained commercial significance. The final product resembles plaster but is more chip resistant.

Castings solidify in less than 5 min and are removed from molds in 8 to 15 min. These resins are handled somewhat differently from conventional polyesters. From 40 to 65 percent water based on resin weight is used. The higher proportions of water reduce mechanical strength and increase the mix viscosity. A typical formula is:

WEP resin: 100 g

Cobalt octoate: 1.25 g (12 percent reactivity)⎱ promoter blend
Dimethyl aniline: 0.50 g   ⎰

Water: 100 g

Peroxide blend: 1.0 g

The peroxide blend is a mixture of 50 percent hydrogen peroxide and water-soluble MEK peroxide[1] in a ratio of 3:1.

Procedure

1   Mix promoters together and add to resin *first*.

2   Add water *slowly*, using a high-speed mixer (2 to 3,000 rpm) until a creamy white emulsion develops. *Never add resin to water*.

3   Stir in catalyst blend. Emulsion turns green, providing a clue to mixing. Stir several minutes.

The emulsion should be poured immediately. Molds used for conventional polyester casting are adequate for WEP. Release agents are normally unnecessary. If used, they should be water repellent, such as polyvinyl chloride film-forming types. Certain silicones and waxes may reduce surface tension, causing pitting.

Safety

Use protective clothing when handling promoters and catalysts. Cobalt octoate is not considered toxic, but dimethyl aniline is poisonous and is absorbed through the skin. On accidental contact, wash with soapy water. Hydrogen peroxide, 50-percent volume, causes serious burns.

---

[1] Lupersol DSW is the registered trade name of Lucidol Division, Wallace-Tiernan Company.

CASTING WITH PHENOLIC RESIN

Phenolic casting resins have long been used in transparent form to produce models of gears, pump parts, and similar products for stress evaluation. A tendency to darken upon aging limits more extensive use. Opaque casting grades are used to produce novelties, plaques, knobs, and art objects. Industrially, it is used for tooling and patterns for laminating and thermoforming. Its advantages are lower cost than epoxy tooling resins, low exotherm in massive castings, and good machining characteristics.

The phenolic has a syrup consistency, activated by addition of an acid catalyst in amounts of 10 to 12 percent. Curing is slow and may require several days. Molds may be of any material not affected by the acid catalyst. The only precaution in handling is to avoid contact with catalyst by using proper protective clothing and face shield.

CASTING WITH EPOXY RESIN

The introduction of epoxy casting resins has been instrumental in developing whole new concepts in mass producing automobiles, aircraft, and other high-volume metal products. The volume of epoxy resins used for stamping, bending, checking, and assembling steel and aluminum sheets has reached such proportions that a separate professional division, the Tooling Division, has been established by the SPI to guide other industries.

The use of epoxy resin has been successful for the following reasons:

1 It takes less time to produce tools compared with the same tool in metal. Lead time from design to production is greatly reduced, often to weeks instead of months.

2 Tools are easier to repair or modify.

3 It is lighter weight, tough and durable.

4 It is dimensionally stable for long periods.

5 It has low, predictable shrinkage during cure.

6 It does not corrode in storage.

7 Tool-building skills are learned more readily than metal-working skills.

8 Variety of colors are available for identification or specific use.

9 It is easily duplicated from a master pattern in case of loss, or for multiple tools.

Epoxy resins for casting art objects, statuary, and other competitive products prove too expensive when compared with phenolic, WEP, and other materials. The extremely high impact strength, stiffness, and compressive strength available in hundreds of formulations place epoxy resins highest on the list for tools and molds.

The viscosity of epoxies can be varied to suit the details of a pattern or for incorporation of fillers. Fillers such as aluminum or iron powder are used for heat-conducting molds for thermoforming; expanded volcanic rock is used to reduce overall weight. (See Chap. 5 for curing conditions.)

ENCAPSULATING WITH EPOXY RESINS

Electronic units requiring protection from heat, moisture, shock, vibration, or tampering are protected with a single application of epoxy encapsulating resin. Molds can be of the reusable type (silicone, epoxy, urethane rubber, steel, or fluorocarbon), or they can remain as the outer shell. Compression-molded alkyd or phenolic, or injection-molded resins are used as shells, as are metal forms. Thermoformed polyethylene molds can be used for mass pouring of many small components.

Although many resins have been used for encapsulating and potting, epoxies are chosen when low shrinkage is required to prevent damage or displacement of delicate encapsulants.

SLUSH CASTING

A process useful to production of hollow rubber, plaster, ceramic, and chocolate candy items is adaptable to various plastics. Particularly useful in casting flexible and semirigid shapes from special formulations of liquid polyvinyl chloride, slush casting is also performed with finely powdered resins, notably polyolefins.

Polyvinyl chloride dissolved in a plasticizer (*plastisol*) is poured into a one-piece shell mold or a multisectioned sealable mold made of good heat-conducting metal such as aluminum, brass, or electroformed copper alloy. The mold is preheated from 350 to 400°F, filled with plastisol or powder, held for several minutes, and inverted. Heat transfer from mold to plastic fuses the material to the depth required, depending on temperature, heat-transfer rate, and fusion characteristics of the resin. A sufficient coating is built up on mold walls in a matter of seconds, after which excess is poured out for reuse. Molds are placed in an oven at 350 to 450°F for several minutes to complete the fusion, then cooled by water immersion or spray mist at the ejection station.

Many products are made on a fully automated sequence, except for manual

28-4 Encapsulated solenoid coil. (*Reprinted by permission of Hysol Div., Dexter Corp.*)

stripping of parts, attention being paid to maintaining original plastisol viscosity, color, and batch temperature as discharged resin is fed back to the fill tank. Hollow articles (such as a ball) with no openings require accurate weight and mold rotation to develop uniform wall thickness.

DIP MOLDING

When a design dictates the use of a male mold, the technique is similar to slush casting. A preheated form is dipped into the liquid for a predetermined amount of time, withdrawn at a predetermined rate, and placed in an oven for fusion. Gloves, rain boots, and coin purses are made by stripping fused resin from male forms.

FLUIDIZED BED COATING

A modification of dip molding permits use of finely pulverized thermoplastics such as nylon, cellulosics, polyolefins, and thermosets, particularly epoxy and polyester. Metal parts to be coated are preheated above the melt point of the resin (or the cure temperature), dipped into the aerated bed, and withdrawn for final fusion or curing. Thickness is variable. Because powders never reach a fluid state, problems associated with liquid systems, such as runs, drips, and viscosity changes, are avoided.

Applications are numerous. Resins are selected on the basis of performance in the environment. Outdoor furniture, automotive and appliance parts, marine hardware, food-handling equipment, and electrical units are protected by coating.

ROTATIONAL CASTING

Products too large to produce by other methods such as blow molding or multiple units, when economically sound (see Table 28-4), are made by a process known as rotocasting, rotomolding, or rotational molding.

28-5 Excess powder being removed by air as pipe section is withdrawn from fluidized bed. (*Reprinted by permission of 3M Co.*)

28-6   Multiple molds shown in fusing oven, with single mold in cooling section. (*Reprinted by permission of McNeil Akron Corp.*)

Anyone who has visited an amusement park or a carnival has watched or ridden an amusement that spins in a circle on a horizontal plane, gradually changing to vertical, while at the same time spinning a car on an extension arm in a clockwise direction. This, in essence, is what takes place when molds charged with liquid or powdered resin are fastened to arms and rotated in an oven heated to 900°F. Rotation occurs slowly in two planes, the major axis turning at 0 to 12 rpm, while spinning the mold at 0 to 40 rpm on the minor axis. Variations of spinning are required for complex or unsymmetrical parts.

TABLE 28-4   Comparison of rotocasting with competitive methods

Factor	Rotocast	Injection molding	Blow molding
Equipment cost	Lowest	Highest	
Mold cost		Highest	Lowest
Operating cost	Lowest	Highest	
Maintenance cost	Lowest	Highest	
Labor cost/hour	Highest	Equal	Equal
Production rate	Lowest	Highest	
Molded-in inserts	Practical	Practical	Limited
Uneven wall thickness	Practical	Not recommended	Difficult (except with programmed parison)
Weld lines	Nonexistent	Occur frequently	Nonexistent
Large area, thin wall	Practical	Impractical	Practical
Color change	Rapid	Slower	Slowest
Raw material cost	Highest	Lower	Equally lower

Closed molds of cast or fabricated aluminum, nickel, or beryllium copper are charged with an exact amount of material and moved into the oven, while continuously rotating. A cooling station receives the unit where water spray and air rigidize the part for manual removal.

Some methods involve cored molds with heat-transfer fluids and coolant for faster cycles, which now require from 2 to 17 min.

Tanks, business-machine housings, toys, chair backs and seats, automotive and trailer parts are molded by rotocasting. Polyolefins, polyvinyl chloride (PVC), nylons, ABS, impact polystyrene, polycarbonate and acetal (glass-filled and natural) are used in various applications.

One great advantage of rotational molding is the concept of multilayered products, for example, a nylon outer shell for abrasion resistance surrounding a polyolefin core for chemical resistance. Other applications include inner and outer shells with a foamed core for insulation.

CASTING SILICONE ELASTOMERS

Flexible molds, seals, gaskets, cushioning devices, and encapsulations are readily cast with liquid silicone rubber. These elastomers, known as RTV (room-temperature vulcanizable), are cured by addition of up to 10 percent catalyst[1] normally supplied by the manufacturer.

An advantage to using RTV is the long working time (pot life) after mixing. Low temperature and high humidity retard curing. Disadvantages are high material cost and the fact that RTV is not reclaimable when cured.

[1] The catalyst used is stannous octoate or dibutyl tin laurate.

28-7   Polyethylene industrial containers made by rotational casting. (*Reprinted by permission of Rotodyne, Inc.*)

Silicone rubber is cast into molds and patterns of intricate detail. Parting agents are unnecessary in making the mold or in casting most materials in the RTV. Casting epoxy or polyester, requiring heat cure in RTV, or pouring molten metal up to 600°F, has no effect on the material.

CASTING POLYURETHANE ELASTOMERS

Costing between RTV and flexible polyvinyl chloride, urethane elastomers are also cast to produce flexible molds, tires for forklift trucks, rollers, metal forming pads, rod, tube, and bar stock.

Polyurethane rubber can be cast into or around any material sealed and treated with parting agents, preferable silicone. The pattern to be duplicated should be pre-heated to 150°F and resin poured from one corner. A hot air blast will help dissipate air ahead of the "pour front" and prevent condensation on mold surfaces.

Like silicone elastomers, urethanes require 12 to 16 h to demold. Polyurethane and silicones may be cast in any thickness, without exotherm.

PREPARATION OF HOT-MELT VINYL MOLDS

For many years, the movie industry has pioneered the use of hot-melt polyvinyl chloride molds to produce movie sets of fine detail. Novelties, vibration dampening pads, and shock-arresting shapes are also produced commercially. The reasons for the development and continuing use of this material are : relatively simple techniques ; comparatively low raw-material cost ; faithful reproduction of detail ; extreme under-cuts easily demolded because of elasticity ; scrap and obsolete molds are reclaimable ; low fire hazard (self-extinguishing) ; no size limitation ; and low order of toxicity.

Hot-melt PVC is obtained as a soft, irregular extrudate that melts easily at 325 to 360°F, depending on grade. Heating with an electric pot such as a turkey roaster, or in a double boiler surrounded by glycerine, is desired because of poor heat

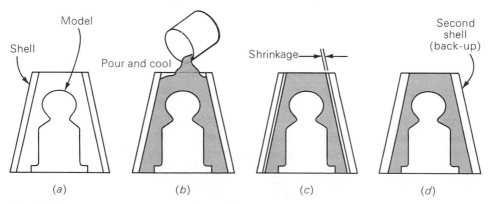

28-8   Steps in preparing a hot-melt PVC mold.

conductance and possible charring when heat is applied at the bottom only. Decomposition is rapid unless contents are constantly stirred away from the sides of the pot.

Faster melting of large quantities can be accomplished by first covering the bottom of the melter with a thin layer of dioctyl phthalate plasticizer, then adding a small amount of plastic. As the resin liquefies, continue adding material until all lumps disappear.

The model to be duplicated can be plaster or wood (thoroughly dry), clay, metal, or heat-resistant plastic. Heating a model from 200 to 300°F prevents premature chilling when melt contacts. Use of a heat lamp above the poured material sustains the melt until air rises, and exerts a hydraulic pressure on the pour.

Vacuum applied at the base aids in pulling the melt against the pattern and

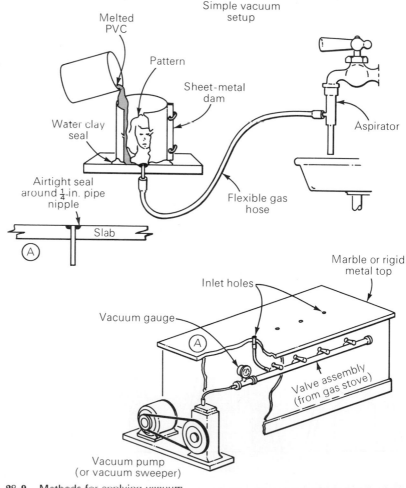

28-9 Methods for applying vacuum.

aids air removal. The vacuum is applied prior to pouring. Care should be taken to prevent drawing vinyl into the pump. Felt, tissue, or cloth can be used under the pattern.

Several hours are required to cool the casting. When cool, remove the PVC mold from the plaster shell, but not the pattern from the mold. Prepare a second plaster shell against the outside of the mold, with a flat top and access hole. This second shell is required to take care of the high shrinkage that occurs as the melt cools. If this is not done, particularly on large objects, the elastomer distorts as weight forces it outward to the original shell.

Small molds, weighing as much as several pounds, are reclaimed by cutting into cubes with shears. Large mold segments can be reduced in a power-operated meat grinder.

As plasticizer evaporates through repeated heating, flexibility decreases and melt point increases. Virgin material and/or plasticizer is added as needed.

Hot-melt molds are used to cast plaster, ceramic clay, WEP resins, phenolic, and epoxy. Release agents are unnecessary because of the high plasticizer content.

The movie and television industries use fire-retardant or general-purpose polyester resins with strips of glass cloth to produce replicas of log cabins, western towns, trees, walls, ornate fireplaces, columns, and many other realistic-appearing stage props.

GLOSSARY

Core   A male section of a mold or pattern; also called plug or punch.

REVIEW QUESTIONS

1 Define the term *casting*.
2 What is the difference between encapsulating, embedding, and potting?
3 What type of product is produced by slush casting?
4 How does a polyester resin used for casting differ from one used for laminating?
5 What problems may be encountered when embedding metallic objects?
6 What factors influence the percentage of catalyst selected for a polyester casting?
7 What causes cracking of a polyester casting?
8 What is the purpose of wax additives in a polyester casting resin?
9 What is the cause of a tacky surface?
10 What methods can be utilized to reduce exotherm?
11 How can a tacky surface be cured to a hard state?
12 Which materials inhibit the cure of a polyester?
13 Explain thixotropy. Which materials are used to impart this characteristic to a liquid resin?
14 What danger should be recognized in handling dimethyl aniline? In working with hydrogen peroxide?

15 What may happen if a promoter comes in contact with a catalyst?

16 For what purposes are phenolic casting resins used?

17 What benefits has the automotive industry received through the use of epoxy casting resins?

18 Why are epoxy resins a logical choice for encapsulating delicate electronic components?

19 Describe the procedure used to produce slush cast or dipped products from polyvinyl chloride resins.

20 What is a plastisol?

21 Which form of resin is used for fluidized bed coating, liquid or powder?

22 Compare the rotomolding process with blow molding.

23 What advantages can be cited for using RTV silicone molds?

24 In preparing a hot-melt PVC mold, what is the purpose of:
    (*a*)   using vacuum
    (*b*)   placing a heat lamp above the pour opening
    (*c*)   making a second plaster backup shell

25 Why is virgin material added to reclaimed PVC?

FINISHING AND ASSEMBLING THERMOPLASTICS

29

Thermoplastic products made by injection, extrusion, blow molding and thermoforming require some degree of finishing, although in many instances the procedure is carried out at the machine by the press operator or an assistant. Assembly of components, or postmolding decoration, is a function of a separate department, except where production cycles are slow enough to permit the machine operator to do the work.

The greatest variety of finishing, decorating, and assembling is performed on injection-molded and thermoformed products. This chapter is concerned primarily with these aspects.

INJECTION MOLDING

Flash removal

Normally, injection-molded parts have no flash when molded correctly. Parts that flash excessively are reground or discarded. With some molds, either through minor distortion, overheated material, or because of movable mold cores, a slight fin or ridge can develop. This is removed by scraping with a sharp blade, either by the press tender or in the finishing room when presses are automatic.

Gate trimming

Edge-gated parts are clipped at the press by using diagonal cutters or similar tools. Thick gates are cut with a band saw. Multiple-cavity molds may contain as many as 200 parts, each with a runner and gate. To enable the press operator to keep up with the machine, manual clipping gives way to a steel-rule shearing die attached to an arbor press or air cylinder. Individual parts fall through holes into receptacles in a single operation. Sprue-gated parts are clipped above the part surface and spot faced on a bench-mounted drill press.

Gate trimming, if done incorrectly, can create stresses that show up later as cracks radiating outward from sprue gates or fanlike from edge gates. Brittle plastics like polystyrene should be cut while warm.

29-1 Complete recycling system can be used when parts require no finishing. Runners and sprue are conveyed to scrap grinder. Reground particles are fed back to hopper. (*Reprinted by permission of Beloit Corp.*)

Machining

Drilled holes, slots, and other features impractical to mold are machined as a secondary operation. In all machining processes, speeds and feeds are adjusted according to the degree to which the thermoplastic softens. Unlike metals, plastics do not help to dissipate the frictional heat generated. As a result, tools become dull and gummed, and the work may scorch or warp.

Drilling Standard twist drills commonly used for metal are adequate for many of the tough thermoplastics such as nylon, polycarbonate, acetal, and cellulosics. Special drills, although recommended for all materials, are particularly effective with polystyrene, acrylics, and soft cellulosics. These drills, of high-speed steel, have deeper, wider flutes with a longer spiral to facilitate chip removal. In addition, drill points are ground to an included angle smaller than 118 deg as with metal drills.

For large or deep holes, drills should be fed slowly at reduced rpm and removed periodically to cool. An air jet or soapy water is used to help cool the work. Jet drills, which feed coolant through the drill to exit at the bottom, are also used.

Drilling thermoplastics must be learned by experience. An adequately sharpened drill at proper feed and speed (see Table 29-1) should produce a continuous ribbon, rather than chip particles.

Products that crack during drilling are those highly stressed by molding. Acrylic and polystyrene may require annealing to relieve stresses prior to machining. Annealing by immersion in warm water (to 190°F) or in air-circulating ovens is common practice.

TABLE 29-1 Recommended drilling speeds

Drill type	rpm
Numbered:	
1–16	2,500
17–32	3,000
33+	5,000
Fractional:	
1/16	5,000
1/8	3,000
3/16	2,500
1/4	1,700
3/8	1,300
1/2	1,000
Lettered:	
A–D	2,500
E–M	1,700
N–Z	1,300

Sawing   Band saws having some "set" in the teeth and 6 to 8 teeth per inch are used for gate trimming and for normal cutting operations.

Circular saws should be hollow ground to run cooler and prevent gumming. A slight set can be tolerated, but is not necessary. Waxing the blade with paraffin or using silicone or fluorocarbon sprays also contributes to faster cutting. These aids should be avoided with parts that are later painted or solvent-welded.

Assembling

Injection-molded parts are designed according to the method by which components are to be assembled. In many cases, the press operator trims and assembles components at the press station as part of his molding duty. Products requiring multiple assemblies such as toys, appliances, and electrical circuitry are put together on an assembly line.

Solvent welding   Mating surfaces are placed on a pad soaked in solvent and immediately brought together. Products can be fixtured on a rotary table when necessary. This practice is hazardous in the molding room because of the combination of solvent vapors and heating devices. Solvent guns are used to inject solvent between mated edges where bonding occurs because of capillary action.

Bodied solvents, or dopes, consisting of small amounts of the material to be cemented dissolved in an appropriate solvent, are used to fill irregular surfaces caused by mold defects, or resulting from sawing. Evaporation is slower, necessitating clamping for as long as 24 h.

As with machining, stresses may cause cracking from the effect of solvent vapors unless parts are annealed.

29-2   Rotary table used for ultrasonic welding bottom section. (*Reprinted by permission of Branson Sonic Power Co.*)

Ultrasonic Welding   Special units that generate sound waves at 20,000 cycles per second (Hertz, abbreviated Hz) eliminate the need for solvent assembly. By means of a transducer, electrical impulses are converted to mechanical energy. The oscillations are transmitted through specially designed "horns" machined from titanium and scientifically "tuned" for optimum performance.

Ultrasonic welding is applicable to thermoplastics of sufficient rigidity to transmit energy waves from the horn to the weld area. Polystyrene, acetal, polycarbonate, cellulosics, and ABS are among those to which sonic power can be applied. Soft vinyls, polyolefins, and urethane elastomers are too flexible to transmit vibrations.

Silicone release agents may interfere with welding since the contact surfaces may not generate friction because of sliding. For this reason, fluorocarbons and thermoplastics with antifriction additives such as molybdenum di-sulfide are also impractical. Best results are achieved by designing the part to have "energy directors," which provide leakproof joints. (See also Fig. 29-3.)

Common examples of ultrasonically welded products are camera flash cubes, ballpoint pens, toys, and houseware items. Bonding occurs in a fraction of a second, while solvent bonding requires days for joints to attain maximum strength.

Electronic heat sealing

One method of fabricating and assembling films and sheets, as well as extruded profiles, is high-frequency welding or electronic heat sealing. Polyvinyl chloride is used primarily, although other resins find applications, notably polyurethane foam.

In this process, aluminum or brass dies fixed to either platen or positioned manually transmit electrical impulses at 20 to 30 megahertz (abbreviated MHz) to the thermoplastic. Frictional heat is generated between the two sheets because of the insulating characteristics of the material. Welds are stronger than the material because of the formation of a small bead.

Air mattresses, water beds, inflatable toys, cushions, flotation devices, and automotive parts are examples of products made by high-frequency assembly methods.

Thermal welding   Although ultrasonic welding generates sufficient heat at the bond line to fuse sections together, it is considered apart from thermal welding. Other methods have long been in use to heat mating surfaces.

Circular parts can be spin welded against each other until sufficient heat softens the thermoplastic. When surfaces have melted, a slight pressure is maintained to provide a leakproof weld when cool.

A lathe, drill press, or other rotational machine can be used, with proper chucking and nesting devices. Peripheral speed, pressure, and spin time are determined by trial. Rotational speeds vary from 5 to 150 ft/sec linear velocity with pressures from 10 to 200 psi. In general, smaller diameters require faster rotation; softer plastics melt adequately at lower pressures.

Spin-welded parts are designed to have interlocking joints such as V, tongue and groove, or modified butt surfaces. Plain, flat surfaces tend to move off-center.

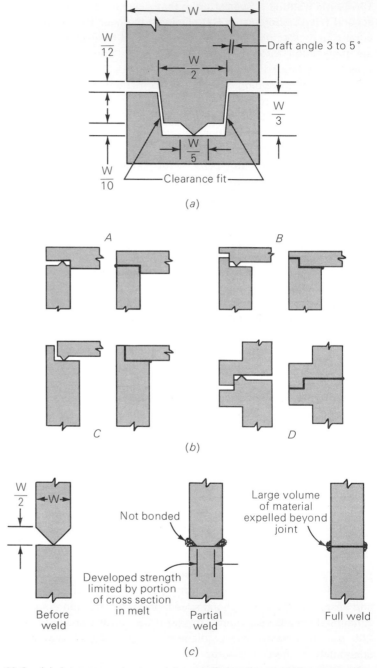

29-3 (*a*) A tongue and groove joint usually has the capability of providing greatest strength. The need to maintain clearance on both sides of the tongue, however, makes this more difficult to mold. Draft angles can be

29-4 High-frequency heat sealer. (*Reprinted by permission of Los Angeles Trade-Technical College.*)

Parts that cannot be solvent welded or attached by ultrasonic or spin techniques can often be melted and fused together by heated tool welding. This method is particularly effective for polyolefins, vinyls, and cellulosics. A hot plate covered with a thin silicone or fluorocarbon film and thermostatically controlled is used. Surfaces to be joined are brought in contact with the plate, held briefly to soften, then joined. Very light pressures (below 3 psi) are required.

Wide surfaces can be put together with a sliding motion to insure good distribution of the molten plastic across the interface. Molded pipe fittings are attached to flanges and pipe sections, often during actual installation, by hot-plate welding.

---

modified concurrent with good molding practices, but interference between elements must be avoided. (*b*) Basic joint variations suitable for ultrasonic welding. These are suggested guidelines for typical joint proportions. Specific applications may require slight modification. Practical considerations suggest a minimum height of 0.005 in. for the energy director. Where height greater than 0.020 in. is indicated, two or more directors should be provided with the sum of the heights equalling the formula dimension. (*c*) The result of beveling one joint face at 45° angles. This is to be avoided. Note also that joints designed for solvent sealing can generally be modified to meet ultrasonic welding. (*Adapted from illustrations supplied by Branson Sonic Power Company.*)

A special unit designed for rapid assembly of injection-molded parts is used by many molders. An operator places the sections to be welded in nesting fixtures attached to air cylinders. When the cycle starts, the weld surfaces contact a heated plate, retract to allow the plate to move out of position, then move together to effect the weld.

Mechanical fastening   The yield characteristic of many thermoplastics is an advantage in designing products for snap fits, press fits, cold heading, and stretch-over. For snap fits, an undercut is machined or molded into one surface, which engages a lip to form the assembly. Selection of the material is based partly on its ability to be molded with undercuts. Nylon, polycarbonate, cellulosics, and ABS are used for snap-fit applications.

Knobs, gears, handles, and similar parts are attached to shafts by press fitting. Acetal, nylon, acrylic, and polyolefins are pressed on metal or plastic shafts for radio and television components, tool handles, gear trains, and pump parts. During pressing, the plastic yields to conform to the slightly oversize shaft diameter. Metal shafts can be chilled and the plastic heated to reduce the amount of interference.

29-5   Cosomatic horizontal thermal sealer showing polyethylene mallet head after welding. (*Reprinted by permission of Rodgers Plastics Equipment, Inc.*)

Cold heading is a method of assembling metal or wood to plastic, dissimilar plastics, or plastics of the same type. Because of their toughness, polycarbonate, ABS, nylon, and acetal are most widely used. In this procedure, plastic rivets are headed with an arbor press, punch press, or similar pressure devices, and a suitable die.

Stretch fits are useful for assembling soft grades of polyolefins, vinyls, and some styrene copolymers when parts are designed to utilize the elasticity of these materials. The most common examples are coffee-can lids, containers for small tools, and pill boxes.

Decorating

Many of the processes used for decorating thermoset plastics are used for thermoplastics. Spray painting, silk-screen printing, roller coating, vacuum metallizing, and electroplating are applied to injection-molded products. Paint or plating applied to the underside of transparent products provides a depth of surface not possible with opaque thermosets.

Although applicable to laminated and compression-molded parts, hot-leaf stamping is more widely used to decorate thermoplastics. A roll of pigmented foil is fed under a hot die, which presses the design firmly into the heat-softened plastic. Production rate is rapid and the process economical.

BLOW MOLDING

Blow-molded containers, toys, and industrial products require trimming of fins and necks. A shearing die trims the fin automatically in most operations. Necks are trimmed with a lathe tool or rotating knife. Containers are printed by offset printers or by hot stamping.

EXTRUSION

Extruded products are decorated by transfer rollers that apply designs, wood-grain effects, or information items as the extrudate is being cooled. Profile shapes, particularly cellulose acetate butyrate, are vacuum metallized for table edging, signs, and frames.

THERMOFORMING

Scrap factors are greater in thermoforming processes than in any other. Products are trimmed by band saw, routing, or die cutting. Edges are smoothed as required by the same techniques used for injection molding.

Decoration is not as widespread because of textures available in sheet stock or embossing during forming. However, any process applicable to other products can be used to decorate or assemble thermoformed products.

GLOSSARY

Capillary action   The phenomena whereby a liquid coats a surface because of cohesive and adhesive forces acting at minute distances.

Megahertz   One million electrical cycles per second.

Set   Configuration of saw teeth in which one tooth is pointed slightly to the left, the next slightly to the right.

Transducer   A device that converts electrical energy into mechanical energy.

REVIEW QUESTIONS

1   How is a thin ridge or fin of plastic removed when it appears as flash on molded parts?
2   How are sprues finished off when they form a part of the molded item?
3   What factor creates a problem in drilling thermoplastics that is not present when drilling metal?
4   How is cracking avoided when machining highly stressed molded parts?
5   What is a bodied solvent?
6   Which thermoplastics are more readily joined by ultrasonic welding?
7   What is an energy director?
8   What shape of products can be spin welded?
9   What types of common household products lend themselves readily to assembly by snap-fit methods?
10   What is meant by cold heading?
11   What types of products are sealed by high-frequency electronic methods?
12   List some of the methods for decorating thermoplastics.

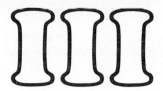

TEST PROCEDURES

The growth of any industry is reflected in public acceptance of its products. If an item is properly designed, made correctly from a suitable material, and sold at a reasonable price for a definite need, it can compete in the marketplace. Plastic products are increasingly competing with traditional materials and continue to do so on the basis of quality.

This situation did not always exist. There was a period during and immediately following World War II when plastics were considered inferior substitutes for many products formerly made from metal. Because of careless selection of material, or because of improper mold design or faulty processing methods, plastic items were made that warped, cracked, shrunk, faded, and failed in many other ways. In some cases, substitution of plasticizers or solvents during compounding created unforeseen problems for the manufacturer.

These difficulties have been largely overcome. Unfortunately, a few inferior products occasionally appear to blight the image of the industry. In virtually each instance the fault can be attributed to poor quality control at the processing level. If raw materials are not stored properly, are not used within a certain period, or become contaminated, then quality is affected. When machine cycles are not held constant or not changed to meet changing conditions, the product will vary. Selection of a material that has not been thoroughly evaluated for the service condition and poor product or mold design occasionally plague an unwary customer.

The quality control department should be considered pivotal to other operations. Quality control personnel should have the final judgment on:

1   Rejecting raw materials and ancillary products (such as inserts) before they are committed to production

2   Interrupting a process during any production phase when standards are lowered

3   Evaluating finished goods

4   Establishing acceptable standards and deviations from the norm

Control of product quality should be the function of a separate department (or even one well-qualified individual), not of the production department. The tendency exists, in the interests of production quota, to overlook minor, correctable defects. As a result, some products offered for sale may appear satisfactory to the buyer when, in reality, they are inferior. Many defects are subtle and result in product failure only after a period of time when minor stresses have been applied.

Many causes of defects in molding and laminating have been cited in previous chapters. The following chapters explain some of the important tests performed for evaluation of incoming raw materials, as well as in-process controls and final acceptance standards.

Hundreds of different kinds of tests are used throughout the industry. They are classified according to the requirements that must be served.

AMERICAN SOCIETY FOR TESTING AND MATERIALS (ASTM)

This organization is composed of professional people who draw up and publish standard methods of classifying and evaluating materials used by industry. Standards and specifications include mechanical, physical, electrical, chemical, thermal, and environmental values with which manufacturers should comply.

NATIONAL ELECTRICAL MANUFACTURERS' ASSOCIATION (NEMA)

Specifications and test methods governing laminated materials for use in the electrical and allied industries are published by this specialized professional group.

UNDERWRITER LABORATORY (UL)

Products that bear the seal of this governing body have undergone rigid testing to conform to strict standards of safety in electrical applications.

GOVERNMENT STANDARDS

Test methods for the purpose of meeting specifications established by various federal agencies are set forth in Publication LP 406-B. Products made for U.S. Naval, Air Force, and Ordnance Departments, for example, are evaluated in accordance with federal requirements.

Tests constantly undergo revision to meet changing standards brought on by new materials and new methods. Companies that exercise adequate quality control are conscientious in every phase of their operations from evaluating incoming raw materials to final inspection.

QUALITY CONTROL PROCEDURES

30

Each manufacturing process requires relevant quality control and test measures that will ensure conformance to industry standards. The types of tests depend on the stringency of the factors established to make the product totally acceptable. An item such as a dustpan may require only visual inspection by a press operator, with occasional spot checks by an inspector. An aircraft connector, by comparison, undergoes tests for electrical insulating values at various temperatures, water absorption, dimensions, degree of cure, and insert positioning. Controlling the quality of a high-reliability product can be costlier than the actual molding expense when extensive testing is demanded. Quality control should begin when the raw materials are purchased.

Materials are received from suppliers in bags, cartons, drums, or bins. Large-volume users have bulk shipments delivered by truck or railroad cars, which unload into silos or underground storage tanks.

Thermoplastic resins are fed to production machines without further treatment, although the trend is increasing toward in-plant blending of resins to reduce material inventory. Most processors add color pigments, special additives such as lubricants, and develop "secret" formulations to reduce overall costs. In general, the processor relies heavily on the material supplier to maintain resin quality and performance from shipment to shipment and conducts only those tests necessary to assure uniformity.

After the resin is made into a product, it is the processor's responsibility to ascertain its quality. Because quality control and testing add to product cost, only those tests that prevent loss of production and customer acceptance are considered.

More testing is involved for thermoset resins as used in molding, laminating, preimpregnating, and coating than with thermoplastics because of the intermediate chemical stage in which they are received and used. It should be remembered that thermosets complete their polymerization during the processing steps, but thermoplastics are completely polymerized by the material manufacturer. Certain exceptions occur, as with vinyl fusion, foaming of urethanes, and casting of silicones.

TESTS FOR INCOMING RAW MATERIALS

Thermosetting molding compounds

Percentage of fines   Chapter 2 describes how molding compounds are formulated from resins, fillers, and special additives. Those furnished in powder form must be checked to determine the proportion of "fines" to larger particle sizes. Fines develop in the final crushing and screening operations of compounding, when brittle resins flake off the granules. A small percentage (5 percent) is not detrimental, but an undue amount can create serious problems such as: some loss in strength; uneven heating and mold-filling rate; orange-peel surface; and nonuniform dimensions. Rolling unopened drums or spinning in a drum tumbler redistributes the fines.

ASTM test method D1921-63 outlines a sieve analysis to determine the extent of particle sizes and the proportions of each. A 100-g sample is placed on the top (coarsest) sieve, which is nested on top of two sieves having successively smaller screen sizes, and bottom pan collects the finest particles. A mechanical shaker operated for 10 min separates the coarse, medium, and fine components, and the percentages recorded.

Specific gravity   An important property of plastics is their inherent light weight. Aluminum is a light metal, but it is twice as heavy as the average plastic. Steel is six times heavier than most plastics. From a structural standpoint, however, we cannot make a direct substitution for metals because plastics are less than half as stiff as aluminum. When sufficient strength can be obtained in a plastic product through proper design and correct choice of reinforcement, the lighter weight becomes advantageous to the user. In load-bearing applications, plastics can be made thicker and eliminate the necessity for multipart assemblies, thus actually proving to be stronger at the same weight as a competitive product of metal.

As a quality control measure, specific gravity is useful to determine the quality of laminates, foam densities, and the soundness of molded parts. Specific gravity is the ratio of the weight of a given volume of a substance compared with the weight of an equal volume of distilled water at 73.4°F (23°C). The specific gravity of water is 1 g/cc (metric units); or 0.036 lb/in.^{3} (English units).

IMMERSION METHOD
Equipment required:

Analytical balance

Fine wire

Steel or lead sinker (smooth, regular shape)

Immersion tank for distilled water

Procedure: Test sample should not weigh over 2 oz, preferably under 1 oz. Samples should be such that air cannot be trapped during immersion. The following steps should then be taken:

1   Weigh sample in air and record.

2   Fasten wire to the sample and attach to the hook beneath the balance pan.

3   Suspend sample and immerse to cover fully, but not touching sides or bottom of vessel.

4   Record weight of immersed sample.

Calculation: Specific gravity (g/cc) $= \dfrac{a}{a + w - b}$

where a = specimen weight in air

w = weight of wire (plus weight of sinker, if used) immersed to same depth as when attached to specimen

b = weight of specimen (sinker, if used) and wire immersed

Note: A steel or lead sinker slightly heavier than required to sink the specimen is necessary for plastics that float.

The terms *density* and *molded density* are used interchangeably with specific gravity. When samples can be precisely measured, the formula is

Density (g/cc) $= \dfrac{\text{wt in grams}}{\text{vol in cc}}$

1 in.^{3} = 16.39 cc

Bulk factor   is defined as the ratio of the volume of loose, uncompacted molding material to the volume of the same quantity of material after molding. Bulk factor is also called the compression ratio. For example, if a quantity of molding compound has a volume of 2 in.^{3} that compresses to 1 in.^{3} during molding, the bulk factor (compression ratio) is 2:1.

This physical property is important to the mold designer who must provide adequate loading space in cavities. Design of loading trays, scoops, and storage areas must take into account the bulk factor of materials. Bulk factor dictates whether automatic preforming equipment can be used. Fluffy or high-bulk compounds require costly, manual tablet making.

When a molding charge is measured rather than weighed, incomplete or porous moldings are more apt to occur if any change in bulk factor is not detected. Fluffiness or bulk of a molding compound can be measured to obtain apparent density. By comparing apparent density with molded density, the bulk factor is determined (see Table 30-1).

TABLE 30-1   Specific gravity and bulk factor of plastic materials

Material	Filler or reinforcement	Specific gravity, g/cc	Specific weight, lb/cu in.	Specific volume, cu in./lb	Bulk factor
Phenol formaldehyde	Cellulose	1.32–1.45	0.047–0.052	21.0–19.1	2.1–4.4
	Mica	1.65–1.92	0.059–0.069	16.8–14.4	2.1–2.7
	Glass	1.60–2.20	0.058–0.079	17.3–12.6	2.0–10
	Asbestos	1.45–1.90	0.052–0.068	19.1–14.6	2.0–14
	Macerated fabric	1.36–1.43	0.048–0.051	20.4–19.4	3.5–18
Urea formaldehyde	Cellulose	1.45–1.55	0.052–0.056	19.1–17.9	2.2–3.0
Melamine formaldehyde	Cellulose	1.45–1.55	0.052–0.056	19.1–17.9	2.2–2.5
	Asbestos	1.70–2.00	0.061–0.072	16.3–13.9	2.1–2.5
	Glass	1.80–2.00	0.065–0.072	15.4–13.9	5.0–12
	Macerated fabric	1.50–1.55	0.054–0.056	18.5–17.9	5.0–10
Epoxy (cast, unfilled)		1.11–1.40	0.040–0.050	25.0–19.8	...
Molded	Glass	1.60–2.00	0.057–0.072	17.3–13.9	2.0–7
Polyester and DAP (cast, unfilled)		1.12–1.18	0.040–0.042	24.7–23.5	...
Molded	Clay	1.40–1.60	0.050–0.058	19.8–17.3	2.0–4
	Glass	1.35–2.30	0.049–0.083	20.5–12.0	2.0–10
Silicone	Asbestos	1.60–1.90	0.057–0.068	17.3–14.6	6.0–8
	Glass	1.68–2.00	0.060–0.072	16.5–13.9	6.0–9
Alkyd:					
Powder		1.60–2.30	0.057–0.083	17.3–12.0	1.8–2.5
Putty		1.60–2.30	0.057–0.083	17.3–12.0	1.0–1.2
ABS		1.02–1.25	0.037–0.045	27.2–22.2	1.1–1.2
Acetal		1.41–1.42	0.050–0.051	19.6–19.5	...
Cellulose acetate		1.20–1.34	0.044–0.048	21.9–20.7	1.8–2.6
Cellulose acetate butyrate		1.15–1.22	0.041–0.044	24.1–22.7	1.8–2.4
Cellulose nitrate		1.35–1.40	0.049–0.050	20.5–19.8	...
Cellulose propionate		1.17–1.24	0.042–0.045	23.7–22.4	...
Fluorocarbon:					
PTFE		2.10–2.22	0.075–0.080	13.2–12.5	...
FEP		2.12–2.17	0.076–0.078	13.0–12.8	2.0
Polyamide		1.09–1.15	0.039–0.041	25.4–24.1	1.7
Polycarbonate		1.20	0.043	23.2	1.7–5.5
Polyethylene:					2.0–2.5
Low density		0.910–0.925	0.032–0.033	30.4–30.0	2.0–2.5
Medium density		0.926–0.940	0.033–0.034	29.9–29.5	...
High density linear		0.941–0.965	0.034–0.035	29.4–28.7	...

TABLE 30-1 Specific gravity and bulk factor of plastic materials (*Continued*)

Material	Filler or reinforcement	Specific gravity, g/cc	Specific weight, lh/cu in.	Specific volume, cu in./lb	Bulk factor
Polymethyl methacrylate		1.17–1.20	0.042–0.043	23.7–23.2	1.6–2.0
Polyphenylene oxide		1.06	0.038	26.1	2.0–3.5
Polypropylene		0.88–0.906	0.032–0.033	31.5–30.6	2.0–2.4
Polystyrene		1.04–1.100	0.037–0.040	26.6–25.2	1.6–2.4
Polysulfone		1.24	0.045	22.4	1.8–2.2
Polyvinyl acetate		1.10–1.14	0.040–0.041	25.2–24.3	. . .
Polyvinyl chloride:					
Rigid		1.35–1.45	0.049–0.052	20.5–19.1	2.0–2.3
Flexible		1.16–1.35	0.042–0.049	23.9–20.5	2.0–2.3
Polyvinylidene chloride		1.65–1.72	0.059–0.062	16.8–16.1	2.0
Polyurethane		1.11–1.25	0.040–0.045	25.0–22.2	2.4–2.8
Aluminum		2.40–2.70	0.086–0.097	11.5–10.3	. . .
Glass (plate)		2.50	0.090	11.1	. . .
Maple		0.68	0.024	40.8	. . .
Oak		0.86	0.031	32.2	. . .
Rubber (hard)		1.15–1.25	0.041–0.045	24.10–22.2	. . .
Steel		7.60–7.80	0.270–0.280	3.65–3.56	. . .

Determination of apparent density and bulk factor[1]   Apparatus includes a cylindrical measuring cup (100-cc capacity), gram balance (accurate to 0.1 g), funnel.

Procedure: Close off small end of funnel with a small strip or by hand. Pour 115 ± 5 cc of sample into funnel; allow sample to fall freely into cup. Scrape excess with a straight edge. Do not vibrate, shake, or tamp contents. Calculate net weight of sample.

$$\text{Apparent density (gm/cc)} = \frac{\text{net weight of sample (grams)}}{100 \text{ cc}}$$

$$\text{Bulk factor} = \frac{\text{molded density (g/cc)}}{\text{apparent density (g/cc)}}$$

Note: Thermoplastics vary widely because of the incorporation of plasticizers, release agents, pigments, flame retardants, and other additives. The addition of fiberglass increases base-resin density to 40 percent.

Flow or plasticity   Flow is the amount of mold filling a material exhibits during controlled rates of press closing under exact pressure and temperature conditions. Comparisons of flow are made to compare one resin with another or to determine whether or not a resin has "aged." A knowledge of flow is important to the mold designer and the molder so that the correct grade of material can be selected. For

[1] This is ASTM test method D1895-67.

example, a phenolic compound that shows good mold-filling properties in a shallow mold may be incapable of filling a deep mold.

FACTORS AFFECTING FLOW

1   *Resin types*:   Urea formaldehyde is heat sensitive. Rapid curing reduces flow. As a class, melamine resins exhibit longer flow than urea, when molded at the lower end of the temperature range. Phenolics, because of the variety of resin types, enable the molder to select the flow best suited for a particular design. Polyesters, diallyl phthalates, and epoxies can be tailored for cure rate by adjusting catalysts, accelerators, or inhibitors. Postponing the gel time has the effect of increasing flow.

2   *Type of filler*:   Small-particle size represented by wood flour, mica, cotton linters, and minerals creates less turbulence and less frictional drag during mold filling. Short fibers of glass, nylon, carbon, and rayon present problems in molding fine threads or engraved mold details. Longer fibers, in length to 2 in., create a "log-jam" effect, often resulting in entrapped air, unfilled sections, or resin-rich area. Metal fillers reduce flow by increasing thermal conductivity through the resin matrix.

3   *Degree of resin advancement*:   Resin manufacturers control the degree of resin advancement during the later phases of production. Milling over heated rolls for final blending of ingredients is effective in controlling cure time and flow. Molders can advance resin polymerization with oven or radiant heat, or electronic preheating.

    All resins have a natural tendency to polymerize in storage, causing partial precure that reduces flow. An exception might be a polyester in which catalyst decomposition slows or prevents curing, which increases flow duration.

Flow tests (molding index)

CUP MOLD

A standard laboratory test particularly useful for compression-molding materials is conducted to determine the minimum pressure required to mold a standard cup. Specimens are molded at a temperature deemed realistic for the type of compound at pressures to 6,580 psi. The time required to close the mold fully is also noted.

SPIRAL MOLD

In this test, the compound is forced through a sprue into a spiral channel. Numbers from 0 to 40 are engraved from the center or entrance point. Compounds are classed as low (1 to 10), medium (11 to 22), and high (23 to 40) plasticity. For high-flow epoxies, spiral length is increased to 100. Either the cup test or the spiral can be used to determine flow for compression and transfer grades. A long cup-closing time or short spiral length indicates a poor transfer grade of material. Conversely, a short cup-closing time or long spiral indicates a soft flow material suitable for either transfer molding or deep-draw compression parts.

WILLIAMSON FLOW TEST

A method of evaluating flow of high-bulk-factor materials, particularly diced fabrics, prepreg roving, and premix, meets requirements of the aerospace industry. The flow rating is based on the quantity of compound remaining in the loading well of a split cylindrical mold.

Liquid resins for molding and laminating

Compression molders that concentrate on low-pressure molding and premix manufacturing, and laminators that impregnate fabrics, mats, and roving, conduct tests to insure resin quality.

Solids content   This indicates the amount of solid resin in the liquid. Polyesters and epoxies are available as 100 percent solids or diluted with solvents. Phenolics and aminos are furnished diluted with an appropriate solvent. When viscosity is low enough to combine in correct proportion with colorant, reinforcement, filler, and

30-1   Spiral-flow samples showing the effect of molding pressure on transfer-molding grade of epoxy. (*Reprinted by permission of Furane Plastics, Inc.*)

catalysts, then resins are used without thinning. Liquid materials must be checked when received and checked again during their production life.

Several methods are used to determine solids content:

1   By specific gravity of the liquid

2   By the evaporation method:

Equipment needed:
Balance, analytical type
Evaporating dishes (tin, glass, or aluminum)
Oven, air circulating and vented, 275° ± 5°F
Thermometer
Tongs
Desiccator

Procedure: Accurately weigh three evaporating dishes for tare weight. Weigh 10-g samplings in each to 0.001 g. Place samples in oven close to thermometer. Evaporate solvent, if any. Weigh periodically. When no further weight loss can be noted (sample weight remains constant), remove to desiccator to cool for 15 min.

$$\textit{Calculation}: \quad \text{Percent solids} = \frac{\text{original weight} - \text{final weight} \times 100}{\text{original weight}}$$

Viscosity   Viscosity of a liquid is a measure of its resistance to flow. Increases in viscosity indicate advancing stages of polymerization or solvent evaporation. Figure 30-2 shows equipment used to determine viscosity with a bubble viscometer. In this procedure a test sample is placed in a stoppered glass tube. The rate of bubble rise is compared with a set of calibrated bubble tubes containing liquids of known viscosities. This method is useful for testing liquids at ambient and elevated temperatures.

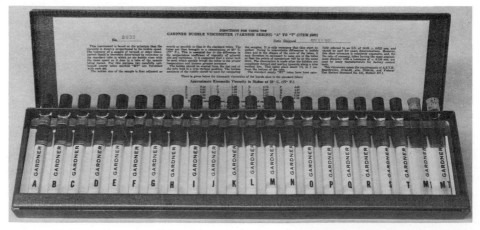

30-2   Gardner bubble viscometer. (*Reprinted by permission of Gardner Laboratory, Inc.*)

Another measure of fluidity involves the use of a Brookfield Viscometer. In this procedure a spindle is inserted in a cup of liquid and caused to rotate at predetermined speed. Readings are taken at standard spindle speeds (see Table 30-2) and stated in poises.[1]

Tests for cure conditions

Gel time   Resins used in liquid form, such as polyester, diallyl phthalate, and epoxy, reach a semisolid, rubbery stage before becoming hard and infusible. This condition is called a gel. In making a wet lay-up or spray-up part, air must be removed by the laminator prior to gelation (kick-over).

The compression molder who molds FRP products adjusts press-closing rate, temperature, plus initial and final pressure to coincide with resin gel time. Catalysts can be checked for deterioration in comparative tests.

Equipment needed:
Balance, accurate to 0.01 g
Water bath, controlled to 180°F (82.2°C)
Pyrex test tubes, 20 × 150 mm, with rack
Recording potentiometer (pyrometer), range 0 to 500°F
Thermocouple wire, Iron/Constantan, 24 B & S gauge, 6 in. long, or thermocouple needle
4-oz glass jar or nonwaxed paper cup
Nylon adapters or vented corks
Spatula

Materials needed:
Polyester resin
Benzoyl peroxide paste (50% BPO in tricresyl phosphate) catalyst

[1]A poise is the standard unit of measurement equal to a force of 1 dyne acting on a moving plane 1 sq cm in area at a velocity of 1 cm/sec, when liquid-film thickness is 1 cm. One centipoise equals 0.01 poise.

TABLE 30-2   Typical viscosities at 77°F

Resin	Centipoise
Styrenated polyester at 100% solids:	
Laminating	400–700
Spray-up	210–400
Matched metal molding	850–8,000
Casting	450–900
Epoxy at 100% solids:	
Filament winding	275–700
Tooling (casting)	7,000–10,000
Laminating	500–700
Phenolic at 62% solids:	
Preimpregnating	200

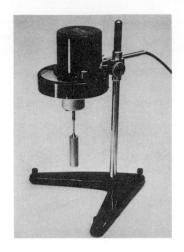

30-3   Brookfield viscometer. (*Reprinted by permission of Brookfield Engineering Laboratories, Inc.*)

Procedure[1] :   Weigh 50 g into container and add 2-percent paste catalyst. Stir for 2 min. Avoid stirring in air bubbles. Transfer 10 g each (or to a depth of 3 in.) into two or more test tubes. Place the adapter with thermocouple in the test tube to a depth of 1.5 in. Place the rack in the water bath so that water level is above resin height. Set recording chart at 2 in./min speed. Record the temperature curve until the peak temperature is obtained.

Calculation :   The SPI gel time is the difference in time between 150 and 10°F above the bath temperature (in this test, 190°F).

Gel-time tests may be conducted at any temperature pertinent to the process for which a resin is to be used. A matched-die molding application may call for bath temperatures to 300°F.

Room-temperature gel time   Determination of "pot life" of polyesters using room-temperature catalysts can be effectively made.

Equipment needed:
Balance, accurate to 0.1 g
Paper cup or metal container, 16-oz capacity
Stopwatch
Stir stick (tongue depressor)
Spoon

Materials needed:
Polyester resin, promoted with 0.5 percent cobalt naphthenate
MEK peroxide catalyst

Procedure :   To 300 g of resin, add 1 percent catalyst (3.0 g). Start timing with stopwatch simultaneously with catalyst addition. Stir well to disperse catalyst for

[1] This method is recommended by SPI and is known as SPI gel-time procedure.

1 min. Remove the spoon. Each 60 sec thereafter stir very slowly with stir stick for 10 sec. Withdraw slowly with a flat side up. As viscosity increases noticeably, stir for 10 sec each half-minute. When resin thickens and tends to crumble, or adheres to the stick and rises as it is withdrawn, stop the timer.

Calculation: The elapsed total time is recorded as the room-temperature gel time. Variations should be expected if greater or lesser quantities of resin are used, or when temperature of the test area is raised or lowered appreciably from normal 77°F.

Peak exotherm   Resins that undergo addition-polymerization generate heat (exotherm) as a function of the curing mechanism. With polyesters, the heat rise and subsequent dissipation occur after gelation. The epoxy systems exotherm before gelling, which can create problems in laminating. As the temperature climbs, resin viscosity lowers, and the resin may drain off vertical mold surfaces, unless thixotropic agents are used.

High exothermic temperatures are generally undesirable. As heat is generated, thermal expansion occurs, while at the same time shrinkage takes place when the thermosetting, three-dimensional network forms. This simultaneous opposition creates stress or causes crazing, cracks, or porosity. The exotherm peaks for rigid polyesters range from 250 to 470°F; for flexible resins, 190 to 290°F.

Peak-exotherm data is derived along with gel-time determinations.

Degree of cure   Thermosets never cure completely during the initial processing operations. The terminal cross-linking of a few molecular chains may not occur for many days and may even require heat treatment to speed the reaction. Fortunately, molded, cast, and laminated products attain 90 to 95 percent of their ultimate strength when removed from the mold and cooled, providing a realistic cure time has been established. Laminates designed for high reliability, such as rocket nozzles, aircraft components, and molded parts manufactured to close tolerance, are often postcured in ovens as a precaution against failure.

The degree of cure is determined by analyzing the percentage of resin that has not reacted. When fully cured, thermosets are insoluble in solvents, acids, or alkalies. Methods to determine the extent of undercure take into account the fact that unreacted chemicals are soluble.

Acetone extraction test for phenolic resins   In this procedure, a sampling is obtained from a cast, molded, or laminated part by drilling, milling, planing, grinding, or filing. Care should be taken to ensure that frictional heat generated by dull tools or fast cutting does not cause additional curing. Particle sizes are reduced until they pass through a number 40 sieve but are retained on a number 140.

The test is conducted in triplicate. A 3-g sample is weighed into a filter paper or filter thimble and placed in a glass siphon. The siphon is suspended below a condensing coil and placed into an assay flask containing 50 ml of acetone. An electric hot plate is adjusted so that the acetone evaporates, condenses, fills the siphon, and empties between 15 and 20 times per hour. Extraction is conducted for 4 h, after which the contents in the flask are poured into a tared dish. Any residue is washed out with fresh acetone and added to the extract. All solvent is evaporated in a well-ventilated oven to obtain a constant weight.

Calculation: Percent acetone extractable matter $= \dfrac{W - D}{S} \times 100$

where W = weight of dish and extract
D = weight of dish
S = weight of original sample

Chloroform test for diallyl phthalate   In this test, a small piece of molded material, preferably with one sawed surface exposed, is placed in the bottom of a wide-neck flask and covered with chloroform. The solvent is slightly boiled for 3 h. Whitening or slight swelling indicates a minor undercure; cracking or disintegration is evidence of serious undercure.

Undercure tests for urea-formaldehyde   A review of the characteristics of urea resins will show that even when given an excellent cure, some deterioration occurs in boiling water. A comparative test involves boiling for 15 min and noting the degree of chalking of the surface resin, or minute cracking. Thick sections should be sawed prior to boiling in order to examine the central portion. Undercured sections appear granular, indicating a case hardening of the exterior without sufficient heat penetration to complete the cure.

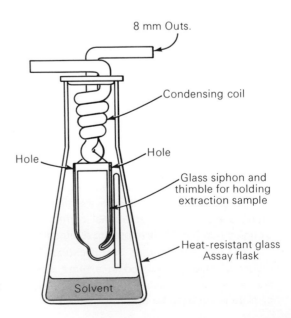

30-4   Acetone extraction apparatus.

Undercure test for melamine-formaldehyde   Boiling for 10 min in a 1% aqueous solution of sulfuric acid, followed by rinsing and drying, will show evidence of undercure by dulling or chalking of surface.

Barcol hardness test for reinforced plastics   The progress of a cure is effectively checked with a Barcol indenter (needle penetrometer). Readings are taken at various locations and averaged. Pressing the instrument against the surface for a direct dial indication (0 to 100) provides an instant clue. Specifications vary upward from a minimum reading of 45, depending on the type of resin. Most polyesters and epoxies harden to a Barcol of 65 and up. Care must be exercised to avoid placing the indenter on starved or void areas and on fibers.

Shrinkage   All plastics shrink to some degree during molding and laminating cycles. Initial shrink enables the molder to remove parts from cavities or causes parts to stick to cores for mechanical ejection. The greatest shrinkage develops during initial cure of thermosets, but it is completed only by an afterbake or upon long-term aging. Thermoplastics, by comparison, shrink by thermal contraction as they cool from the hot-melt condition. This can be more troublesome in maintaining part-to-part uniformity since control of melt temperature, pressure, rate of fill, and cooling must be closely maintained. Then, too, as molded-in stresses are relieved, or plasticizers migrate, original dimensions change.

   The molder compensates for shrinkage by building a mold oversized to the extent that the material shrinks. Shrinkage manifests itself in several undesirable ways unless the effects are understood:

1   Warpage, especially with uneven wall thickness.

2   Cracking of plastic around inserts.

3   Variations in shrink factors, depending on molding method:
   (*a*)   Transfer-molded parts shrink less than compression moldings.
   (*b*)   Parts molded in a flash-type mold shrink more than when material is confined in the cavity.
   (*c*)   The degree of packing an injection mold influences shrink.

30–5   Barcol impression test. (*Reprinted by permission of Los Angeles Trade-Technical College.*)

Specimens usually 4-in. diameter, or 0.50 × 5 in. bars are molded in 0.125-in. thickness for thermoplastic or 0.50 in. for thermosets. Any molding method can be used, including compression molding of thermoplastics.

Calculation (to nearest 0.001 in.) :

$$\text{Dimensions of } \frac{\text{cold mold} - \text{cold piece}}{\text{cold mold}} = \text{shrinkage (in./in.)}$$

Measurements are reported at initial 24- and 48-h intervals. With cast resins, readings are taken up to 7 days.

Tests for quality of reinforcements

Laminators, premix manufacturers, coaters, and filament winders must exercise quality control on reinforcing materials as well as on resinous materials. Not only must fabrics, roving, and mats be of uniform quality, but the final product must contain the specified ratio of reinforcement to resin matrix so that end-use properties are realized (see Table 30-3).

Tests for pre-impregnated materials

Preimpregnated materials (prepreg) are produced to meet the end-user's specifications. The factors that control the properties of the final laminate must be held to close limits since each affects all others. For each laminate there is an optimum fabric-to-resin ratio that governs mechanical strengths, water absorption, weight and electrical values (see Table 30-4).

To control final resin content in the laminate, it is necessary to control volatile content, resin solids, and flow at various cure conditions. Preimpregnation is carried out by passing the web (fabric, felt, or mat) through a dip tank containing solvated resin. Percentage of resin solids, solution viscosity, and the clearance of the metering device determine the amount of wet resin picked up on the web. Resin solutions may range from 20 to 95 percent, depending on the nature of the web being coated. As the wet web passes through the evaporating zone of a heated tunnel or tower, the major portion of solvent is flashed off. Further passage advances the resin to a stage of partial precure (*B* staging). A small but variable retention of solvent is desirable in the prepreg to influence the following factors.

Drape   A desirable quality in a prepreg is its ability to permit laying the material over complicated curves and contours. Conversely, producers of flat laminates prefer stiff, boardy material to facilitate handling at the press. No standard test exists for drapability, but this characteristic is largely influenced by type and amount of resin, volatile content, and fabric style.

Tack   The stickiness of a prepreg is of concern to the laminator. For flat laminates,

TABLE 30-3 Tests for fibrous glass products

Test*	Method or equipment
Woven fabric:	
Thread count (per in.)	Thread counter
Weave type	Magnifying glass
Thickness	Micrometer
Weight (per sq yd)	Template 6.76 × 6.76 in. to cut sample. Gram weight of piece equals fabric weight in ounces per square yard.
Selvedge	Visual or micrometer. Check thickness for machine-coating purposes. Wide variations between selvedge and weave cause incorrect resin pickup.
Quality	Visual checks for broken yarns, dirty loom splices, slubs (balled or fuzzy area), tight or loose selvedge, and other noticeable defects.
Glass mat:	
Weight (oz/sq ft)	Cut template to 6.76 × 6.76 in. $$\frac{\text{weight in grams}}{9} = \text{oz/sq ft} \quad \text{or}$$ Take square-foot samples across the roll width; weigh and average.
Thickness	Use micrometer-thickness gauge to measure to nearest 0.001 in. at no pressure.
Ignition loss (binder content)	Heat square-foot sample in a muffle furnace at 1,157 ±45°F until all organic binder is removed. $$\text{Binder }\% = \frac{\text{original weight} - \text{final weight}}{\text{original weight}} \times 100$$
Tensile strength	See Fig. 30-6, which shows machine set up to test tensile strength of square-foot samples.
Solubility in styrene monomer	See Fig. 30-7, which shows test for solubility of mat binder to determine wet-handling features when impregnated with polyester. Note elapsed time from immersion to break.
Quality	With aid of a fluorescent panel box, visually examine mat for holes, thin spots, tangles, fluffy areas, dirt, or excess binder concentrations.
Roving:	
End count	Cause a length of roving to separate into individual strands by shaking, and count manually. End counts of +4, −0 are typical for 60-end rovings and higher.
Ribbonization	Visually check the integrity of the roving bundle for strand-to-strand bond.
Catenary	Check uneven lengths of individual strands by suspending 50 ft of roving between uprights. Ripple the roving between thumb and forefinger until strands separate. Maximum distance of 0.50 in. is permitted from uppermost and lowest strands, measured at the midway point.

*Complete test details are found in ASTM Standards D2408 to D2410 and D2660.

absence of any tack (dry tack) is preferred for ease of stacking. For lay-ups on vertical or inclined planes, medium to high tack is specified so that plies remain where placed. As with drape, definition of tack is a matter of agreement between purchaser and coater.

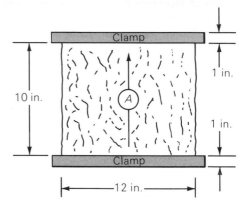

30-6 Setup to test tensile strength of glass mat. *A* arrow indicates lengthwise direction of roll of mat.

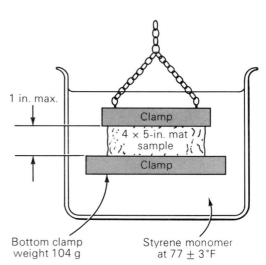

30-7 Test setup for mat-binder solubility.

TABLE 30-4 Resin content vs. laminate properties

Excessively low	Excessively high
High water absorption	Low mechanical strengths
Poor chemical resistance	Higher manufacturing costs
Poor surface appearance	Higher shrinkage
Inferior electrical insulation	
Rougher machined surfaces	

Resin Flow   Controlling resin flow is necessary to insure that the laminate contains the specified resin content. In press laminating, flow is minimal (0.5 to 5 percent) when pressure up to 1,500 psi is available. Lagging, vacuum-bag, low-pressure molding and autoclave methods require higher flow conditions to aid in air removal. Flow continues until resin gels or precures; flow is contingent upon volatile content, resin solids, and degree of resin advancement. Heat-transfer rate, thickness of lay-up, and catalyst systems must also be considered.

Procedure: Bias-cut specimens measuring 4×4 in. are stacked to a weight approximating 20 g. These are placed between sheets of release paper (cellophane or other) assembled between thin caul plates and cured for 15 min at pressures from 15 to 1,000 psi. Cure temperatures are 275°F for polyesters, 325°F for phenolics, and 350°F for epoxies. Upon cooling, the laminate is deflashed by scraping carefully to avoid removing any fibers, and then it is reweighed.

Calculation: Percent resin flow $- \dfrac{\text{orig. wt} - \text{final wt}}{\text{orig. wt}} \times 100$

Volatile content   Percentage of residual solvent is controlled in amounts to 20 percent. Volatiles influence flow by retarding cure until they are eliminated. This, in turn, governs flow at specific temperature-pressure arrangements, as well as contributing to tack and drape.

Procedure: A bias-cut sample is weighed prior to suspending it for 10 min in an air-circulating oven at temperatures prescribed above for flow.

Calculation :   Percent volatiles $= \dfrac{\text{orig. wt} - \text{final wt}}{\text{orig. wt}} \times 100$

Resin-solids content   Prepreg manufacturers and users are interested in the amount of resin solids contained within the web. Fabric-thickness variations or inconsistencies in the coating operation result in out-of-specification yardage.

Procedure: A sample from which volatiles have been completely removed is placed in a muffle furnace heated to 1050°F for 30 min or until resin has carbonized and disappeared.

Calculation (for fibrous glass webs) :

Percent resin solids $= \dfrac{\text{wt of devoled sample} - \text{wt after burn-off}}{\text{wt of devoled sample}} \times 100$

For high-silica fibers, a muffle temperature of 1550°F is used for resin burn-off. For asbestos, burn-off is conducted at 1550°F, and calculation is changed to include a water-loss factor as the asbestos is pyrolyzed.

Wet Resin Pickup   The total of volatiles plus resin solids contained in the impregnated material provides a further check on coating uniformity. Procedures are similar to those used to determine solids content, except that the volatile test is bypassed.

Procedure :   Specimens 4 × 4 in. are cut on a bias from across the roll width, weighed, and placed directly in the muffle furnace until volatiles and resin solids are burned away.

Calculation :   Percent wet resin pickup $= \dfrac{\text{orig. wt} - \text{burn-off wt}}{\text{orig. wt}} \times 100$

Combustible fabrics such as nylon, Orlon, or Dacron[1] are immersed in a suitable solvent until all resin is removed. The original 4 × 4 in. specimens are weighed before immersion and after forced air drying.

Calculation :   Percent wet resin pickup $= \dfrac{\text{orig. wt} - \text{final wt}}{\text{orig. wt}} \times 100$

Final resin content   Cured laminates are tested for resin to reinforcement ratio by burn-off tests conducted on any suitable size, usually 2 × 2 in. Obviously a laminated product cannot be ruined by cutting test pieces from it. To provide test coupons, representative areas in the trim sections are selected. When practical, a special test panel is made at the same time and under identical conditions as the production part. Burn-off is conducted under the same conditions as for solids content of prepreg.

Calculation (for glass, high silica, quartz) :

Percent resin content $= \dfrac{\text{orig. wt} - \text{final wt}}{\text{orig. wt}} \times 100$

Calculation (for asbestos) :

Percent resin content $= \dfrac{\text{orig. wt} - \text{final wt} \times 1.16}{\text{orig. wt}} \times 100$

Tests for thermoplastic materials

Processors of thermoplastic resins are not required to perform tests on incoming raw materials to the extent that compression molders and laminators do. Reliance on manufacturers' data suffices for the most part since chemical reactions are not essential to the conversion from pellet or sheet to finished item. There are exceptions, as in viscosity controls and plasticizer content of vinyl plastisols for coating, dipping, or rotomolding. When in-plant compounding becomes more widespread, quality control methods will expand in scope.

Melt index   All thermoplastics are sensitive to minor temperature changes in a

[1] Orlon and Dacron are Du Pont registered trade names for acrylic and polyester fibers.

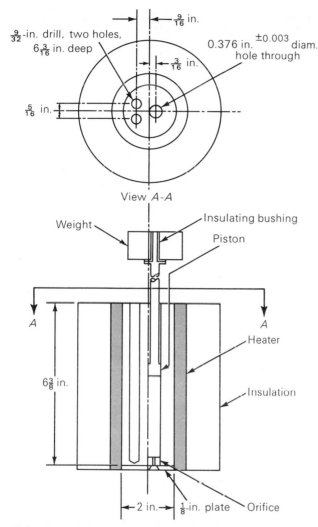

30-8 General arrangement of extrusion plastometer.

narrow range; these changes affect flow. Flow, in turn, affects machine output and part quality. One useful test[1] for comparing flow rates is conducted by using an extrusion plastometer. When evaluating polyethylene resins, it is referred to as melt index; for all others, the value obtained is designated as flow rate.

Procedure: A sample weighing 3.5 to 4.0 g is charged to a preheated cylinder and a piston is placed in position. Extrudate is cut off after 5 min, again 1 min after the start, then discarded. The next portion is collected and weighed (Table 30-5), and

[1] This is ASTM test method D1238-65T.

TABLE 30-5   Standard flow-rate conditions

Flow range, g/10 min	Time interval, min	Factor to obtain flow rate, g in 10 min
0.15– 1.0	6	1.67
1.0 – 3.5	3	3.33
3.5 –10.0	1	10.00
10.0 –25.0	0.5	20.00

flow rate or melt index is obtained by multiplying weight of extrudate by the appropriate factor.

Temperatures and piston weights are adjusted for different grades of one resin. For polyethylene the most common test condition is 374°F (190°C) with 2,160-g piston weight.

Melt-index determinations are primarily for classification of resin types, but they are useful in segregating unidentifiable scrap for purposes of reprocessing. The values obtained by laboratory analysis may not correlate with the flow performance in production. Two polyethylenes, made by different processes, may be completely different in their response to heat and pressure conditions in the extrusion or molding process. For this reason, it is not prudent to mix resins based solely on the melt-index rating. Melt viscosity, shear rate, and frictional drag can change drastically with changes in the machine. Cooling and shrinkage rates between two blended materials may vary, thus leading to stress formation, warping, or future cracking.

These are only a few of the hundreds of tests that serve to control quality of products and provide the basis for comparing materials. As new materials and processes are developed, or new markets explored, additional tests will be standardized.

GLOSSARY

Bias cut   Material cut at 45 deg from the normal weave pattern.

Caul plates   Metal plates between which laminates are pressed.

Desiccator   Glass vessel that has an airtight seal to prevent moisture absorption by products placed in it.

Muffle furnace   Electrically heated furnace capable of attaining temperatures of several thousand degrees. Oxygen is excluded in order to prevent combustion.

Resin solids (solids content)   The amount of resin deposited onto a web of roving, mat, or fabric.

Selvedge   The edge of woven fabric created when the shuttle forms loops as it traverses to weave fill yarns.

Starved area   That portion of a laminate which has little or no resin in support of the reinforcement.

Void   Empty space within a molded or laminated product, usually as a result of trapped volatile substances.

Volatile content   The percentage of solvent (water, alcohol, acetone) remaining in the resin portion of a preimpregnated material.

Wet resin pickup   The sum of the resin solids plus volatile content in a prepreg, expressed as a percentage of the total weight.

REVIEW QUESTIONS

1   Why is a greater amount of in-plant testing required for thermosetting plastics than for thermoplastics?

2   How are molding compounds checked for percentage of "fines"?

3   What benefits are derived from using a material with a low specific gravity?

4   What other term is used for bulk factor?

5   Why is a knowledge of bulk factor important?

6   Why are comparisons of flow rates important to a molder?

7   What does plasticity mean?

8   How is the flow rate of polyester and DAP resins adjusted?

9   Explain how fillers affect the flow of molding compounds.

10   What does a slow cup-closing rate tell a molder about mold-filling properties?

11   What other tests (besides cup-mold tests) are used to determine flow?

12   What is meant by the term *solids content*?

13   What does high viscosity in a liquid indicate?

14   What procedures are used to determine viscosity?

15   Of what value is a gel-time test?

16   What type of molded or laminated products requires a postcure? What happens to the resin during the postcure period?

17   What test is performed to determine the extent of cure for phenolic products?

18   How does the test of cure for diallyl phthalate differ from that used for phenolic?

19   What is the purpose of a Barcol test?

20   How does a molder compensate for shrinkage in order to mold parts to close dimensional tolerance?

21   What tests are performed to determine if a fabric meets specifications?

22   What is catenary?

23   What is prepreg?

24   Why is control of resin flow in prepreg critical?

25   What influence does residual volatile content have on flow?

26   How is resin solids content determined?

27   Of what value is a melt-index test?

28   A thermosetting molded product with a volume of 200 cc weighs 264 g. What is the specific gravity of the item? Which material (phenolic, urea, or melamine) has this specific gravity?

29   A thermoplastic product weighs 70.4 g and contains 80-cc volume. What is the specific gravity? Which material (polystyrene, polypropylene, or polyethylene) has this specific gravity?

30   A flexible thermoplastic weighs 425.25 g and has a volume of 315 cc. Which material is it?

MECHANICAL PROPERTIES

31

When a new product is considered for the market, studies are undertaken to select the best possible material. Questions must be answered as to how the product will perform under mild or rugged conditions of service. Tests are contrived to approximate as nearly as possible the actual service environment. The purpose of mechanical tests is to learn how a material will react when forces are applied.

To provide a basis for understanding mechanical properties (see, too, Table 31-2 at end of chapter), it becomes necessary to introduce terms that are involved in testing procedures and calculations:

Strain: The amount of deformation that takes place as a material yields under an applied force (load). This value is expressed in inches per inch.

Stress: The force applied to the test specimen, expressed in pounds per square inch or kilograms per square centimeter.

Elastic limit: The strain point at which material under stress can still recover to its original dimensions if stress is removed.

Permanent set: The permanent change in shape (deformation) that occurs beyond the elastic limit, but before ultimate failure.

These factors come into play in most types of mechanical tests and should be recognized as specific tests are described. For a test to have value and provide a basis of comparison between materials and products, it is imperative that it be conducted under identical conditions by each laboratory (see also Table 31-2). One laboratory device, the universal testing machine, so named because it can be used for many kinds of mechanical tests, is to be found in every well-equipped laboratory. A typical machine shown in Fig. 31-1 is capable of applying force at a constant rate, recording the load throughout the test and plotting the strain on a chart.

TENSILE STRENGTH

The ability of a material to withstand forces tending to pull it apart is called tensile strength. Fishing line, tote boxes, luggage handles, and pressure vessels are examples of applications of tensile stress.

Sample preparation

Although tensile tests can be conducted on strips, monofilaments, rods, and tubes, standard shapes (called dumbbell or dog bone) are preferred, as designated by ASTM standard D638-67T.

Care must be exercised in preparing specimens to eliminate scratches, nicks, or irregularities. Smooth surfaces prevent stress concentration and premature failure. Molded or cast samples are used, as are pieces sawed, milled, routed, or die-cut from

31-1 Specimen undergoing tensile testing. (*Reprinted by permission of Composites Div., Ferro Corp.*)

sheet stock. Fabric laminates are tested in warp, fill, and bias directions. Extruded sheet is tested in both machine and transverse directions. Tensile tests can also provide information on the strength of weld areas in injection- or transfer-molded items. A minimum of five specimens should be tested, discarding and replacing any that prove below or above the average.

　　　　Speed of testing is determined by the rate at which the fixed and movable crossheads separate when the testing machine is "on idle," that is, no specimen is

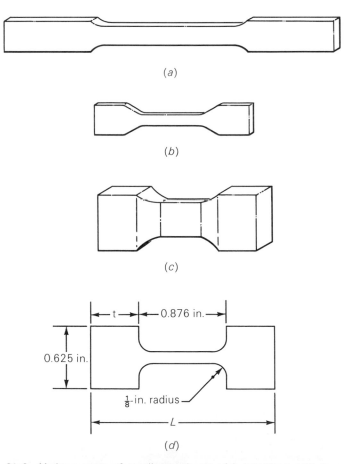

31-2　Various types of tensile specimens. (*a*) Injection molded or machined from cast, extruded, or laminated sheet; (*b*) die-cut from thin (0.16 in. or less) sheet or film; (*c*) compression-molded specimen; and (*d*) suitable for compression- or injection-molded thermoplastics or thermosets; or die-cut, or machined from sheet of film. (Tab length *t* should be a minimum of 0.312 in. Overall length *L* should be a minimum of 1.50 in.)

being stressed. Depending on the material and rate of straining desired, speeds are variable from 0.01 to 5.0 in./min. When speed is unspecified, 0.04 to 0.05 is used.

Procedure: Specimen is measured for width and thickness of the "necked-down" portion to the nearest 0.001 in. and clamped in the jaws perpendicular to the machine base.

Solution:

$$\text{Ultimate tensile strength (psi)} = \frac{(L) \text{ load to cause failure (lb)}}{(A) \text{ min. cross-sectional area (sq in.)}}$$

Sample problem: A tensile specimen measuring 0.495 in. wide and 0.060 in. thick fails at 240-lb stress.

Solution: Ultimate tensile strength $= \dfrac{L}{A} = \dfrac{240}{0.495 \times 0.06} = \dfrac{240}{0.0297} = 8{,}080$ psi

Practice problems: A test sample of polystyrene measures 0.500 in. wide × 0.125 in. thick. If failure occurs at 620 lb, what is the ultimate tensile strength?

Answer: 9,920 psi

A phenolic-glass laminate measures 0.509 in. wide × 0.119 in. thick. If failure occurs at 3,355 lb, what is the ultimate tensile strength?   *Answer*:   55,000 psi

COMPRESSIVE STRENGTH (ASTM D695–63T)

The ability of a material to resist forces that tend to crush or compress it is called compressive strength. Examples of products that are subjected to compressive forces are chair seats, furniture legs, toys, and washing-machine agitators.

The test piece is a cylinder or prism, with diameter or principal width half its length to prevent buckling. Specimens should be carefully made to absorb stress uniformly. Particular attention must be paid to the direction of load according to the way in which a sample was molded or laminated. In compression testing rigid materials, injury from flying fragments can occur if the machine is not equipped with a wire enclosure or heavy plate.

Crosshead speed is fixed at 0.05 in./min.

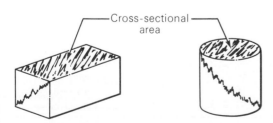

Cross-sectional area

Diagonal fracture indicates proper break   **31-3**   Compressive test specimens.

Procedure: Specimen is measured for width and thickness to the nearest 0.001 in. and located on the base of the testing unit.

Solution:

$$\text{Ultimate compressive strength (psi)} = \frac{(L) \text{ load to cause failure (lb)}}{(A) \text{ min. cross-sectional area (sq in.)}}$$

Sample problem: A compressive specimen measuring 0.490 in. wide × 0.492 in. thick × 1.020 in. long fails at 3,000 lb.

Solution:

$$\text{Ultimate compressive strength} = \frac{3,000}{0.490 \times 0.492} = \frac{3000}{0.241} = 12,448 \text{ p.s.i.}$$

Practice problem: A test sample of molded phenolic measures 0.50 × 0.50 × 0.75 in. long. If failure occurs under a compressive load of 8,000 lb, what is the compressive strength?   *Answer*:   32,000 psi

FLEXURAL STRENGTH (ASTM D790-66)

The ability of a material to withstand bending forces is called flexural or transverse strength. Fishing rods, diving boards, benches, and archery bows undergo flexural stresses. Flexural stressing involves a combination of tensile and compressive forces. Specimens are loaded as a simple beam supported near each end, with load applied at center. Specimens may be molded or cast, or machined from laminates or extrusions. For rigid specimens, crosshead speed is 0.20 in./min.

Calculation: Ultimate flexural strength (psi) $= \dfrac{3\,pl}{2\,bd^2}$

where p = load, in pounds, to cause failure
$\quad l$ = span
$\quad b$ = breadth (width) of specimen
$\quad d$ = depth (thickness) of specimen
Span distance is fixed at a ratio of 16 times the sample thickness.

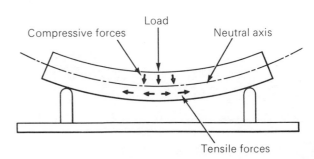

31-4   Forces involved in bending a simple beam.

Sample problem: A test bar measuring 0.496 × 0.125 × 5 in. long fails at 120-lb flexural stress. Calculate flexural strength.

Solution: Span distance = 16 × 0.125 = 2 in.

$$\text{Ultimate flexural strength} = \frac{3 \times 120 \times 2}{2 \times 0.496 \times 0.125^2}$$

$$= \frac{720}{0.992 \times 0.0156}$$

$$= \frac{720}{0.015} = 48{,}000 \text{ psi}$$

Practice problem: An acrylic test bar measuring 0.125 in. thick × 0.500 in. wide fails in flexure at 40 lb. What is the ultimate flexural strength? *Answer*: 15,384 psi

IMPACT STRENGTH

The ability of a material to withstand being dropped or struck a sharp blow is termed impact strength. Safety helmets, luggage, auto bodies, boat hulls, and dinnerware are typical products that must have good impact strength.

Impact tests standardized under ASTM D256-56 prove useful for testing bar stock molded by any method or machined from cast or extruded resins. During the test, the energy expended by the pendulum is measurable, based on the energy

31-5 Conducting a flexural test. (*Reprinted by permission of Composites Div., Ferro Corp.*)

absorbed by the specimen. Impact strength is expressed as foot-pounds to cause failure. Thus, if a glass laminate is rated at 18 ft-lb/in. of width, it can withstand shock equivalent to an 18-lb weight dropped from a height of 1 ft, or 1 lb dropped from 18 ft.

Izod impact test (ASTM D256–56)

The test bar is clamped as a vertical cantilever beam. Dimensions of the bar are standard except that thickness cannot exceed 0.50 in. A notch is milled across the narrowest dimension to provide an area of stress concentration and initiate failure. The specimen is positioned with the notch on the striking surface. A direct reading is taken from the machine, based on the amount of follow-through by the pendulum. Impact value is derived by dividing the energy expended in foot-pounds by the notch dimension.

Charpy impact test (ASTM D256–56)

The test bar is placed horizontally against end supports and struck in the center. Specimens can be unnotched. If notches are used, as in Fig. 31-6, the notched side is positioned opposite the impact side.

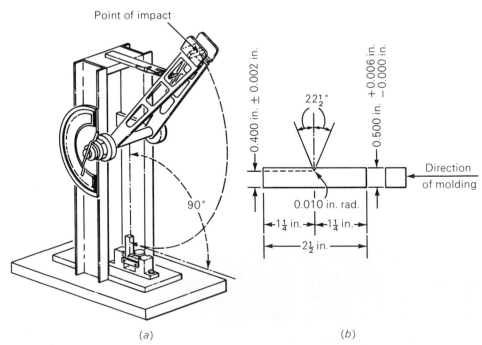

(a)   (b)

31-6   (a) Cantilever beam (izod type) impact machine; (b) dimensions of test specimen.

Miscellaneous impact tests

Other tests have been devised for situations where irregular shapes or large structures make it necessary to test the actual abuse of a product. A boat manufacturer maneuvers the craft against a dock or over submerged logs. A helmet maker uses a falling steel ball dropped from successively higher points. Other shapes such as V point, chisel edge, or square may simulate service conditions when dropped. Housings for portable tools and appliance handles are fastened to a cord attached to a concrete wall, and then allowed to fall against the wall.

Elongation (ASTM D638–67T)

The amount of stretch a material can undergo before taking a permanent set or failing otherwise is called elongation. Knowledge of this property is important to producers of articles such as blow-molded packaging, trash containers, and other flexible items.

Cut strips, extruded filament, and tensile specimens are used for testing. Gauge lines are marked 1 or 2 in. apart and their extension measured at various stresses during a tensile test until failure occurs.

Calculation :

$$\% \text{ elongation} \quad = \frac{\text{final gauge length} - \text{original gauge length}}{\text{original gauge length}} \times 100$$

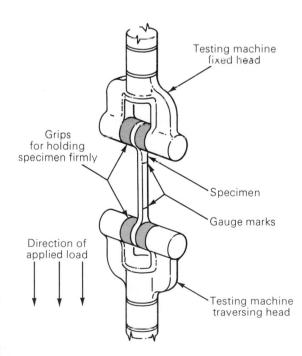

Testing machine
fixed head

Grips
for holding
specimen firmly

Specimen

Gauge marks

Direction of
applied load

Testing machine
traversing head

31-7   Setup for elongation test.

Sample problem: A sample of linear polyethylene elongates to 5.50 in. from an original gauge length of 2.00 in. Calculate elongation.

Solution: $\dfrac{5.50 - 2.00}{2.00} = \dfrac{3.50}{2.00} = 1.75 \times 100 = 175\%$

Plastics that stretch a minimum of 200 percent are classified as elastomers. Those that stretch 600 percent or more and recover are termed elastic. Materials such as polyolefins are classified as flexible, while others such as urethanes can be elastic or flexible.

Practice problem: A calendered vinyl sheet stretches from an original gauge distance of 2 to 4.50 in. between gauge lines. What is the percentage of elongation?

Answer: 125 percent

SHEAR STRENGTH

The ability of a material to resist shearing when one portion is moved from a stationary section is calculated as shear value. Gear teeth, threads, hammer handles, and glue lines exemplify applications of shear stresses.

Shear strength is expressed in pounds per square inch calculated on the basis of the load to cause failure in the area of the sheared section. ASTM standard D732-46 utilizes a punch type shear fixture for testing flat specimens from 0.005 to 0.500-in. thick at a speed of 0.05 in./min.

Calculation:

$$\text{Shear strength (psi)} = \frac{\text{pounds to cause failure } (L)}{\text{area of sheared edge } (A)}$$

(Area of sheared edge = circumference × thickness.)

Another shear test, called *interlaminar shear*, is widely used to evaluate the degree of adhesion between plies of laminates. A hardened steel ball, 0.50 in. in diameter, is forced into the edge of a laminate.

A direct load reading is observed as pounds to delaminate. Area is not considered because of spring back when the ball is removed.

MODULUS OF ELASTICITY[1] (STIFFNESS)

The proportion of deflection, deformation, or elongation (strain) to applied stress is known as modulus. There is a law that states the relation between the amount of stress applied to a body and the amount of strain produced:

Hooke's law: The amount of change in the shape of an elastic body is directly proportional to the applied force, providing the elastic limit is not exceeded. Stated

[1] Methods for determination of modulus are outlined in ASTM test methods: D695–63T (tensile modulus), D638–67T (compression modulus), and D790–66T (flexural modulus).

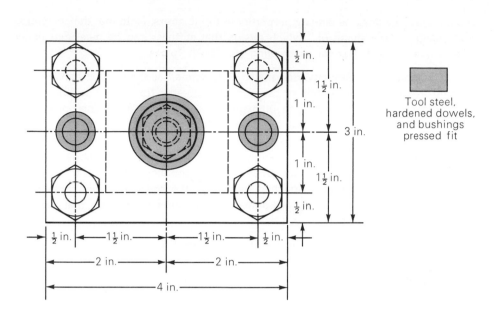

Tool steel,
hardened dowels,
and bushings
pressed fit

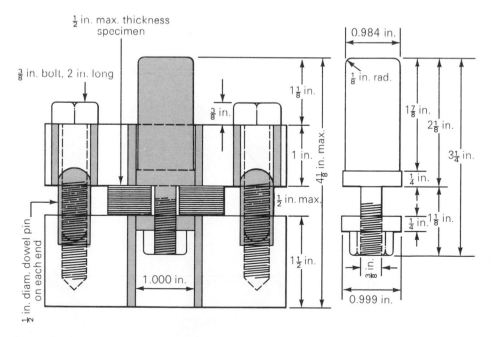

31-8   ASTM punch to test shear strength.

another way, the strain is directly proportional to the stress. Note that the key words are "change in the shape of," which means that stiffness can be measured under tensile, compression, flexural, or twisting (torsional) forces.

The modulus of elasticity, designated E, is defined as the relation between the amount of compression, stretch, or bending of a material (in inches) and the stress (per square inch). As in testing to ultimate failure, stress is determined at various intervals within the elastic limit by use of the appropriate formula.

Figure 31-7 illustrates a test to determine modulus in tension (tensile modulus). At any load increment, for instance every 50 lb, the amount of stretch is noted or recorded automatically on a graph. Strain gauges containing transducers can also be used for greater accuracy. Attached directly to the specimen, they convert the strain resulting from molecular realignment to electronic impulses that activate recording devices.

Sample problem: A polycarbonate material is tested for ultimate tensile strength and tensile modulus. The specimen is 0.500 in. wide and 0.130 in. thick (0.065-sq in. area). Strain-gauge readings are plotted for 50-lb load increments. The stress-strain curve is shown in Fig. 31-9.

Select any point on the curve, for example at load 250 lb. At this increment,

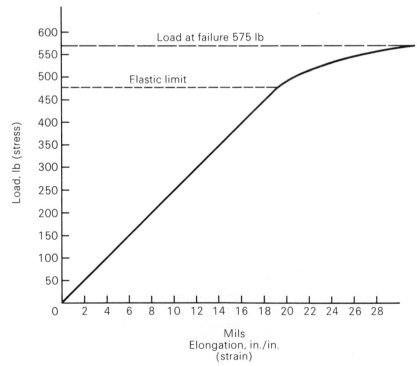

31-9   Stress-strain curve.

stress equals 250/0.065 or 3,847-psi tensile stress. Note that elongation (strain) is 0.010 in. Therefore tensile modulus (E) is 3847/0.010 or 384,700 psi.

HARDNESS

Hardness can be considered in several ways. One kind has to do with the manner in which a material resists penetration when the entire mass is involved. Another kind involves only surface characteristics and would include scratch or abrasion resistance.

Rockwell hardness

A Rockwell number is directly related to the resistance of a rigid material to penetration by a steel ball. Rockwell testers register a number based on the amount of impingement as a load is applied; higher numbers indicate harder materials. In this test, the scale symbol represents ball-indenter size and load (see Table 31-1).

The scale is a progression of increasing severity, and it is desirable to use the smallest ball and highest load factor. The proper scale is found by trial, with readings over 125 indicating that the next smaller ball or higher load should be used.

Durometer

Soft plastics and rubbers are tested for resistance to penetration by the Shore hardness method, which is similar to Barcol hardness for thermosets. As a quality control test, durometer readings are useful in checking calendered, extruded, or dipped vinyl products, and other plasticized resins.

Scratch resistance

An item such as a phonograph record can be tested by means of a diamond stylus drawn across the surface at a constant speed under specified load, and then measuring the width and depth of the groove produced.

Abrasion resistance

Plastic materials used in counter tops, floor tile, sinks, and windshields are subject to the scouring action of grit and cleansers. Transparent products are measured for loss of clarity; opaque products are compared by weight or volume losses.

TABLE 31-1 Rockwell-hardness scales

Scale symbol	Major load, kg	Ball indenter (diameter)
R	60	0.5000
L	60	0.2500
M	100	0.2500
E	100	0.1250
K	150	0.1250

One type of abrasion test makes use of a Taber abraser. Test specimens 4.50 ×
1.75 in. are weighed and attached to a mounting plate. Abrasive, such as aluminum
oxide, is discharged from a hopper to the top surface of a revolving plate, passing
under the specimen, which is also rotated. Weight loss is calculated after 1,000
revolutions.

Formula :

$$\text{Vol. loss (cc)} = \frac{\text{orig. wt (g)} - \text{final wt (g)}}{\text{specific gravity}}$$

Abrasion resistance is also expressed as

$$\text{Av. thickness loss} = \frac{\text{orig. vol.} - \text{final vol.}}{\text{area of specimen}}$$

Bouncing-ball hardness test

Releasing a steel ball and noting the initial height of rebound and the total number
of bounces determines the hardness of resins that are used in applications like
coatings, mouthpieces for musical instruments, or playground surfaces.

When materials are evaluated for hardness, it should be recognized that resins
may rate as excellent in one kind of test and poor in another. The evaluator must
choose the test that most realistically simulates service.

CREEP AND COLD FLOW

Any material subjected to any kind of stress tends to relieve the stress by some sort
of molecular reorientation, which is evidenced by change of shape. Metals, ceramics,
and concrete exhibit extremely small magnitudes of creep or cold flow, while plastics
and rubbers demonstrate a more pronounced effect. Materials with high elongation
generate more of this undesirable change than rigid types.

Creep and cold flow can best be explained by referring to Fig. 31-10. The
moment stress is applied to a material, it undergoes some elastic deformation. A
polypropylene tow rope stretches, a fluoroplastic gasket compresses, or a paper-base
laminate deflects under the weight of a relay it is supporting, to mention a few examples.
Elastic deformation is depicted as occurring the moment stress is applied and continuing until the load is removed. Once relieved, there is an immediate elastic recovery
substantially equal to elastic deformation. Complete recovery is not immediate, and
the material continues to creep back toward its original size or length. If it does not
recover fully, it has undergone a permanent set, the magnitude depending on the
stress applied, the length of time, and the temperature of the plastic.

Thus creep and cold flow is defined within this frame of reference :

Creep   The rapid change of dimensions in a material that develops immediately
after it is stressed. During the creep period, *the rate* of deformation continues to
change—rapidly at first, then decelerating.

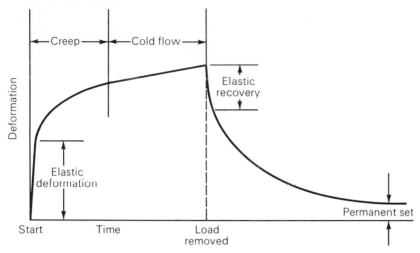

31-10   Diagram illustrating creep and cold flow.

Cold flow   The change in dimensions occurring in a material during application of stress, the rate of which deformation is substantially constant.

The incorporation of fibrous reinforcements improves dimensional stability, and addition of plasticizers or selection of low-molecular-weight resin grades generally result in higher orders of creep and cold flow.

FATIGUE FAILURE

Plastics, as well as all other materials, can fail when subjected repeatedly to stresses below their elastic limits. Vibrations from sonic or mechanical sources induce low-level stresses at a high rate to cause failures.

The manufacturer of a foam cushion may conduct a test where the pad is exposed to 100,000 compression cycles to establish information on cell-wall collapse. Fishing rods are examined for breaks after repeated bending through a 360-deg arc. Coated fabrics for draperies are exposed to cycles of folding in order to determine adhesion to the base fabric.

Endurance tests are designed to provide data for guarantees of the life expectancy of the product. Aside from visual observations, other methods of evaluating fatigue compare tensile, flexural, or other mechanical properties after the prescribed cycles have been endured, with the original values, even though failure has not occurred.

A prominent manufacturer of testing machines has the slogan "One test is worth a thousand opinions." This might be paraphrased to "One reliable and reproducible test is worth a thousand guesses." The point here is that tests performed by sophisticated machines that record all data are more reliable for a series of com-

TABLE 31-2 Typical mechanical properties of plastics

Plastic	Tensile strength, psi	Compressive strength, psi	Flexural strength, psi	Tensile modulus, psi	Impact strength, ft/lb	Hardness, Rockwell or Shore
Thermoplastics:						
ABS	4–8,000	7–12,000	5–13,000	3.5×10^5	1.0–10	R75–115
Acetal	10,000	16–18,000	14,000	5.2×10^5	1.2–2.3	M78–R120
Acrylic	5–11,000	4–18,000	8–19,000	4.5×10^5	0.3–0.5	M70–R120
Cellulose acetate	4–9,000	2–36,000	6–16,000	$0.65–6.0 \times 10^5$	0.4–5.2	R34–R120
Cellulose acetate butyrate	2–7,000	2–22,000	2–9,000	$0.50–2.5 \times 10^5$	0.8–6.3	R30–R115
Fluoroplastics:						
CTFE	4–6,000	4–7,400	7–9,000	$1.5–3.0 \times 10^5$	2.6	R75–R95
TFE	2–5,000	1,700	...	0.58×10^5	3.0	D50–D65
FEP	3,000	...	...	0.5×10^5	No break	R25
Nylon	7–14,000	7–13,000	No break	$1.1–4.5 \times 10^5$	1–5	R103–R120
Polycarbonate	8–9,500	12,000	13,500	3.5×10^5	12–17.5	M70–M118
Glass filled	12–20,000	16–19,000	17–30,000	$10–18 \times 10^5$	4.5–6.5	M88–M95
Polyethylenes:						
Low density	1–2,000	...	...	$0.14–0.38 \times 10^5$	No break	D41–D46
Medium density	1–3,500	...	6,000	$0.25–0.55 \times 10^5$	0.5–16	D50–D60
High density	3–5,500	2–3,600	10,000	3.8×10^5	0.8–20	D60–D70
Polyphenylene oxide	11,000	13,000	15,000	$1.6–2.2 \times 10^5$	1.5–1.9	R118–R120
Polypropylene	4–5,500	5–8,000	6–8,000	$1.6–2.2 \times 10^5$	0.5–2.0	R85–R110
Polystyrene:						
General purpose	5–12,000	11–16,000	9–14,000	$4–6 \times 10^5$	0.3	M65–M80
Impact	1–7,000	4–9,000	5–10,000	$1.5–5.0 \times 10^5$	0.5–11.0	R50–R100
Glass filled	9–15,000	13–18,000	10–20,000	$8–13 \times 10^5$	0.4–5.0	M70–M95
Polyvinyl chloride:						
Rigid	5–9,000	8–13,000	10–16,000	$3.5–6 \times 10^5$	0.4–20	D65–85
Flexible	1–4,000	900–1,700	...	...	Varies	A–40–100

Material						
Thermosets:						
Epoxy:						
Cast resin	4–13,000	15–25,000	13–21,000	3.5×10^5	0.2–5.0	M85–M120
Glass laminates	35–100,000	35–90,000	40–100,000	$20{-}35 \times 10^5$	5–25	M105–M120
Melamine:						
Molded:						
Cellulose filler	7–13,000	40–45,000	10–16,000	11×10^5	0.2–0.35	M115–M125
Glass filler	5–10,000	20–35,000	15–23,000	24×10^5	4.0–4.6	M115–M125
Laminated						
Paper base	10–25,000	30–48,000	12–20,000	. . .	0.3–1.5	M110–M125
Glass base	25–40,000	25–85,000	35–85,000	$20{-}25 \times 10^5$	5–15	M115–M125
Phenolic:						
Molded:						
General purpose	6–10,000	22–36,000	8–12,000	$8{-}17 \times 10^5$	0.2–0.6	E64–E95
Asbestos filled	5–7,500	20–35,000	8–14,000	$10{-}30 \times 10^5$	0.3–3.5	E75–E97
Mica filled	5–7,000	25–30,000	8–12,000	$25{-}50 \times 10^5$	0.2–0.4	E85
Glass filled	5–18,000	17–70,000	10–60,000	33×10^5	0.3–18.0	E54–E101
Laminated:						
Paper base	11–16,000*	25–34,000†	18–21,000	. . .	0.5–0.6	M75–M118
Glass base	13–37,000*	37–70,000*	23–72,000	. . .	12–15*	M70–M125
Polyester:						
Cast resin (rigid)	6–13,000	13–36,000	3–23,000	$3.0{-}6.4 \times 10^5$	0.2–0.4	M70–M115
Glass laminates	30–40,000	39–53,000	35–90,000‡	. . .	25–32*	M70–M125
Urea	5–13,000	25–45,000	10–18,000	$10{-}15 \times 10^5$	0.2–0.4	M110–M120

*Tested edgewise.
†Tested flatwise.
‡Dependent on type of reinforcement and resin content.

parative tests, with reproducibility assured. However, nonstandard, simple tests have a definite place in the scheme of things when they are meaningful and reproducible.

GLOSSARY

Bias direction   The fabric direction of 45 deg in relation to warp and fill directions (90 deg).

Crosshead   Thick steel plate to which test fixtures are attached.

Transverse direction   The width of the sheet being extruded, as opposed to the machine or takeoff direction.

REVIEW QUESTIONS

1   What is the purpose of mechanical tests?
2   Define the following terms:
   (*a*)   strain
   (*b*)   stress
   (*c*)   elastic limit
   (*d*)   permanent set
3   What is meant by tensile strength?
4   What products must exhibit high tensile strength?
5   Why is extruded sheet tested in both the machine direction (lengthwise) and the transverse (width) direction?
6   Why should tensile specimens be prepared carefully?
7   What is meant by compressive strength?
8   What type of products must exhibit good compressive strength?
9   What danger exists in conducting a compressive test?
10   What is meant by flexural strength?
11   Which forces are combined in a flexural test when it is conducted on a beam?
12   Name the two types of impact tests conducted on standard bar stock.
13   What is the difference between an elastomer and an elastic material?
14   What two tests can be conducted to determine the shear strength of a laminate?
15   Define modulus of elasticity.
16   What property would be involved as a needle is dragged across a phonograph record?
17   What property is determined by a Rockwell test?
18   What is the difference between creep and cold flow?
19   What causes fatigue failure?

ELECTRICAL PROPERTIES

32

The continuing expansion of the electrical and electronic industries closely parallels the growth of the plastics industry. In many ways, these giant industries serve together, with each contributing to the development of the others. This has been true since the earliest days, when phenolic resin varnishes replaced shellac as an electric insulation. The close alliances between the plastics manufacturer and the electrical trades are destined to continue. As more houses and buildings are constructed, more plastics are required for wire insulation, conduit, meters, circuit breakers, switches, and appliances. Conversely, when new electronic devices are perfected, such as computers and satellites, plastics play an important role.

When a new polymer is introduced, it is evaluated against the record of current materials. If it can be demonstrated that additional safeguards from electrical failure can be realized, or performance of an insulator can be improved under adverse conditions of heat, cold, humidity, or vibration, then a large and ready market is assured. Not infrequently a new resin development leads to a new or improved electrical device. In one instance, the advent of silicones and fluorocarbons permitted engineers to design better motors a smaller size, thus maintaining original horsepower ratings. Because these resins are superior as insulation, particularly as motors become overheated, they can be applied in thinner coatings to decrease space required for motor windings.

Plastics enjoy an enviable position in electrical applications because they offer structural advantages and improved design capability and at the same time provide manufacturing economy. Thousands of common electrical products rely on some form of plastic to fulfill many general-purpose uses, and some highly specialized problems are solved by special formulations of resins. The task is to select the correct plastic for the specified purpose, taking into account the method by which the product must be made (molding, casting, lamination, extrusion) and the cost factor (see also Table 32-3 at end of chapter).

Previously the mechanical properties of plastics were compared with those of metals, and it was stated that plastics often fall short as structural load-bearing members except on a strength-to-weight basis. As temperatures increase or decrease, this becomes more severe.

In electrical terms, metals are classified as conductors of electricity; plastics are considered as nonconductors or dielectrics. In certain instances, plastics semiconductors are made by using graphite or other conductors in highly filled resin systems. To study the electrical properties of plastics (see, too, Table 32-3 at end of chapter) it is necessary to compare them with other insulating materials, notably ceramics, mica, glass, rubber, and vulcanized fiber.

KINDS OF ELECTRICAL ENERGY

Electrical energy is transmitted in several different ways, such as radio transmission vs. wire conductors. The energy can be classified in a general way according to wave frequency, although in some instances the categories overlap.

Power-line frequencies

Energy supplied at 50 to 60 Hz is commonly used for house current, small industrial requirements, and general-purpose uses.

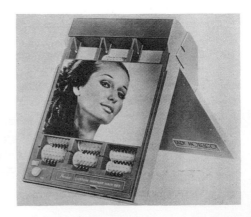

32-1 Tray for heating and storing curlers molded with Noryl ®. Selection of material is based on inherent resistance to electric current at elevated temperature. (*Reprinted by permission of General Electric Co.*)

Audio frequencies

Telephonic and television equipment operates in the audio-frequency range of 20 to 20,000 or more hertz.

Radio frequencies

Radio signals are generated in a frequency band ranging from 1,000 KHz (1 MHz) and more.

Radar frequencies

Transmission and reception of radar impulses are normally carried by wavelengths of 1,000 MHz and above.

The properties most frequently investigated to determine the relative value of an insulator at different frequencies form the basis for this chapter.

POWER FREQUENCIES

Dielectric strength

The maximum voltage an insulator can withstand before arc-through (puncture) occurs is a measure of dielectric strength. This value is derived by dividing the maximum applied voltage at failure by specimen thickness, and it is expressed in volts per mil.[1]

A single, perfect insulator that can perform under any combination of circumstances involving heat, chemicals, moisture, contamination, and stress has yet to be found. Even the best insulation cannot prevent small current leakage around or through it. As leakage progresses, heat is generated, resulting in accelerated leakage, increased heating, etc., and when voltage is great enough, failure occurs. The effects of leakage are cumulative, therefore insulation breakdown is measured with time-dependency factor in mind.

Test method ASTM D149-64 describes several procedures in applying voltage to compare dielectric performance. In the short-time test, the rate of voltage increase is set variously from 100 to 3,000 V/sec, with 500 V/sec more widely used to simulate common service conditions. Voltage at the outset is 0. Another test, called step-by-step, applies an initial voltage equal to 50 percent of the breakdown voltage as determined by short-time values. Voltage is then increased at rates agreed upon by customer and manufacturer. A typical test procedure for a glass-filled diallyl phthalate molded sample, 0.125 in. thick, would have initial voltage at:

  0 sec = 10,000 V held for 60 sec

 60 sec = 20,000 V held for 60 sec

120 sec = 40,000 V held for 60 sec

180 sec = 50,000 V and failure

[1] 1 mil = 0.001 in.

Solution: Dielectric strength (V/mil) $= \dfrac{\text{total voltage at failure}}{\text{thickness (mils)}}$

Dielectric strength $= \dfrac{50{,}000}{125} = 400$ V/mil

Values reported for dielectric strength can be misleading unless specimens tested are of comparable thickness. Dielectric strength, when expressed as volts per mil, *decreases* as specimen thickness *increases*. A 1-mil film of polycarbonate has a dielectric strength of 1,500 V/mil, while the same material 0.125 in. thick demonstrates a dielectric value of 400 to 475 V/mil. What then accounts for this seeming discrepancy common to all materials? As voltage increases to puncture the specimen, heat genera- tion causes thermal expansion and reduction of intermolecular bonds, which is more pronounced in thick sections. Thicker sections may have internal voids, occluded moisture, or poor weld areas. Undercure of thermosets is more likely in a heavy section than a thin one, leading to earlier failure.

Dielectric strength tests are important in selecting a material for a particular electrical use, but it is also a useful control tool to ascertain, nondestructively, the relative mechanical properties. There is often a close correlation between dielectric value and some mechanical property such as compressive or flexural strength. This is indicative not only of the degree of cure, density, and distribution of constituents in a nonhomogeneous compound, but also of the quality and correctness of the fabricating process.

Electrical properties are reduced by exposure to elevated temperature. One of the major drawbacks in the use of many polymers as insulators is the reduction of electrical ratings when the part operates at high temperatures. Silicones and fluoro- carbons are useful to 500°F, but many other thermoplastics are not recommended for usage over 200°F, even on an intermittent basis. Thermosets often show dielectric strengths at room temperature slightly below many thermoplastics, but they retain more of their values as temperatures increase (Table 32-1).

In addition to the normal electrical operating conditions, organizations such as Underwriter's Laboratory and the National Electrical Manufacturers' Association (NEMA) require a margin of safety. Insulation must pass proof-test voltages defined by electrical codes. In some instances, proof-test voltage is equal to 900 V plus twice the rated voltage of the electrical appliance.

TABLE 32-1   Dielectric strength, V/mil, 0.125-in. thickness

NEMA-grade laminates	25°C (77°F)	100°C (212°F)
X, XX	700	500
A	225	100
LE	500	400
CE	500	400

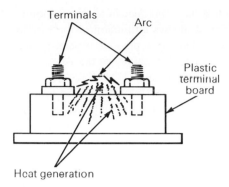

Terminals

Arc

Plastic terminal board

Heat generation

32-2 Process by which failure occurs because of arcing.

Arc resistance

Terminal boards, switch elements, and commutator segments are often exposed to arcing conditions. Arc resistance is defined as the time required for surface failure to occur in the presence of a given electrical current, which produces a high-intensity arc. An arc is generated when two conducting elements are brought into close contact, as in closing a switch. The severity of the arc is determined by the voltage across the terminal as well as the amperage. Home appliances such as vacuum sweepers; electric shavers, sewing machines, and portable power tools may arc if stalled, or when a switch opens. ASTM standard D465-61 outlines the procedure for testing arc resistance. An interruptor in the electrical circuit provides for voltage to be applied intermittently for the first 3 min, after which current is continuous but increased in severity each succeeding minute.

From Table 32-2, note that during the first minute the arc is applied for only one-eighth of that period, or 7.50 sec, allowing time for the test piece to cool and stabilize. In each succeeding sequence, heat is generated at more frequent intervals, then continuously. Materials exhibit different kinds of failure, such as softening, cracking, or melting. In such cases, arc resistance is not a time function. Some materials, notably phenolic resins, fail when the intense heat of the arc carbonizes the surface, thus creating a conductive path that renders the part useless as an

TABLE 32-2 Sequence of 1-min current steps

Step	Current, milliamps	Time cycle	Approx. rate of heat generation, watts	Total time, sec
0.125–10	10	0.25 sec on, 1.75 sec off	3	60
0.25 –10	10	0.25 sec on, 0.75 sec off	6	120
0.50 –10	10	0.25 sec on, 0.25 sec off	12	180
10	10	Continuous	24	240
20	20	Continuous	34	300
30	30	Continuous	45	360
40	40	Continuous	56	420

insulator. Phenolics containing a heat-conductive filler can sustain an arc for longer periods before carbonizing, but carbon formation is always a function of the exterior resin surface, regardless of filler. Amino and alkyd plastics excel in arc resistance, and although charred, they do not carbonize to the extent that phenolics do. As a result, electrical components do not require as frequent a replacement (Table 32-2).

The majority of plastic components in relays, starters, contactors, and terminal boards used in circuitry are made from phenolic compounds because of cost and mechanical properties. When severe arcing conditions are anticipated, alkyds, melamine, or cold-molded cement-asbestos compounds are selected.

Dielectric constant

To ensure that there is a minimum of voltage fluctuation in electrical circuits, such as is found in radio or television receivers, the property of dielectric constant is important. Dielectric constant refers to the ability of an insulator (condenser) to store electrical energy. This characteristic is involved in the design of condensers and capacitors used in equipment operating at power, audio, and radio frequencies. One of the simplest types of condensers is of parallel-plate construction.

If a condenser is the device for storing electrical energy, we can use the formula

$$\frac{\text{Energy stored in condenser, plastic as insulation}}{\text{Energy stored in condenser, air (or vacuum) as insulation}} = \text{dielectric constant}$$

The dielectric constant of air is 1.0 at any frequency. The greater the dielectric constant, the greater the capacitance (energy-storing capability).

$$\text{Capacitance of a parallel-plate condenser} = \frac{KA}{4\pi L}$$

where K = dielectric constant
π = 3.1416
A = plate area in square centimeters (1 cm² = 0.155 in.²)
L = distance separating plates in centimeters (or thickness of dielectric material)

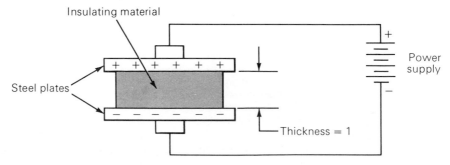

32-3   Schematic diagram for dielectric constant.

From the above equation it follows that the thinner the insulation, the greater the capacitance of the condenser. The limiting factor is, of course, the dielectric strength of the insulation.

Plastics in thin-foil form are used extensively as the insulation. In selecting the proper material, if dielectric strengths of two materials are equal, then the material having the highest dielectric constant would be the logical choice. Plastics most suitable for condensers at all frequencies are cellulose acetate, cellulose acetate butyrate, polyethylene terephthalate, polyethylene (low density), polystyrene, poly-urethane, and fluorocarbons.

Ozone resistance

A serious drawback to natural rubber insulation is its poor resistance to ozone. Ozone is created when atmospheric oxygen breaks down in the presence of high voltage. Because ozone is a powerful oxidizing agent, insulations must be selected for stability in this situation. Polyvinyl chloride, fluorocarbons, and silicone elastomers have found wide usage as wire and cable insulation because of ozone resistance.

AUDIO FREQUENCIES

Insulation resistance

Insulation resistance is a characteristic that may be classified as two distinct effects:

Volume resistance   This is the resistance to leakage of electrical current through a specific volume of material.

Surface resistance   This is the resistance to leakage of electrical current on the surface of an insulator.

In insulation resistance there is a factor that is somewhat controllable by the molder or fabricator of the plastic material. If the surface of a product is dirty, for example, because of a film of oil or release agent, leakage of current can be greater. Surface resistivity is improved when highly polished molds impart a high luster to insulators, which in turn provides cleaner surfaces.

Even though insulation resistance is of particular importance in audio- and radio-frequency applications, it cannot be overlooked in designing electrical appliances and power tools operating at power frequencies. High-current leakage could produce a dangerous electrical shock when insulation resistance is inadequate.

ASTM standard D257-66 outlines procedures for determination of direct-current insulation resistance. In many cases the molded part can be used as the test piece. Flat sheet, tapes, tubing, and laminates are also tested.

When practical, electrical leakage is used as a control or research device to measure moisture content, degree of cure, percentage of voids, or deterioration caused by chemical or other environmental factors. Direct measurements are taken with a galvanometer, voltmeter, or other methods.

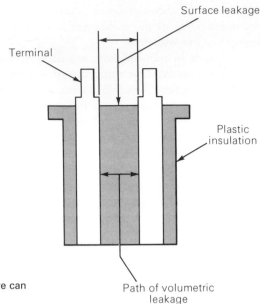

32-4 Two ways by which electrical failure can occur.

RADIO FREQUENCIES

Loss and power factors (dissipation)

In addition to electrical properties of interest at audio and power frequencies, a knowledge of additional properties is important to designers of high-fidelity equipment. The expression *low loss* is sometimes used to describe an insulating characteristic. The loss factor is the product of the dielectric constant and the power factor. The power factor becomes important in electrical circuits operating at high or ultrahigh frequencies found in radio, television, and other wireless equipment. For example, if a television broadcast once tuned in is to stay tuned in, then the constancy of the

32-5 Terminal blocks must exhibit good arc resistance, volume, and surface resistance under a variety of operational conditions. Material is glass-fortified Noryl ®. (*Reprinted by permission of General Electric Co.*)

circuit must be maintained. Constancy is sustained under these conditions when:

1 Changes in capacitance are negligible within the operating temperature range of the insulation. If the power factor or loss factor is high, then insulation becomes heated, altering initial values.

2 Electrical values do not change with changes in humidity.

3 The dimensional stability of the insulation can be depended upon to prevent changes in spacing between components.

One or more of these conditions may be present, as in walkie-talkie units used in military operations. Power factor determines the power loss through the insulation, which appears as heat. As mentioned in the discussion of dielectric strength, the loss of current (power) is cumulative. When insulation is heated, power is lost through it, leading to further heating, increasing loss, and so on.

Among nonplastic materials, fused quartz and mica demonstrate minimum loss at radio frequencies. Low-loss plastics include polystyrene, methyl pentene, polyphenyl oxide, polypropylene, and fluorocarbons. Because of low cost and moisture resistance, polystyrene is widely used for insulating terminal strips, antenna bushings, coil frames, and condensers when high precision is required. If high temperature is to be encountered, mica-filled phenolics are molded to required shapes for coil forms, bushings, and spacers. For temperature environments above the limitations of organic plastics, ceramic or cold-molded refractory products (cement-asbestos) should be specified.

Standard procedures exist for testing plastics to determine loss and power factors. Accurate measurement requires a high degree of skill in the use of special electrical apparatus.

It is characteristic of any organic insulation that resistance falls off appreciably with temperature. Consequently, tests should be performed at temperatures of operation. Different materials vary widely regarding insulation resistance at elevated temperatures. Most plastics are replaced by inorganic insulators at temperatures above 300°F. In selecting a plastic for insulation, the designer must not only be aware of the electrical limitations, but he must also consider the mechanical forces that can affect the useful life of the equipment. Relays, starters, and housings are subjected to impact, flexural, and compressive loads that can have an adverse effect over a period of time.

GLOSSARY

Cold-molded plastics   Products made by compression molding at room temperature, followed by baking. Binders consist of phenolic resin, bitumen, or cement. Fillers are asbestos or other inorganic material.

Galvanometer   An electrical instrument that indicates the presence of small currents or is used to measure them.

TABLE 32-3 Electrical properties of major plastics

Plastic	Arc resistance, sec ASTM D495	Dielectric strength ASTM D149 short-time V/mil 0.125 in. thick	1- to 2-mil foil	step-by-step 0.125 in. thick	Dielectric constant ASTM D150, cycles 60	10^3	10^6	Dissipation (power) factor ASTM D150, cycles 60	10^3	10^6	Volume resistivity, ohm/cm ASTM D257 at 50% relative humidity, 23°C
ABS	50–85	350–500	...	350–450	2.4–5.0	2.4–4.5	2.4–3.8	0.003–0.008	0.004–0.007	0.007–0.015	1.0–5.0×10^{16}
Acetal	240	500	...	400	3.7	3.7	3.7	...	0.004	0.004	6.0×10^{14}
Acrylic	No track	450–550	400	350–500	3.5–4.5	3.0–3.5	2.2–3.2	0.05–0.06	0.04–0.06	0.02–0.03	2.0×10^{14}
Allylic	120–250	380–450	...	320–410	3.4–5.0	3.3–5.0	3.6–4.5	0.006–0.019	0.01	0.028–0.060	4.0×10^{14}
Cellulose acetate	50–310	250–600	3–5000	200–520	3.5–7.5	3.4–7.0	3.2–7.0	0.01–0.06	0.01–0.07	0.01–0.10	...
Epoxy:											
Cast resin	45–120	400–550	...	380	3.5–5.0	3.5–4.5	3.3–4.0	0.002–0.010	0.002–0.020	0.03–0.05	...
Glass laminate	15–180	400–750	...	300–600	4.2–5.3	...	4.2–5.3	0.003–0.015	...	0.010–0.025	1×10^{15}
Fluorocarbons:											
Polychlorotrifluoroethylene	>360	500–600	1–3700	450–500	2.2–2.8	2.3–2.7	2.3–2.5	0.0012	0.023–0.027	0.009–0.017	1.2×10^{18}
Polytetrafluoroethylene	>200	480	4300	430	<2.1	<2.1	<2.1	<0.0002	<0.0002	<0.0002	1.0×10^{18}
Fluorinated ethylene propylene	>165	500–600	5000	...	2.1	2.1	2.1	0.0003	0.0003	0.0003	2.0×10^{18}
Polyvinylidene fluoride	50–70	260	3500	...	8.4	8.0	6.6	0.049	0.018	0.017	2.0×10^{14}
Melamine:											
Alpha-cellulose filler	95–135	270–300	...	240–270	7.9–9.5	7.8–9.2	7.2–8.4	0.030–0.083	0.015–0.036	0.027–0.045	8.2×10^{12}
Asbestos filler	120–180	410–430	...	280–320	6.4–10.2	9.0	6.1–6.7	0.07–0.17	0.07	0.041–0.050	1.2×10^{12}
Glass-fiber filler	180	170–300	...	170–240	9.7–11.1	...	6.6–7.5	0.14–0.23	...	0.013–0.015	2.0×10^{11}
Phenolic:											
Wood-flour filler	Tracks	200–400	...	100–375	5.0–13.0	4.4–9.0	4.0–6.0	0.50–0.30	0.04–0.20	0.03–0.07	...
Asbestos filler	120–200	200–350	...	150–300	7.5–50.0	6–30.0	5–10.0	0.10–0.30	0.10–0.40	0.40–0.80	...
Mica filler	Tracks	350–400	...	250–390	4.7–6.0	4.4–5.5	4.2–5.2	0.03–0.05	0.03–0.04	0.005–0.010	...
Glass-fiber filler	4–190	140–400	...	120–270	7.1	6.9	4.6–6.6	0.05	0.02	0.012–0.026	...

Polyamides:										
Nylon 6	130–140	310–530	1500	300–440	3.9–5.5	4.0–4.9	3.5–4.7	0.04–0.06	0.01–0.06	0.03–0.04	...
Nylon 6/6	...	385–470	850	340–410	4.0–4.6	3.9–4.5	3.4–3.6	0.014–0.040	0.02–0.04	0.04	...
Nylon 6/10	...	400	...	...	3.9	3.6	3.5	0.04	0.04	0.04	2.1×10^{16}
Polycarbonate	5–120	...	1500	364	2.9–3.1	3.02	2.96	0.0006–0.0009	0.0021	0.009–0.010	...
Polyester:										
Film	...	...	7500	...	3.2	3.0	2.8	0.005	0.016	0.008	...
Cast resin	125	380–500	...	170–420	3.0–4.3	2.8–5.2	2.8–4.1	0.003–0.028	0.005–0.025	0.006–0.026	...
Glass laminate	60–240	350–500	...	220–600	3.8–7.3	4.0–6.0	3.5–6.4	0.01–0.04	0.01–0.06	0.008–0.030	...
Polyethylene:										
Low density	135–160	450–1000	4700	420–700	2.2–2.3	2.2–2.3	2.2–2.3	<0.0005	<0.0005	<0.0005	10^{16}
Medium density	200–235	450–1000	500?	500–700	2.2–2.3	2.2–2.3	2.2–2.3	<0.0005	<0.0005	<0.0005	10×10^{16}
High density	...	450–500	500	440–600	2.3	2.3	2.3	<0.0005	<0.0005	<0.0005	10×10^{16}
Polyphenylene oxide	20–75	400–500	...	...	2.5–2.9	2.5–2.9	2.5–2.9	0.0003	...	0.0009	10^{18}
Polypropylene	136–185	500–560	3000–4500	450–650	2.2–2.6	2.2–2.6	2.2–2.6	<0.0005	0.0005–0.0018	0.0005–0.0018	10^{16}
Polystyrene	20–150	300–700	2500	300–600	2.4–4.7	2.4–4.6	2.4–3.8	0.0001–0.003	0.0001–0.002	0.0001–0.002	1.4×10^{16}
Polysulfone	75–122	425	...	430	3.14	3.13	3.10	0.0008	0.001	0.003	5×10^{16}
Silicone:										
Cast resin	115–330	550	...	550	2.7–3.0	...	2.6–2.7	0.0007–0.001	...	0.001–0.002	2×10^{15}
Glass reinforced	240–420	180–480	...	125–300	3.3–5.2	3.2–5.0	3.2–4.7	0.004–0.030	0.003–0.020	0.002–0.020	$1 \times 10^{10-14}$
TPX methyl pentene	...	...	...	700	2.1	2.1	2.1	0.00007	0.00003	0.000025	1×10^{16}
Urea-formaldehyde	80–150	300–400	...	220–300	7.0–9.5	7.0–7.5	6.0–8.0	0.035–0.043	0.025–0.035	0.25–0.35	$1 \times 10^{12-13}$
Urethane	122	330–900	600–900	450–500	4.0–7.5	4.0–7.5	6.5–7.1	0.015–0.017	0.05–0.06	...	$2 \times 10^{11-15}$
Polyvinyl chloride and	...	...	...	...	...	...	...	...	...	...	...
Polyvinyl chloride-acetate:										
Rigid	60–80	400–1300	425–1300	375–450	3.2–3.6	3.0–3.2	2.8–3.1	0.007–0.020	0.009–0.017	0.006–0.019	1×10^{16}
Flexible	...	300–1000	250–1000	275–900	5.0–9.0	4.0–8.0	3.3–4.5	0.080–0.150	0.070–0.169	0.040–0.140	$1 \times 10^{11-15}$
Flexible, filled	...	250–800	...	225–750	5.0–6.0	4.0–5.0	3.5–4.5	0.100–0.150	0.090–0.160	0.090–0.100	$1 \times 10^{11-14}$
Fused quartz	...	...	...	...	...	4.1	4.1	...	...	...	...
Natural rubber	...	...	...	...	2.8	2.9–3.1	3.1	...	...	...	...
Vulcanized rubber	110–150	125–225	...	...	...	...	...	...	...	...	...

Voltmeter   An electrical instrument used for measuring potential difference.

Vulcanized fiber   Product made by treating cellulose with a gelatinizing agent such as zinc chloride, followed by pressing, leaching, and drying. Used for electrical insulation, tubing, and mounting panels.

REVIEW QUESTIONS

1   State, in your own words, the reasons for the close alliance between the plastics industry and the electronic and electrical fields.
2   What is meant by dielectric strength?
3   List the categories of electrical energy and the applications for each category.
4   How are plastics made conductive?
5   How does thickness affect dielectric strength?
6   What factors affect dielectric strength of thermosets? Of thermoplastics?
7   How can dielectric tests be used to define mechanical properties?
8   What is meant by "proof-test" voltage?
9   What is arc resistance?
10   What common products found in the home or the workshop are insulated against electrical shock?
11   Which plastics exhibit outstanding resistance to arcing?
12   Is phenolic suitable for applications where severe arcing can occur? Why?
13   What does dielectric constant mean?
14   What types of products have condensers?
15   In what ways can electrical leakage occur?
16   How is loss factor determined?
17   Which organic polymers exhibit low-loss factors?
18   How does elevated temperature affect
    (*a*)   dielectric strength
    (*b*)   loss factor
19   Which thermoplastics rank highest as insulators at high temperature?
20   A dielectric test specimen 4 in. in diameter and 0.120 in. thick fails at 60,000 V. What is the dielectric strength?

THERMAL PROPERTIES

33

Because of the effects of heat, the organic nature of plastics leads to problems un-common to metals or minerals. When properly selected, designed, and fabricated, plastics perform exceptionally well under adverse conditions of extreme heat or cold.

Heat, or thermal influence, which permits us to mold and form a plastic easily and economically, may be the cause of failure sometime during a plastic's service life. Many plastics are affected by even moderate temperatures to the extent that mechan-ical, electrical, chemical, and optical properties are impaired. Molding and thermo-forming operations are governed by the way materials react during exposure to the heating sequence.

In this chapter, the effects of a thermal environment are discussed to provide a basis for the intelligent selection of materials that must function above or below normal temperatures. It should be recalled that the thermal behavior of thermoplastics (as a class) is contrary to that of thermosets. Thermosetting resins are softened briefly during the processing phase before they polymerize. Once hardened, they cannot be heat softened. Thermoplastic polymers can be softened as many times as required, providing thermal degradation is prevented. This does not mean that thermosets are immune to the influence of heat. When sufficient heat is applied, charring and de-composition are inevitable. Most thermoplastics soften beyond the point of utility at temperatures that have little or no effect on thermosetting resins. (See Tables 33-1 and 33-2.)

THERMAL CONDUCTIVITY

The rate of heat transfer through a material is called thermal conductivity. A major reason for the popularity of plastics in industrial and commercial applications is their relatively low rate of heat conduction. Molded items such as pot handles, electric iron handles, and hair driers are commonplace products that demonstrate this property. Less widely known is the outstanding role plastic plays in the fabrication of reentry heat shields, rocket nozzles, and flame skirts that are indispensable to vehicles employed in outer-space missions.

Using a pot handle as an example, one can compare the rate of heat transfer of the plastic (usually general-purpose phenolic) with that of the metal component. At a temperature of 212°F the metal would cause a burn, but the handle could be touched and not cause an injury. This is because the rate of heat transfer in metal is much higher, and even though the plastic is at the same temperature, it is not as hot to touch. Hot-molded plastics can be handled safely without protective gloves, although gloves are used as a protection against the hot steel mold.

Data on thermal conductivity (referred to as K factor) is obtained on flat panels or loose powders by a method known as the guarded hot-plate test. In this procedure (ASTM C117-63) samples are placed upright between an electrical heat source and a cooling plate. The test is carried on for the length of time required to stabilize input and output temperatures, which is usually several hours. Thermal conductivity is then calculated

$$K = \frac{QL}{A\,(T_1 - T_2)}$$

where K = thermal conductivity in BTU/in./h/sq ft/°F

Q = time rate of heat flow, in BTU/h*

*BTU (British Thermal Unit): The amount of heat required to raise temperature of 1 lb of water 1°F. Thermal conductivity is also expressed in calories, the amount of heat necessary to raise 1 g of water 1°C. 252 cal = 1 BTU.

33-1 Rocket nozzle capable of performance at 5,000°F. (*Reprinted by permission of of United Technology Corp.*)

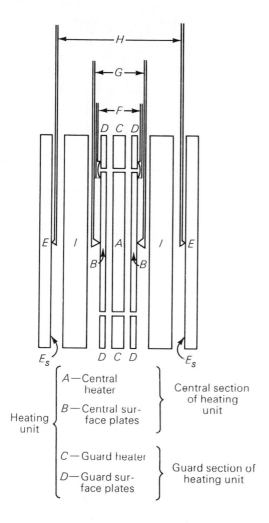

A—Central heater } Central section of heating unit

B—Central surface plates }

Heating unit

C—Guard heater } Guard section of heating unit

D—Guard surface plates }

33-2 General features of the metal-surfaced hot plate to determine thermal conductivity. Code: E = cooling units; E_3 = cooling-unit surface plates; F = differential thermocouples; G = heating-unit surface thermocouples; H = cooling unit surface thermocouples; and I = test specimens.

L = thickness of specimen along path of heat flow, in.
A = area under test, sq ft
T_1 = temp. of hot surface, °F
T_2 = temp. of cold surface, °F

At this point, the difference between heat conductivity and heat resistance should be emphasized. To illustrate, let us take the example of a molded electric iron handle with a metal base that attains a maximum operating temperature of 400°F. The first choice of material for the handle might be asbestos-filled phenolic, since thermal properties charts indicate thermal stability on continuous exposure to 500°F. General-purpose phenolic, by comparison, is limited to 360°F for continuous operation. However, cellulose-filled phenolic is used for electric iron handles because heat

TABLE 33-1 Thermal properties of thermosetting materials

Material	Thermal conductivity 10^{-4} cal/sec/cm², 1°C/cm	Deflection temperature, °F fiber stress, 264 psi	Heat resistance, °F (continuous)	Flammability	Thermal expansion, 10^{-5}/°C
Phenol formaldehyde and phenol furfural					
Molding compounds:					
Unfilled	3–6	240–260	250	Very low	2.5–6.0
Wood flour or cotton flock filler	4–7	260–340	360	Very low	3.0–4.5
Asbestos filler	8–22	300–400	350–500	None	0.8–4.0
Mica filler	10–14	300–350	250–300	None	1.9–2.6
Glass-fiber filler	0–14.5	300–600	350–550	None	0.8–1.6
Melamine formaldehyde					
Molding compounds:					
Alpha-cellulose filler	7–10	350–370	210	None	4.0
Asbestos filler	13–17	265	250–400	None	2.0–4.5
Glass-fiber filler	11.5	400	300–400	None	1.5–1.7
Urea formaldehyde					
Molding compound:					
Cellulose filler	7–10	260–290	170	SE*	2.2–3.6
Polyesters					
Cast:					
Rigid	4	140–400	250	SE to 1.1	5.5–10
Flexible	...	...	200	Slow to SE	...
Molding compound:					
Glass premix	10–16	400–450	300–325	Slow to SE	2.0–3.3

Diallyl phthalate

Molding compound:				
Glass filled	5–15	350–500	300–450	None to slow	2.0–5

Alkyd

Molding compounds:				
Mineral filled				
(granular or putty)	15–25	350–500	300–450	None to slow	2.0–5.0
Asbestos filled	10–20	315	450	SE	3.0
Glass filled	15–25	400–500	450	None to slow	1.5–2.5
Synthetic fiber filled				
(Orlon, Dacron, nylon)	7–12	245–430	300–430	None to SE	3.0–5.5

Epoxy

Casting grades:				
No filler	4.5	115–550	250–550	SE to slow	4.5–6.5
Aluminum filler	...	190–600	200–500	None to slow	5.5
Flexible fillers	...	75–250	250–300	Slow	2–10
Molding compounds:				
Glass filler	4–10	250–500	300–500	SE	1.1–3.5
Mineral filler	4–30	250–500	300–500	SE	2.0–5.0

Silicone

Molding compounds:				
Glass filler	7.5	900+	600+	None to slow	0.8
Mineral filler	11–13	900+	600+	SE	2–4

*SE = self-extinguishing.

511

TABLE 33-2 Thermal properties of thermoplastics

Thermoplastic	Thermal conductivity, 10^{-4} cal/sec/cm² 1°C per cm	Deflection temp., °F fiber stress, 66 psi	Deflection temp., °F fiber stress, 264 psi	Heat resistance, °F continuous	Flammability burning rate, in./min	Thermal expansion $10^{-5}/°C$	Vicat softening point, °F (method B)	Specific heat
ABS (Acrylonitrile-Butadiene-Styrene):								
Extrusion grades	4.5–8	210–230	200–224	140–220	Slow	7–13	215	...
Molding grades:								
High impact	↓	205–225	200–218	160–210	→	9.5–10.5	218	...
Medium impact		210–230	199–225	160–200		5.0–8.5	215	...
Heat resistant		230–250	200–245	190–230		6.0–9.0	253	...
Self-extinguishing		205	150–210	150–200	SE*	...	211	...
20–40% glass filled		220–250	210–240	200–230	Slow	2.9–3.6	...	...
Acetal:								
Homopolymer	5.5	338	255	185	1.1	8.1	...	...
Copolymer	5.5	316	230	220	1.1	8.5	324	...
20% glass filled	...	331	315–345	185–220	0.8–1.0	2.0–4.7	324	...
Acrylic:								
Cast	4–6	165–235	160–215	140–200	1.0–1.3	5–9	...	...
Molded	4–6	165–225	155–210	140–190	0.9–1.2	5–9	208–255	...
Cellulosics:								
Cellulose acetate:								
Molded and sheet	4–8	120–209	111–195	140–220	Slow to SE	8–18	...	...
Cellulose acetate butyrate:								
Molded and sheet	4–8	130–227	113–202	140–220	Slow	11–17	...	0.3
Cellulose propionate:								
Molded	4–8	147–250	111–128	...	1.0–1.3	11–17	...	0.3
Cellulose nitrate:								
Sheet	5.5	...	140–160	...	Very fast	8–12	...	0.3
Ethyl cellulose:								
Molded and sheet	3.8–7	...	115–190	...	Slow	10–20	...	0.3–0.7

Fluoroplastics:								
Polychlorotrifluoroethylene (CTFE)	4.7–5.3	253	...	350–390	None	4.5–7.0	414	0.22
Polytetrafluoroethylene (PTFE)	6	250	...	550	None	10	...	0.25
Fluorinated ethylene propylene (FEP)	6	...	...	400	None	8.3–10.5	487–540	0.28
Polyvinylidene fluoride (PVF$_2$)	3	300	195	300	SE	12	313	0.33
Nylons:								
Type 6:								
Molded	5.85	300–365	150–175	175–250	SE	8.3	...	0.38
Cast	...	400–425	200–425	180–250	→	9.0	...	...
Type 6/6	5.85	360–470	150–220	180–300	→	8.0	518	0.4
Type 6/10 (unfilled)	5.16	300	...	180–250	→	9.0	...	0.4
20–40% glass filled	5.15	400–510	320–502	300–400	→	1.2–3.2	420–440	0.3
Polycarbonate:								
Unfilled	4.3	270–290	285–300	250	SE	6.6	302–325	0.3
10–40% glass filled	2.5–5.2	265–280	295–300	275	SE	1.8–3.8	...	...
Polyolefins:								
Polyethylene:								
Low density	8.0	100–121	90–105	180–212	1.04	10–20	...	0.55
Medium density	8–10	120–165	105–120	220–250	1.0–1.04	14–16	...	0.55
High density	11–12.4	140–190	110–130	250	1.0–1.04	11–13	...	0.55
Ethylene-ethyl acetate (EEA)	...	...	...	190–200	Very slow	16–23	...	0.55
Ethylene-vinyl acetate (EVA)	...	140–147	93	...	Very slow	16–20	...	0.55
Polypropylene:								
Unmodified	2.8	200–230	125–140	250–320	Slow	5.8–10	...	0.46
Glass filled	...	305–310	230–300	300–320	→	2.9–5.2	...	...
Rubber modified	3.4	160–200	120–135	200–250		6.0–8.5	...	0.5
Polysulfone	...	368	345	300–345	SE	5.2–5.6	406	0.3
Polystyrenes:								
Unfilled:								
General purpose or heat resistant	2.4–3.3	...	220	150–170	Slow	6.0–8	...	0.32
Impact resistant; heat resistant	1.0–3.0	...	210	140–175	Slow	3.4–21	...	0.34
Special heat and chemical resistant	1.9–3.0	...	180–235	170–200	Slow	6.0–8	...	0.33
20–30% glass filled	...	207–236	195–220	180–200	Nonburning to 1.0	1.8–4.5	...	0.26

*SE = self-extinguishing.

TABLE 33-2 Thermal properties of thermoplastics (*Continued*)

Thermoplastic	Thermal conductivity, 10^{-4} cal/sec/cm² 1°C per cm	Deflection temp., °F fiber stress, 66 psi	Deflection temp., °F fiber stress, 264 psi	Heat resistance, °F continuous	Flammability burning rate, in./min	Thermal expansion $10^{-5}/°C$	Vicat softening point, °F (method B)	Specific heat
Styrene-acrylonitrile copolymer	2.9	· · · (Subzero)	190–220 (Subzero)	140–205	Slow	3.6–3.8	· · ·	0.33
Styrene-butadiene copolymer	3.6	−120	−150	130–150	Slow	13.0–13.7	· · ·	0.47
Methyl pentene (TPX)	4.0	· · ·	· · ·	250–320	1.0	11.7	· · ·	0.52
Polyurethane:								
Cast	5.0	Varies over wide range		190–250	Slow	10–20	· · ·	0.43
Molded	1.7–7.4	· · ·	· · ·	190	Slow to SE*	10–20	· · ·	0.45
Vinyl polymers and copolymers:								
Polyvinyl chloride:								
Flexible	3.0–4.0	· · ·	· · ·	150–175	Slow to SE	7–25	· · ·	0.4
Rigid	3.0–7.0	135–180	130–175	150–175	SE	5–18	· · ·	0.25
Polyvinyl chloride acetate	3.0–4.0	140–185	130–175	150–180	Slow to SE	7–25	· · ·	0.4
Polyvinylidene chloride	3.0	· · ·	130–150	160–200	SE	19	· · ·	0.32
Steel						.6–7		
Aluminum						1.2		

*SE = self-extinguishing.

conductivity is one-third that of the mineral-filled grade, making it more pleasant to touch. Intermittent exposure to heat above the recommended limit for continuous operation does not affect it.

Table 33-1 shows that resins containing inorganic fillers (asbestos, glass, clay, powdered metals) have higher rates of heat flow than organic-filled (walnut shell, chopped fabrics, wood flour) resins.

THERMAL EXPANSION AND CONTRACTION

Serious failures can occur when plastics are combined with metals unless compensation is made for differences that exist when temperatures change. The ability of plastics to be molded around a wide variety of inserts made from metals, plastics, rubber, or glass, or to be cast around electrical components, has led to greater simplification of design and assembly.

Table 33-2 indicates that plastics expand from 2 to 10 times as much as steel does, and that aluminum and fiberglass-reinforced plastics approach comparable values. Conversely, plastics contract much more than metals as they cool, which can be either an advantage or a disadvantage.

Several examples of typical applications in which metal is combined with plastic provide a broader understanding of the effects of expansion and contraction.

Example 1:   A phenolic is to be molded around a flat steel strip for a skillet handle. As the utensil is heated, both metal and plastic expand, but at different rates. Upon cooling, the plastic shrinks at a faster rate, creating stresses in localized areas. If the thickness has not been designed to withstand stress, cracking results.

As mentioned earlier, a small amount of polymerization may occur after a part is placed in service. This creates minute shrinkage, which generates additional stress.

Example 2:   Encapsulated electrical products can fail unless the resin system is formulated to withstand differences in expansion when heat is generated as a function of the circuit.

Example 3:   Designers must take into account the expansion of large steel molds operating at a high temperature when close tolerance is required on part dimensions. Mold designers must also know the shrinkage value for the compound and build molds oversized to compensate for contraction caused by cure.

Thermoset resins are often molded to exact dimensions. When no additional shrinkage can be tolerated after the unit is in service, molders subject cured parts to a postcure or afterbake schedule. Parts are baked in ovens for periods to 7 days. Low temperatures (150 to 200°F) are employed during initial phases to avoid blistering, and increased periodically until the operating temperature is reached. The oven is then cooled at a predetermined rate to avoid thermal shock and warpage. Many molded items undergo a postcure shrinkage to 50 percent of the molded amount.

Shrinkage results from various causes, depending on the type of material.

Because of severe temperatures, these characteristics may come about:

1   Further contraction caused by polymerization of traces of unreacted chemicals. All thermosets and thermoplastics that may contain traces of monomeric constituents are susceptible.

2   Loss of water as with nylon, polyvinyl alcohol, and water-extended polyesters.

3   Migration of plasticizer commonly used with flexible vinyls, cellulosics, and modified ABS.

4   Molecular changes caused by stress relief or orientation. Thermoformed and molded products, particularly polyethylene and polystyrene, may warp if not processed correctly.

Thermal expansion can be measured directly by micrometer if the piece is sufficiently large. More accurate measurements are derived through the use of a dilatometer, which defines linear expansion over a narrow temperature range. Immersion of a fused quartz tube containing the sample in a controlled temperature bath permits close control of temperature rise and an accurate plot of expansion coefficient over a useful thermal range of -30 to $+30°C$.

Formula:

$$\alpha = \frac{\Delta L}{L\,I}$$

where α = coefficient of linear expansion °C
 ΔL = average of change in length of test specimen caused by heating and cooling
  L = length of specimen at room temperature
  T = temperature difference in °C over which length changes are measured

FLAMMABILITY

Some plastics support combustion readily. Others burn slowly or are self-extinguishing, but a few do not support burning at all. One reason cellulose nitrate is not more widely used is that of fire hazard, since it is rapidly consumed after ignition. Safety

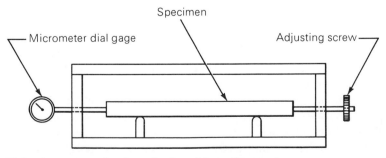

33-3   Arrangement for determination of thermal expansion.

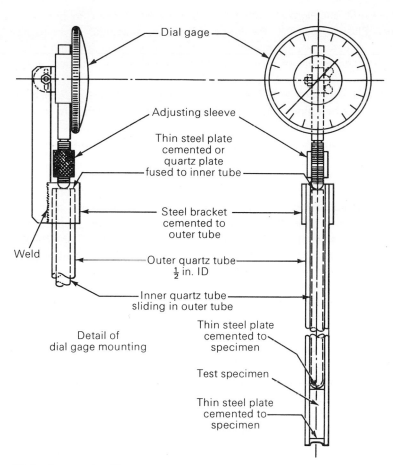

Dial gage

Adjusting sleeve

Thin steel plate
cemented or
quartz plate
fused to inner tube

Steel bracket
cemented to
outer tube

Weld

Outer quartz tube
$\frac{1}{2}$ in. ID

Inner quartz tube
sliding in outer tube

Detail of
dial gage mounting

Thin steel plate
cemented to
specimen

Test specimen

Thin steel plate
cemented to
specimen

33–4 Quartz-tube dilatometer.

film for movie projection was introduced through the use of cellulose acetate, reducing fire hazard to a minimum. Storage requirements for x-ray film also dictated the early use of cellulose acetate.

With the growing use of plastics in transportation, public buildings, homes, and recreational vehicles, research in flame-retardant and nonburning plastics is continually expanding. The release of toxic fumes when certain plastics decompose (fluorocarbons, vinyls, and polyurethanes) indirectly creates difficulties especially when industrial, aircraft, or building codes are restrictive.

Plastics continue to be maligned for flammability and toxicity by competitive industries that rebel against the inroads of plastics into previously sacrosanct territories. Scare tactics, completely without foundation, are occasionally used to thwart a new use for plastics. A blow torch played on a section of rigid vinyl pipe, for example, of course will burn it as long as flame is applied. But the same test on wood shows that the wood will not only ignite much faster, but it will continue to burn and smolder for hours. Plastics found in the home are not as hazardous as materials that have been used for centuries, namely, wood, cotton textiles, paper, and flammable liquids. The ignition temperature of plastics is almost twice that of products made from cellulose.

When particular resins such as polyester, epoxy, acrylic, and ABS are specified to fulfill an application where fire could prove disastrous, such as in aircraft, special grades are formulated to meet rigid specifications. Incorporating chemicals such as bromine, chlorine, and fluorine, plus inert fillers (antimony) renders an otherwise combustible resin self-extinguishing.

As a class, thermoplastics are more susceptible to flame than thermosets. Some, however, are inherently self-extinguishing because of their molecular structure, rather than because of the addition of flame-arresting chemicals. Notable among them are polyvinyl chlorides, fluorocarbons, polycarbonate, polyphenylene oxide, and certain nylons. Thermosetting resins of the phenolic and amino types are nonburning. Exceptions to this characteristic are paper- or cotton-base laminates that have high filler content and thin cross section. Though difficult to ignite, they burn slowly as the resin chars.

There are many types of tests that compare the burning rates of products depending largely on physical form and intended end use. For plastics under 0.050 in. thick, method ASTM D568-61 is utilized. Samples 18 in. long are fixed vertically after marking a 12-in. gauge length 3 in. from each end. Timing begins when flame from a bunsen burner reaches the lower gauge mark and ends when 12 in. of the strip has been consumed. Values are recorded as inches/minute burning rate.

A similar test, ASTM D635-63, is conducted for plastics over 0.050 in. thick, except that the specimen is mounted horizontally and inclined 45 deg edgewise. Another test for flammability brings a test specimen in contact with a silicon carbide rod electrically heated to approximately 1,750°F to measure burning rate or degree of char.

Building materials are tested for flame spread according to ASTM method E84-61, also known as Underwriters' tunnel test. A rectangular duct 25 ft long, 17.50 in. wide, and 12.50 in. high is used, with the test piece forming the "roof." Burning rates are compared with two arbitrary values: red oak, 100; cement-asbestos board (transite), 0. Materials with comparative values to 25 are rated noncombustible. Chlorinated polyester attains this favorable rating. Materials that rate between 75 and 200 are considered combustible; these include all nontreated woods and general-purpose polyesters.

HEAT-DEFLECTION TEMPERATURES

A practical test designed to arrive at the value of a material operating at elevated temperature under stress is outlined by ASTM D648-56. Suitable for thermoplastics as well as thermosets, the test procedure applies to molded or sheet materials normally rigid at room temperature. A molded or machined sample 5 in. long, 0.50 in. wide,

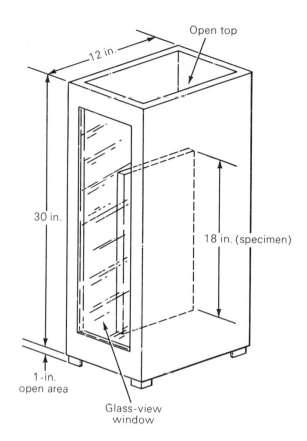

Open top

12 in.

30 in.

18 in. (specimen)

1-in.
open area

Glass-view
window

33-5 Apparatus for burning-rate test.

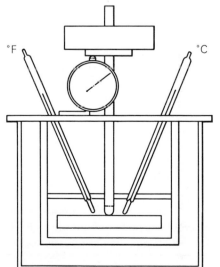
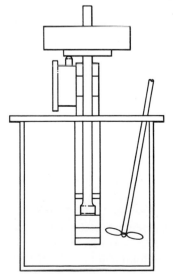

°F

°C

33-6 Apparatus for deflection-temperature test.

by any thickness from 0.125 to 0.50 in. is mounted on supports 4 in. apart, as in a flexural test. Weight calculated to induce a fiber stress of either 66 or 264 psi is applied at the center. The lighter load (1.375 lb for specimens 0.50 in thick) is used for thermoplastics that exhibit deformation at a relatively low temperature, while a heavier stress of 264 psi (5.5 lb for 0.50-in. thickness) is used for thermosets and such thermoplastics as polycarbonate, acetal, and some polyamides. The test is conducted in a controlled temperature bath of mineral or silicone oil and increased at a constant rate of 2°C (35.6°F)/min. Heat-deflection temperature is recorded when a deflection of 0.010 in. is indicated on the spring-actuated dial.

Heat-deflection temperatures are useful in establishing maximum, safe operating temperatures under no load, as well as providing data for safe annealing of thermoplastics.

VICAT SOFTENING POINT

A similar test to that for heat deflection of bars, vicat tests demonstrate the ability of a thermoplastic to resist penetration to a depth of 1 mm by a flat needle having a circular cross section of 1 mm² (0.045-in. diameter). Under total load of 1000 g, bath temperature is raised at controlled rates of 50°C (method A) or 120°C (method B)/h, depending on heat resistance of the material. The vicat softening point is important in evaluating products suspected of creep and cold flow at relatively low temperatures, such as gaskets, seals, and mounting pads.

MELTING POINT

In production, molders and fabricators rely on visual observation to determine when a resin has melted sufficiently for processing. A sophisticated procedure, ASTM D2117-64, is used when a high degree of precision is required by the basic polymer manufacturer. Semicrystalline polymers (polyamides, polystyrene, polyolefins) exhibit a double refraction of light rays that can be viewed through a polarizing medium.[1] When the polymer melts, the ordered crystallinity of the atoms is interrupted, and the optical property of double refraction (birefringence) disappears. The temperature at which this phenomena occurs, as viewed under a microscope, is termed melt point.

Until now only the effect of elevated temperature on plastics have been discussed. Changes caused by temperatures below normal must also be considered. While most products are not subjected to extremes of either heat or cold, certain applications expose a material to one extreme or both. An aircraft door gasket, for example, must retain flexibility over a temperature range of −65°F to +140°F; valve seals in piping systems carrying liquid oxygen sustain temperatures several hundred degrees below zero; cable insulation must withstand frigid conditions without failure.

The comparative effects of heat and cold are problems that confront the engineer responsible for selection of appropriate materials and challenge the designer who gives the product form and function (Table 33-3).

[1] The effect of polarized light is discussed in Chap. 35.

33-7 Weather balloon made of DuPont Mylar must function at 30°F below zero. (*Reprinted by permission of G. T. Schjeldahl Co.*)

BRITTLENESS

Wire coatings, coated fabrics, refrigerator parts, battery cases, and ice chests exemplify familiar products that must withstand subnormal conditions. Aerospace, military, and other governmental agencies have thousands of applications in which plastics must meet rigid specifications for performance in below-freezing environments. ASTM method D746-64T establishes the temperature at which 50 percent of the specimens tested fail because of embrittlement. Specimens die-cut from sheets, measuring 0.25 in. wide and 0.075 ± 0.010 in. thick, are clamped in order to have approximately 1 in. extending in space. The apparatus is immersed in a coolant to drop the temperature to the desired level. Coolants such as methyl alcohol and dichlorodifluoromethane are

TABLE 33-3   Effect of temperature on general properties

Effect of high temperature	Property	Effect of low temperature
Decrease	Tensile, compression, flexural, shear	Increase
Increase	Impact	Decrease
Increase	Elongation	Decrease
Increase	Creep and cold flow	Decrease
Decrease	Surface hardness	Increase
Decrease	Electrical insulation values	Increase
Increase	Solubility in solvents	Decrease
Increase	Attack by acids and alkalies	Decrease
Increase	Water absorption	Decrease
Decrease	Brittleness	Increase

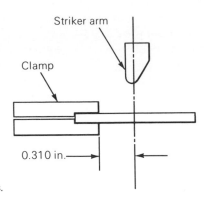

33-8   Method for determining brittleness.

used for −90°C and −120°C, respectively. A striker is permitted to impact at a speed of 6 to 7 ft/sec. Temperatures are lowered in increments of 2°C until 5 of 10 specimens fail at a particular time.

Coated fabrics and extruded or calendered elastomers are tested by bending them around a 0.125-in. diameter mandrel after removal from a dry ice conditioning chamber. Cracking or flaking of the resin is cause for rejection.

Brittleness tests determines the ability of elastomers to retain plasticizers or estimate performance. Plastics can operate below their brittleness ratings when stress is not a condition.

ABLATION

No discussion of thermal properties would be complete without mention of the ablative properties of synthetic polymers. In one decade, man has been successful in placing satellites in orbit, and in returning from the moon. Plastics continue to have an important role in space exploration and in development of strategic ballistic weapons. By assembling performance data on thermal expansion, heat conduction, and decomposition effects, the plastics industry provides structural and protective components for missile and rocket hardware.

Ablation means the loss of material in a sacrificial but controlled manner by the action of gaseous substances impinging on the surface at hypersonic velocity. A meteorite entering our atmosphere is usually consumed by the ablation process before it reaches earth. Those that manage to penetrate the air layer are larger at the outset, or of more heat-resistant composition. Examination of a meteorite shows its substance to be mainly iron, with other inorganics such as silica.

Heat-resistant metals and ceramics used for protective shielding of space vehicles present several limitations: (1) many times heavier than plastics; (2) difficult to fabricate into complex shapes; (3) inability to withstand thermal shock; (4) low impact strength; and (5) greater heat conductivity.

Specific plastics and composites, on the other hand, have fulfilled the demands of space agencies, but not without exhaustive trial and error in the development of new resins and techniques.

The purpose of an ablator is to act as a heat sink and at the same time slow or prevent the transfer of heat to more heat-sensitive elements.

Incongruously, properties charts reveal maximum heat-deflection temperatures of our best materials to be in the order of 600 to 900°F and continuous operating temperatures well below these values. Still we can use synthetic resins in a composite to function at temperatures above 10,000°F. Organic resins are useful because they do decompose, rather than evaporate or sublime as do inorganics. In Fig. 33-9, a sequence of events occurs that makes plastics an ideal ablator. Remember that the same ablation sequence occurs repeatedly in cycles of microseconds as a result of temperature and velocity as frictional heat increases through areas of denser atmosphere.

Ablation sequence for phenolic–high-silica-fiber composite

1 Heat is absorbed at the surface and penetrates into the outer region.
   (a) Surface (skin) temperature rises rapidly, resulting in decomposition of organic particles.
   (b) Gaseous by-products of the decomposition process are forced into the boundary layer, lowering surface temperature by transpiration.

2 Char layer is formed by remaining carbonaceous material.

3 Char layer is blown away, exposing fibrous reinforcement.

4 Fibrous reinforcement vitrifies, boils, and evaporates. Evaporation reduces temperature at the boundary layer.

5 Material directly beneath the ablating layer is repeating phases 1 and 2, thus adding to the cooling effect.

6 The entire mass is gradually increasing in temperature.

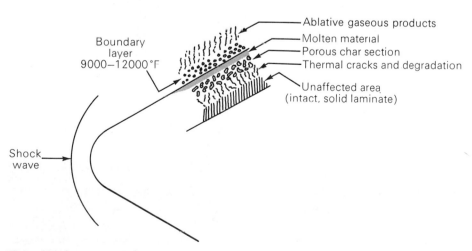

33-9 Ablation sequence for reentry nose cone.

Permissible back-wall temperatures are controlled by a sufficient thickness that would last through the ablative process. By studying data from thousands of laboratory and full-scale firings, design of rocket nozzles and heat shields to 156-in. diameter and design of reentry nose cones is practical.

Quality control information is gathered through a series of meaningful tests conducted on flat, laminated panels, molded blocks, machined specimens, subscale prototypes, and full-scale models. Of interest to the rocket designer are these values: thermal conductivity; temperature rise (back-wall temperature); area and depth ablated; weight loss (g/sec); depth of char layer; type of char (hard; soft); strength retention; gaseous by-products of decomposition; and chemical analysis of remaining material.

SPECIFIC HEAT

Of interest to rocket designers, molders and fabricators of plastic products, and mold designers is the material's specific heat, which is a measure of the quantity of heat required to raise a unit weight of a substance 1 deg in temperature. Specific heats of plastics do not differ appreciably, requiring from 0.3 to 0.4 cal to raise 1 g of molding material 1°C. For example, if 1 lb (454 g) of molding compound is to be raised from room temperature (25°C) to molding heat of 176.6°C (350°F), the heat required is

Weight of material, g × temperature rise × specific heat

454 × 151.6 × 0.35 = 24,089 cal

Expressed in comparable units, 95.5 BTU of heat or 0.0277 kWh of electrical energy would be required. Calculations of this type are seldom necessary since the heat used in actual molding is small compared with what must be furnished to heat

33-10 Subscale model of rocket nozzle under actual firing test. Model is that shown in Fig. 33-1. (*Reprinted by permission of United Technology Center, United Aircraft Corp.*)

the mold and make up heat losses caused by radiation and conduction. Fillers and fabric reinforcements such as cellulose and carbon increase the specific heat value of of a resin, while minerals such as glass, asbestos, and aluminum lower it.

GLOSSARY

Heat sink   Heat absorber.

Hypersonic   Above the speed of sound, which is 1,087 ft/sec.

Sublime   The process by which a solid substance vaporizes before becoming liquid.

REVIEW QUESTIONS

1   List several products found in the home that utilize plastics because low thermal conductivity is important.
2   What is meant by *K* factor?
3   Why does a plastic material not feel as hot as a metal piece at the same temperature?
4   Which fillers would you select to increase the rate of heat conduction of a resin?
5   What is a BTU?
6   How can differences in the expansion of metals and plastics be used to advantage?
7   When is it possible to use a micrometer to measure expansion?
8   What chemical constituents or additives render a plastic flame retardant?
9   Describe the standard ASTM tests used to determine flammability.
10   How is the load applied to a specimen for a heat-deflection test?
11   Why is the heat-deflection test only applicable to rigid plastics?
12   What test can be performed to determine softening points for soft plastics?
13   How are plastics tested for brittleness?
14   What does ablation mean?
15   Why can organic plastics effectively slow heat transfer when ablative conditions are encountered?

CHEMICAL PROPERTIES

34

One reason for the increasing use of plastics is their ability to withstand the corrosive effects of chemicals. When specific chemical problems are to be solved, designers investigate the possible application of a synthetic resin, either as a coating, or as the completely fabricated item. This does not mean that all plastics stand up under chemical attack equally well, but on a selective basis they can outperform metals, glass, wax, or treated paper.

Figure 34-1 is only a slight indication of the endless variety of uses that rely on specific plastics to contain, transport, convey, or protect products from harmful chemicals.

What is the criteria for selecting the proper synthetic material and then devising test and quality control measures for satisfactory performance in a specific chemical environment? Fortunately, chemicals can be classified into groups, This narrows the amount and types of testing that must be undertaken. If plastics are grouped according to the basic chemical nature of each kind, the process of elimination enables us to select the most likely candidates at the outset.

CHEMICAL GROUPS

Acid

Acid is the name given to a large number of compounds containing one or more atoms of hydrogen that may be replaced by a metal.

Example :

$$H_2SO_4 + 2Al \rightarrow Al_2SO_4 + H_2$$
Sulfuric   Aluminum + hydrogen
  acid + aluminum   sulfate

Strong acids are acetic, hydrochloric, hydrofluoric, nitric. Weak acids are boric, citric, lactic.

Base

This is a class of caustic (corrosive) materials having the common properties of being soluble in water and alcohol, able to neutralize acids to form salts, and able to combine with fats to form soaps; bases are also referred to as hydroxides and alkalies.

Example :

$$NaOH + HCl \rightarrow NaCl + H_2O$$
Sodium   Hydrochloric   Sodium   Water
hydroxide   acid   chloride

Strong bases are ammonia, potassium, sodium. A weak base is calcium.

Solvent

Any material capable of dissolving another to form a solution is considered a solvent. By this definition, a solid, liquid, or gas is capable of being a solvent if it constitutes the larger amount of the solution.

34-1   Fiberglas ®-polyester tanks for underground gasoline storage being lowered into sea water. (*Reprinted by permission of Owens-Corning Fiberglas Corp.*)

Solvents are further classified into chemical types:

Esters   Esters are compounds formed by reaction between an alcohol and an acid: amyl acetate, hexyl acetate, butyl acetate, isopropyl acetate, ethyl acetate, and methyl amyl acetate.

Ketones   Ketones are produced from a hydrocarbon when an oxygen atom is substituted in place of two hydrogen atoms that are attached to a carbon atom other than terminal atoms.

Example: Acetone from propane

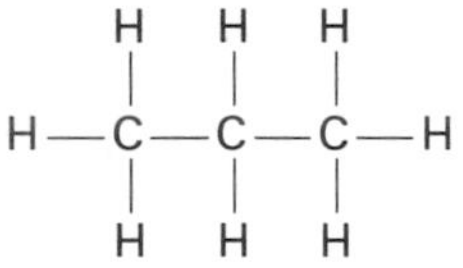

```
    H   H   H                       H       H
    |   |   |                       |       |
H — C — C — C — H           H — C — C — C — H
    |   |   |                       |   ‖   |
    H   H   H                       H   O   H
      Propane                         Acetone
```

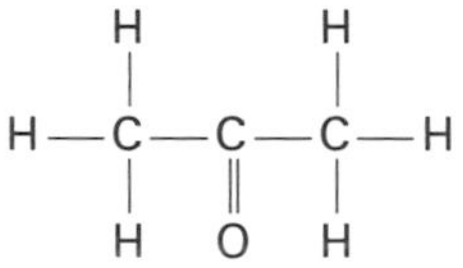

The most widely used ketones are acetone, methyl ethyl ketone (MEK), methyl isobutyl ketone (MIBK), and isophorone.

Ketones are highly active solvents for polystyrene, ABS, acrylics, vinyl elastomers, and cellulosics. Surface coating, impregnation of webs with thermosetting solutions of epoxy and polyester, and premix manufacturing processes utilize ketones to reduce liquid viscosities for compounding and fabricating requirements.

Hydrocarbons   These are compounds containing the elements hydrogen and carbon. They can be further separated into types: aromatic, aliphatic, and halogenated.

Aromatic hydrocarbons are those that contain the benzene ring: benzene, vinyl toluene, toluene, cyclohexanone, xylene, and naphthalene.

Aliphatic hydrocarbons are chain compounds where carbon atoms are combined in the form of a straight chain. They include hexane, tetra hydrofuran (THF), heptane, and dimethyl formamide (DMF).

Halogenated hydrocarbon solvents are derived when one or more hydrogen atoms are replaced with chlorine (chlorinated hydrocarbon); ethylene dichloride, carbon tetrachloride, methylene chloride, trichlorethane, methylene dichloride, and chloroform.

Aromatic hydrocarbons are useful in bonding articles fabricated from polystyrene and ethyl cellulose, and for preparing solutions of synthetic rubber adhesives or cellulose lacquers. Cyclohexanone is used to bond flexible, and some rigid, vinyls. High-molecular-weight vinyls may require THF or DMF.

Aliphatics find wide application in preparing printing inks, contact adhesives, and synthetic varnishes.

Chlorinated hydrocarbons must be handled with the extra precaution of adequate ventilation because of the toxic effects from their vapors. They are sometimes added to other solvents to reduce flammability. These solvents are used with acrylics, vinyls, and synthetic rubber for solvent cementing.

Alcohols   Alcohols are hydrocarbon derivatives in which one or more hydroxy (OH) groups replace a corresponding number of hydrogen atoms:

$$
\begin{array}{ccc}
& \text{H} & \\
& | & \\
\text{H} - & \text{C} & - \text{H} \\
& | & \\
& \text{H} &
\end{array}
\qquad
\begin{array}{ccc}
& \text{H} & \\
& | & \\
\text{H} - & \text{C} & - \text{OH} \\
& | & \\
& \text{H} &
\end{array}
$$

Methane   Methyl alcohol

The alcohols include ethyl, isopropyl, methyl, and butyl. Alcohols are used mainly as a processing aid in the preparation of surface coatings, prepregs, and molding compounds from phenolic, melamine, and phenyl-silane resins.

Monomers   When monomers of thermoplastic resins are available in liquid form, they can be used to dissolve their polymers. Styrene monomer is used to bond molded polystyrene products. It has the advantage of being less volatile and less toxic than solvents, and it minimizes crazing of stressed areas. The monomer must be handled with care to avoid static created when pouring the liquid. Acrylic monomer is used to dissolve its polymer for casting purposes.

What happens when a thermoplastic comes in contact with a solvent that can attack and dissolve it? This can best be explained by what takes place when two mating surfaces are joined for assembly purposes. Initially, a swelling takes place, permitting solvent penetration below the surface. Molecular chains are opened and the polymer becomes soft and sticky. The degree of softening depends on the choice of solvent and the immersion time. The component parts are then joined and clamped

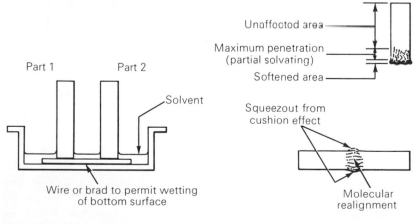

34-2   How solvent bonding occurs.

lightly. As solvent evaporates, molecules from each section regroup into new chains. Complete evaporation requires from 24 h to a week or more, depending on the extent of solvent penetration. Parts may be handled within a few minutes, but they are not at full strength. As residual solvent evaporates, strength of joint increases. An adequate bond is evidenced when the strength at the bond line approaches or meets the tensile or impact strength of the material in an unbonded state.

More serviceable bonds are achieved when mating surfaces are smooth and parallel so that equal penetration and uniform pressure are realized. When irregular surfaces are present, or when more "open-assembly" time is required, solvents are "bodied" by dissolving small amounts (to 25 percent) of the polymer. This technique permits a certain amount of gapfilling and slows evaporation rate enough to allow longer assembly times.

Careful selection of the best solvent is important. When stresses are molded or thermoformed into a product, solvent penetration or vapors will cause crazing or warping as stresses are relieved.

One method of detecting stresses that exist or occur under load is to suspend a sample as a cantilevered beam and to brush a solvent on the top surface. Deflection of the beam under a small weight will disclose stressed areas as a series of craze marks.

The effect of solvents on thermosetting resins is not detrimental when molded or laminated products are fully polymerized. Melamine table and counter tops resist

34-3   Solvent-cemented sign of thermo-formed cellulose acetate butyrate sheet must withstand high-wind loads. (*Reprinted by permission of Ad-Art, Inc.*)

any action by solvents or spilled chemicals. Urea is widely used for cosmetic jars and closures, because it is unaffected by oil and greases found in most cosmetics. Distributor housings in automotive ignition systems molded from phenolic resist attack from gasoline, oil, and cleaning solvents. However, special grades of phenolic are produced for closures of food and pharmaceutical bottles to avoid taste contamination.

Summarizing the behavior of plastics in the presence of solvents, we can note the following:

1 Solvents aid in the identification of plastics by noting their relative behavior.

2 Many thermoplastics can be solvent bonded through use of specific solvents or solvent blends.

3 Incorporation of plasticizers, dyes, and other additives is facilitated through judicious selection of solvent.

4 Stresses in products can be detected by use of coatings or solvent vapors.

5 Preparation of solutions for coatings, impregnants, spray-up, adhesives, and in extrusion of film and fibers is made possible.

6 Slicing thin sheets from softened slabs is accomplished by solvent inclusion.

7 Finishing operations are conducted by using solvent dip or vapor polishing.

8 Thin sheets are made pliable for stretching over cores.

9 Aid in cleaning equipment.

Plasticizers in one polymer can have an adverse affect on another in some instances, especially when migratory. A highly plasticized vinyl object can damage a varnished piece of furniture; a housing containing a volatile plasticizer may cause crazing of a thermoplastic component such as a styrene dial face. Hence, the effect of plasticizer exudation must be thoroughly investigated.

EFFECT OF ACIDS AND ALKALIES

Even though thermosetting resins are insoluble, they are adversely affected by strong alkalis. Molded articles of phenolic immersed in solutions of caustic are degraded beyond usefulness.

Plastics that do not resist solvents often prove exceptionally resistant to strong acids and alkalies. Oxidizing acids (nitric, chromic) have no effect on rigid vinyls, polystyrene, or fluorocarbons. Polyester and phenol-furfural formulations are successfully employed as coatings or laminate binders in chemical and refinery piping to conduct hot sulfuric, phosphoric, and other strong acids.

EFFECT OF OILS

Derived from animal, mineral, or vegetable origins, oils can have a deleterious effect on some thermoplastics and synthetic rubber. Plastics generally exhibit good resistance when in contact with oil. The effect of oils on plastics is evidenced as swelling, cracking, gain in weight, or weight loss, caused by leaching of soluble constituents. Low-molecular-weight polyethylenes, ethylene-vinyl acetate, and cellulose acetate butyrate are examples of thermoplastics having poor oil resistance.

EFFECT OF COMMON HOUSEHOLD CHEMICALS

Quality control and research testing are constantly practiced by resin manufacturers and processors to evaluate the effects of common household chemicals. The average householder seldom realizes the extent of testing carried out to determine the performance of packaging items, food-storage containers, refrigerator components, blenders, mixers, and can openers that come in contact with a variety of chemicals. Kitchen, bath, laundry, garage, and garden provide markets for plastic products that encounter numerous chemicals. Bleach, cleaning fluids, waxes, insecticides, paint thinner, fruit juices, vinegar, olive oil, and spices represent only a few of the numerous chemicals found in the average home.

Major resin suppliers can furnish detailed information on the relative merits of their resins against hundreds of chemicals. The thermoplastics most widely used are the polyolefins because of their general chemical inertness—specifically, polystyrene for low cost and clarity, polyvinyl chloride for low cost and toughness, and methyl pentene for low cost, light weight, and toughness. Dispensers, mixers, blenders,

34-4   Nylon wheels withstand oil and grease without deterioration. (*Reprinted by permission of The Polymer Corp*.)

and appliance housings are molded from ABS, polycarbonate, nylon, and cellulose acetate when impact strength, high gloss, and chemical resistance are required.

TESTING FOR RESISTANCE TO CHEMICAL REAGENTS

Fifty chemicals are listed in ASTM D543-67 to represent the main categories of chemical compounds, solutions, and common industrial products. Specimens are prepared, measured, and weighed prior to immersion in suitable containers for 7 days. When removed, samples are visually examined for change in appearance as shown by dulling, staining, discoloration, decomposition, crazing, cracking, tackiness, and swelling. Weight gain or loss is calculated, as is percentage change in dimensions. Many products can be characterized by percentage loss in some mechanical or electrical value. In other instances, chemical effects are noted at elevated temperatures.

ENVIRONMENTAL STRESS CRACK RESISTANCE (ESCR)

This test is limited to type I (low-density) polyethylenes.

Test samples are scored with a razor blade and bent into a U shape, then inserted into a brass channel. The assembly is placed in a test tube containing the

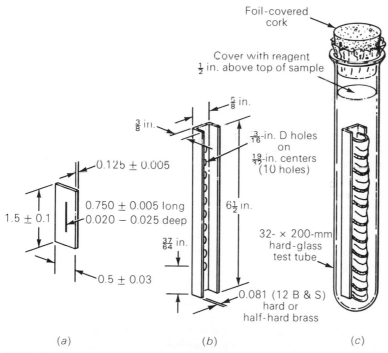

34-5 Test equipment for environmental stress crack resistance. (*a*) Test sample; (*b*) specimen holder; and (*c*) test assembly.

reagent[1] and subjected to a constant temperature bath at 50°C for a maximum of 48 hr. Periodic inspection for visible cracks is undertaken. The evaluation is made of the time to cause failure and the percentage of specimens affected.

The ESCR test is useful to compare various grades and types of polyethylenes, but it is not directly translatable in terms of actual service conditions.

WATER ABSORPTION

Because water exerts a more subtle influence on plastics, the effects of water absorption are discussed separately. Unlike metals or ceramics, many plastics can absorb small amounts of moisture that seriously affect mechanical, electrical, and optical properties. Tests for water absorption are accelerated interpretations that help predict product behavior over long periods in atmospheres of high humidity.

The short-term effect of water is evidenced in several ways, depending on the type of material:

1 Transparent sheets or films may become cloudy (cellulose acetate; polycarbonate).

2 Thin films may wrinkle or curl (polyvinyl alcohol; regenerated cellulose).

3 Products may increase in weight, accompanied by volumetric change (molded thermosets; laminates; polyamides).

4 Electrical insulation values are reduced in relation to the amount of moisture absorbed.

Tests for water absorption

Tests are conducted on standard molded disks of 2-in. diameter and 0.125-in. thickness or on cast, extruded, or calendered sheet 3 in. long and 1 in. wide, by acceptable commercial thickness, when individual, material values are investigated. Sheet thickness of 0.125 in. is required when direct comparison with molded materials is made.

Procedure: Samples are preconditioned by oven drying for 24 h at 122°F (50°C), and then cooled in a desiccator prior to weighing. This is called conditioned weight. After the specified immersion period, samples are wiped dry and reweighed. Percentage increase in weight to the nearest 0.01 percent averaged over at least three specimens is calculated:

$$\text{Increase in wt, \%} = \frac{\text{wet wt} - \text{conditioned wt}}{\text{conditioned wt}} \times 100$$

Sample problem: A cellulose-filled urea disk has a conditioned weight of 9.654 g. After 24-h immersion it weighs 9.703 g. What percent of increase occurred?

[1] The reagent is Igepal CO630 (Antarox A400) alkyl aryl polyethylene glycol.

Solution:

$$\frac{9.703 - 9.654}{9.654} = \frac{0.049}{9.654} = 0.005 \times 100 = 0.5\%$$

A 24-h immersion period is common, but other times and conditions are considered. Heat-resistant plastics are subjected to a 2-h boil, which is considered equivalent to 30 days of immersion at room temperature. Other conditions are specified when long-term effects are desired. For instance, samples are subjected to cyclings of immersion and drying to learn how materials specified for dinnerware, washers, dryers, and similar items would react. Long-term immersion may involve many months, with periodic checks on weight change or dimensions to determine the extent of saturation.

Preparation of samples and interpretation of results must be conducted with some appreciation as to how the material was processed. Compression and transfer-molded parts have a pure resin outer surface that reduces the tendency to absorb moisture. Breaking this skin, as in machining, exposes filler particles for increased moisture pickup. Laminates, particularly glass and cotton reinforced, absorb moisture as a result of capillary action along the fiber. Poor quality laminates show the effect of air entrapment, inferior bonding of resin to the glass interface, and porosity by increased water absorption. Cotton-base laminates absorb moisture in the cellular structure, resulting in swelling. Improper use of canvas laminates for bearings, guides, and support pads results in swelling caused by water absorption. Soaking the item in an oil-graphite solution impregnates the reinforcement, permitting water to be used as a lubricant without damage.

Table 34-1 shows the major plastics and the extent of water absorption. In

TABLE 34-1 Water absorption of common plastics, %

Plastic	Percent
ABS	0.20–0.45
Acetal	0.22–0.25
Alkyd	0.50–0.25
Acrylic	0.30–0.40
Allyl resins:	
Cast	0.03–0.44
Molded:	
Glass filled	0.12–0.35
Mineral filled	0.20–0.50
Synthetic fiber filled	0.20–0.22
Cellulose plastics:	
Cellulose nitrate	1.00–2.00
Cellulose acetate	2.00–7.00
Cellulose triacetate	2.00–4.50
Cellulose acetate butyrate	0.90–2.20
Cellulose propionate	1.20–2.80
Ethyl cellulose	0.80–1.80

TABLE 34-1 Water absorption of common plastics, %
(*Continued*)

Plastic	Percent
Epoxy resins:	
Cast:	
No filler	0.08–0.15
Aluminum filled	0.10–4.00
Laminated:	
Glass fiber	0.04–0.30
Paper base	0.15–0.50
Fluoroplastics:	
CTFE	0.00
TFE	0.00
FEP	0.01
PVF	0.04
Ionomer	0.10–1.40
Melamine:	
Alpha cellulose filled	0.10–0.60
Cotton flock filled	0.16–0.30
Asbestos filled	0.08–0.14
Macerated fabric filled	0.30–0.60
Glass fiber filled	0.09–0.21
Nylon:	
Type 6	1.30–1.90
Type 6/6	1.50
Type 6/10:	0.40
20–40% glass filled	0.20–2.00
Type 11:	
Unfilled	1.10
Glass filled	0.54
Phenol formaldehyde:	
No filler	0.10–0.20
Wood flour filled	0.30–1.20
Asbestos filled	0.10–0.50
Mica filled	0.01–0.05
Macerated or cord filled	0.40–1.75
Glass filled	0.13
Polycarbonate:	
Unfilled	0.15
20–40% glass filled	0.07–0.20
Polyester:	
Cast:	
Rigid	0.15–0.60
Flexible	0.50–2.50
Glass reinforced:	
Preformed	0.01–1.00
Premix	0.06–0.28
Laminated:	
Fabric base	0.20–2.50
Mat base	0.10–0.80

TABLE 34-1   Water absorption of common plastics, %
            (*Continued*)

Plastic	Percent
Polyester (*Cont'd.*)	
Cotton fabric base	1.30–3.50
Paper base	0.10–5.00
Polyethylene:	
Low density	0.015
Medium density	0.010
High density	0.010
Polyphenylene oxide	0.06–0.07
Polypropylene	0.010
Polysulfone	0.22
Polystyrene:	
General purpose	0.03–0.10
Medium and high impact	0.05–0.60
Styrene-acrylonitrile copolymer:	
Unfilled	0.20–0.30
Glass filled	0.08–0.22
Silicone:	
Cast, flexible	0.01
Molded	0.08–0.20
TPX methyl pentene	0.01
Urea formaldehyde	0.40–0.80
Urethane:	
Cast	0.02–1.50
Molded	0.70–0.90
Vinyl polymers and copolymers:	
Polyvinyl butyral	1.00–2.00
Polyvinyl chloride and copolymer:	
Rigid	0.07–0.40
Flexible	0.15–0.75
Polyvinylidene chloride	0.10

general, the chemical composition of the base resin largely influences water resistance. Additives such as plasticizers, fillers, and reinforcements, as well as the fabrication technique, will alter original values of pure compounds. Resins containing chlorine, bromine, or fluorine are water repellent. Polyvinyl chloride, fluorocarbons, and halogenated epoxies and polyesters are good examples. Plastics containing silicone or sulfur are water resistant, as illustrated by silicone elastomers and polysulfide modified epoxies. Resins containing only hydrogen and carbon (polystyrene, polyethylene) are extremely water resistant; resins having an oxy-hydrogen group or oxygen are more susceptible to water absorption. An example of the former is polyvinyl alcohol; the latter, cellulose acetate. Materials that exhibit higher affinity for moisture require predrying before they can be molded or thermoformed to prevent blisters or porosity.

REVIEW QUESTIONS

1   What is the difference between an acid and a base?

2   When is a chemical classified as a solvent?

3   Which solvents are used to dissolve:

    (*a*)   polystyrene

    (*b*)   flexible vinyl

    (*c*)   rigid vinyl

    (*d*)   acrylic

4   What affect does a liquid monomer have on its polymer?

5   What is meant by bodied solvent?

6   What is the advantage of using a bodied solvent?

7   What is the effect of solvent vapors on stressed plastics?

8   Why are thermosetting plastics not affected to the same extent as thermoplastics?

9   Why is a knowledge of solvents important in fabricating and using plastics?

10   Which thermoplastic is chemically inert to most common chemicals?

11   What effect does water absorption have on dielectric strength?

12   What method is used to perform an accelerated test for water absorption?

13   How can water-absorption data be used to determine the quality of a laminate?

14   Which plastics exhibit excellent water resistance?

15   Which plastics exhibit a high percentage of water absorption?

OPTICAL PROPERTIES

35

Plastics have been popular partly because of the exceptional clarity of many resins. Almost all resins are either transparent or translucent, unless modified by additives or fillers. Some are transparent as thin films, becoming translucent as thickness increases. An example is polyethylene.

The basis of comparison of optical properties (other than among different plastics) is plate glass or optical glass. As a group, plastics are less than half the weight of glass, are not as cold to the touch, have higher impact strength, do not shatter into sharp fragments, and are more easily formed, machined, and polished. Glass, on

35-1 Transparent and translucent lamp covers are injection molded in cellulose acetate. (*Reprinted by permission of Eastman Chemical Corp.*)

the other hand, is much more scratch resistant than the hardest plastic; it withstands higher temperature, is more weather resistant, and is a more suitable dielectric at high temperature.

In studying the optical qualities of plastics, it should be noted that lack of hardness, discoloration, shrinkage, moisture absorption, and thermal influences depreciate their original values.

Specific optical properties that influence the selection of any material are as follows:

Index of refraction: The amount of bending a ray of light experiences upon entering a transparent substance obliquely. The refractive index is a ratio of the speed of light in a vacuum (or air) to the speed of light in a transparent substance.

Light transmission as a function of wavelength: The ability of various plastics to transmit various wavelengths of light (colors) is of fundamental importance to designers of optical equipment.

Stress-optical sensitivity: Certain transparent plastics, when observed under polarized light, exhibit a photoelastic phenomenon when stressed. This permits a visual examination of internal stresses present in the molded or fabricated item.

35-2   Weather resistance and dimensional stability are requirements for selection of Acrylite ® acrylic roof panels in Florida shopping center. (*Reprinted by permission of American Cyanamid Co.*)

To be able to examine these properties and apply values to transparent plastics, a brief review of the properties of light is in order.

INDEX OF REFRACTION

The ratio can be expressed as

$$R = \frac{V\hat{v}}{V\hat{s}}$$

where R = index of refraction

$V\hat{v}$ = velocity of light in vacuum

$V\hat{s}$ = velocity of light in the transparent substance (solid, liquid, or gas)

Light travels in the vacuum of space at a speed of 186,285 mps, slowing slightly as it enters earth's denser atmosphere. Since the slowing down is extremely small by comparison, the speed of light in air is regarded as approximately the speed in vacuum. But as light enters a substance of greater density than air, it is slowed further. Thus one can appreciate the difficulty of measuring the variations in the speed of light. A practical method involves the use of a refractometer, which measures the index of refraction by observing the behavior of light as it enters and leaves a transparent material at an angle from its surfaces.

Figure 35-3 shows a beam of light entering a transparent plastic. The right-hand edge R of the beam front (facing in the direction the beam travels) strikes the surface first. From this instant until the corresponding point L in the left-hand edge reaches the surface, R travels more slowly in the plastic than L travels in the air. Consequently, their relative positions R' and L' have changed; the beam is refracted (bent) upon entering the material.

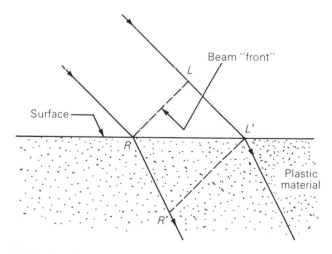

35-3   Light beam entering transparent substance.

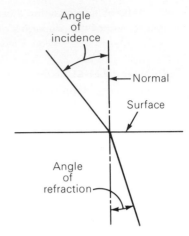

35-4   Angles involved in computing index of refraction.

What is true of a beam of light, that is, a bundle of rays, is true of each individual ray. The path of the ray is readily detected, and its course, upon entering and leaving the surfaces of a transparent substance, is plotted by observing the angles of incidence and refraction. These angles are measured from a line perpendicular to the surface at the point where the ray crosses it. Index of refraction is then calculated

$$\text{Index of refraction} = \frac{\text{sin of angle of incidence}}{\text{sin of angle of refraction}}$$

TABLE 35-1   Refractive index for transparent materials

Material	Refractive index
Allyl resin (cast)	1.50–1.575
Cellulose acetate	1.46–1.49
Epoxy, cast	1.55–1.61
Nylon, type 6/6	1.53
Phenolic, cast	1.50–1.70
Polycarbonate	1.586
Polyester	1.52–1.57
Polyethylene	1.51–1.54
Polymethyl methacrylate	1.48–1.52
Polypropylene	1.49
Polystyrene	1.59–1.60
Polysulfone	1.63
Polyvinyl chloride	1.52–1.55
Urea formaldehyde	1.54–1.60
Water	1.33
Crown glass	1.50–1.55
Flint glass	1.58–1.72
Quartz	1.45–1.53
Diamond	2.42

ASTM method D542-50 describes the technique of determining refractive indices of transparent plastics using the Abbe refractometer.

Calculations based on the index of refraction (see Table 35-1) are important in the design of lenses of glass for cameras, binoculars, telescopes, projectors, corrective glasses, and safety glasses. Contact lenses are ground from special grades of acrylic. Ordinary glasses continue to be made from optical glass because of poor scratch resistance of any suitable plastic.

Light bending

The ability of transparent materials to bend light around curves and to pipe light throughout their structure is a well-known phenomenon. Glass, as well as plastics, is used to advantage, although plastics are more easily fabricated for this purpose. A practical application of light bending is shown in Fig. 35-5, in which an acrylic rod is heated and formed to shape. The light source is at some distance from the tip, providing a more comfortable instrument to handle. A measure of safety is gained over use of a glass rod, which would be equally as good for light piping.

A more recent application of light piping is in the field of fiber optics. Extruded monofilaments are used to transmit light considerable distances in decorative and functional applications.

The ability of a transparent material to bend light is based on internal reflectance as the rays refract from one point of a surface to another. The higher the index of refraction, the greater the amount of light reflected internally with less light escaping at curves. The angle at which total internal reflection occurs is called the critical angle. This can be expressed as a trigonometric function of the refractive index. For example, the critical angle for polymethyl methacrylate is approximately 42 deg. This is determined as follows:

$$\text{Sin of critical angle} = \frac{1}{\text{index of refraction}}$$

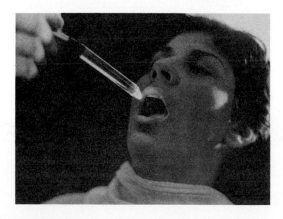

35-5 Medical and dental probes utilize light-bending property of acrylic. (*Reprinted by permission of Los Angeles Trade-Technical College.*)

For polymethyl methacrylate :

$$\text{sin of critical angle} = \frac{1}{1.4885} = 0.67183$$

The angle for which 0.67183 is the sin is 42 deg, 12.50 sec.
For polystyrene :

$$\text{sin of critical angle} = \frac{1}{1.59} = 0.62892$$

The angle for which 0.62892 is the sin is 38 deg, 58 min. This means that when light transmitting through the material strikes the surface at any angle less than the critical angle, it escapes into the air, but rays striking the surface at any angle greater than critical are reflected back into the substance.

This can be shown graphically, as in Fig. 35-6. By considering the 90 deg of any quadrant, it becomes apparent that the smaller the critical angle (or the higher the index of refraction), the greater is the difference between the critical angle and 90 deg, thus the greater number of rays reflected internally.

The practical limits for a good light-bending application are :

1   Radius of curvature should be limited to a minimum of 3 times the sheet thickness or rod diameter to preclude light escaping at the bend.

2   Light source should be well ventilated or nonheat producing.

3   Surfaces should be highly polished and free of scratches.

Edge lighting

Another application that depends upon internal reflection is that of edge lighting. Light introduced at the edge of transparent material is transmitted between highly

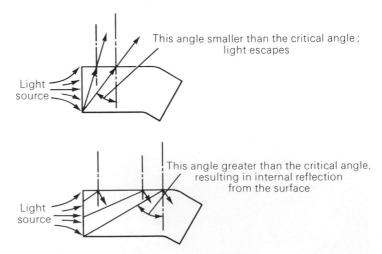

35-6   Light piping caused by internal reflectance.

polished surfaces until it reaches a roughened or recessed area, where it escapes. Automotive and aircraft instrument dials, clock faces, and novelty items are molded or machined to provide illumination.

LIGHT TRANSMISSION AS A FUNCTION OF WAVELENGTH

Optical characteristics of transparent plastics may be further understood if we examine the behavior of white light as it passes through them. Radiant energy, as from the sun or other source, exists as vibrations that are propagated through space in a wave pattern much as waves travel along a rope when one end is moved up and down rapidly.

Our eyes are sensitive to a portion of these radiations, which we call visible light. The physical difference between visible light, and invisible radiations that usually accompany it, is a matter of wavelength. Our senses can register radiations whose wavelengths range from 4,000 to 7,000 Angstroms (Å)*. We are most sensitive to rays of approximately 5,600-Å wavelength. Stated another way, our eyes are most sensitive to green light. The different wavelengths of light produce the different color sensations that we register through our eyes. A blend of all colors appears white to us. We call a certain band of light "red," but the optical designer states it as a wavelength of 6,700 Å.

Figure 35-9 illustrates the following:

1   Color-response curve for the human eye (visible spectrum)

2   The visible spectrum expanded, showing the approximate color sensation caused by each band

3   A spectrum of electromagnetic radiations (waves), including the narrow band of visible light

*Angstrom = 0.00000001 centimeter, or 10^{-8} cm.

35-7   Selector switch is housed in injection-molded acrylic for edge-lighting application. Light appears only through unpainted areas. (*Reprinted by permission of Rohm and Haas Co.*)

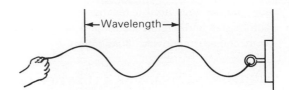

35-8 Propagation of light as a function of wavelength.

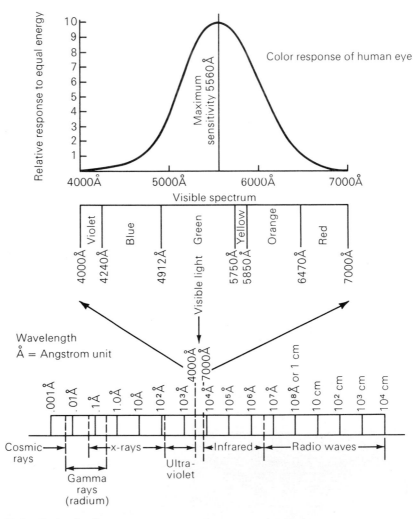

35-9 Scale of radiant energy.

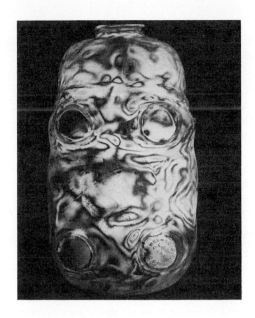

35-10 Internal stresses viewed under polarized light. Toy is molded in clear polystyrene. (*Reprinted by permission of Los Angeles Trade-Technical College.*)

The influence of various wavelengths of light is of interest to designers of optical equipment, as stated previously, but it is of primary concern also to manufacturers of equipment or users of plastics when protection from the harmful effects of radiation is desired.

Ultraviolet light is recognized as beneficial to humans, enabling us to manufacture vitamin D within our bodies. Acrylics have the ability to transmit at least a portion of these rays, whereas plate glass does not. For this reason, acrylic panels are installed in solariums of hospitals and for sun decks. Acrylic panels are too costly and too easily scratched by ordinary cleaning methods to be practical in the home. Clear polyethylene film has proven successful for greenhouses and for protecting young seedlings in outdoor planting. By allowing passage of some ultraviolet light, and at the same time retaining ground moisture, crops are assured a healthier start.

Few clear plastics remain unaffected from exposure to ultraviolet light, which tend to discolor, yellow, or darken them. By incorporating traces of blue colorant, yellowing is masked to a more acceptable green when deterioration occurs. Adding carbon black is common practice with polyethylene and vinyls. Other ultraviolet absorbers used when transparency or pastel colors must be maintained are benzophenones,[1] benzotriazoles,[2] and monobenzoates,[3] among others.

The use of lead powder as fillers for phenolic-molding compounds or cast phenolic for x-ray applications was mentioned previously, as was the use of gamma-ray exposure to crosslink polyolefins.

[1] Uvinul is a benzophenone and is a trade name of General Aniline and Film Corporation.
[2] Tinuvin is a benzotriazole and is a trade name of Geigy Chemical Corporation.
[3] Eastman-RMB is a monobenzoate and is a trade name of Eastman Chemical Products.

STRESS-OPTICAL SENSITIVITY

The ability of some materials to exhibit double refraction of light when placed under stress is referred to as stress-optical sensitivity. Certain natural-occurring crystals, such as Iceland spar, exhibit this property.

In Fig. 35-11, two beams of light are shown emerging from the far side of the crystal; they are designated R and S. As the crystal is rotated, spot S remains stationary, and R rotates around S. The single light beam has been doubly refracted, being divided into two beams traveling different paths by the crystal.

Many noncrystalline substances exhibit this same property (birefringence or double refracting) when stressed. This reaction to stress, for example, in bending or twisting, is called stress-optical sensitivity, and it forms the basis for photoelastic examination of structural members or products that must perform with a high order of reliability.

The purpose of stress analysis is to determine by mathematical means the required size, shape, and strength of the component. The object is to gain sufficient strength without any more material than necessary, and thus reduce cost, weight, or space occupied. There are many such parts, especially when shapes are complex, in which stress is not easily predicted or analyzed. For example, the stresses in a bridge section can be calculated, but the manner in which they are sometimes applied is not always predictable. Sharp changes in cross section of components or sharp internal corners are usual locations of serious stress concentrations.

It is impractical to build a prototype of steel or concrete just to perform destruction tests, but transparent scale models of cast phenolic, polyester, acrylic, polystyrene, or cellulose acetate are easily constructed. When placed under stresses to simulate actual loading conditions, the stresses can be examined visually or photographed, and then compared. Weak sections are revealed for design corrections.

To understand the function of a polarizer, let us return to the analogy by which the nature of light was depicted in Fig. 35-8. In this illustration, as the vibrations travel from the source at the left, each individual point on the rope moves in an up-and-down direction (plane). The vibrations could just as well be in a horizontal plane or at any angle between vertical and horizontal. Actually light emitted from common sources of illumination consists of individual rays vibrating in random planes.

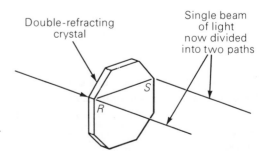

Double-refracting crystal

Single beam of light now divided into two paths

35-11 An example of double refraction.

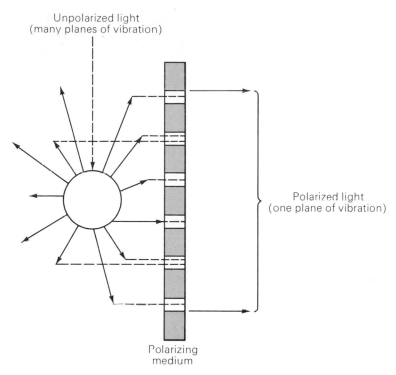

Unpolarized light
(many planes of vibration)

Polarized light
(one plane of vibration)

Polarizing
medium

35-12   The function of a polarizer.

A polarizer can be thought of as a grating with slots in parallel alignment. The slots permit rays vibrating in the corresponding parallel plane to pass through unhindered, but they block the rays vibrating in any crosswise direction. The light passing through is said to be polarized; its vibrations are all parallel.

In examining the plastics for stress, the sample is placed as shown in Fig. 35-13.

Many products benefit from photoelastic analysis whether they are to be made from plastic or metal. Gears, aircraft pulleys, struts, bulkheads, valve parts, and tape reels are examples of products examined through stress-optical methods.

As a production quality control measure, injection- or blow-molded parts can be quickly analyzed to determine the extent of flow lines, weld areas, and stress concentrations created by incorrect degating or ejection. Stresses caused by molding at reduced temperatures of melt or mold can be detected, as well as those created by improper mold or product design. As a research device or for quality control, examination by polarized light can be instrumental in scheduling annealing cycles for stress relief or to ascertain stress level caused by machining.

Not all transparent plastics are stress optical. Among those that are stress optical, some are better than others. Cast phenolic is excellent and can be cast to the required shape or made into slabs for machining. In the thermoplastic class, polystyrene,

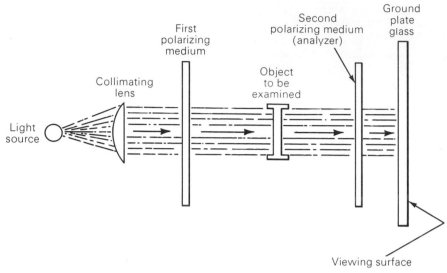

35-13   Setup for examination of stress-optical sensitivity.

cellulose acetate, acrylic, and polycarbonate are used for models. Acrylic and poly-carbonate, although stress optical, show only gray bands in stressed areas, but the others exhibit a variety of multicolored bands of light.

REVIEW QUESTIONS

1   What are the advantages of using plastics for optical applications? What are some of the disadvantages?
2   What is meant by index of refraction?
3   Which thermoplastics exhibit the highest refractive indices?
4   Explain how light bending occurs.
5   What applications for plastics incorporate the light-bending principle?
6   Which wavelengths of light are called visible radiations?
7   How is stress concentration observed in transparent plastics?
8   Which plastics are stress optical?
9   What product-design considerations and molding techniques should be observed to reduce stress concentrations?
10   Of what value is a stress-optical examination to the processor of plastic products?

THE ROLE OF THE
PLASTICS TECHNICIAN

A

If the plastics industry is to become the dominant economic force in the world, as many people predict, it is necessary to attract many thousands more of skilled persons during the decade 1970 to 1980.

A substantial need exists for industrial and mechanical engineers who specialize in plastics processing and related fields. Support for engineering personnel must be provided by plastics technicians or technologists. The ratio of engineers to technicians, in terms of numbers, is difficult to project precisely, but no doubt skilled technicians will be required in far greater numbers.

Critical shortages must also be resolved in the area of tool and die making, if processing is to keep pace with public acceptance of plastics.

Designers of products, tooling, accessories, fixtures, and machinery, plus draftsmen to execute working drawings, are always in demand.

Qualified plastics technicians have unique opportunities to obtain key positions in any facet of the industry because of their specialized training and skill. They know the language of their trade and exhibit proficiency at their tasks.

Exactly what is demanded of the plastics technician? What knowledge and skills should he possess? The answers greatly depend upon the particular area of plastics processing the future technician chooses. Specific processes require specific knowledge. For instance, injection-molding skills are completely different from

551

those used in working with reinforced plastics; preimpregnating processes are foreign to those in unrelated fields such as blow molding or thermoforming.

A career in plastics is like that in medicine. For a medical person to become a heart specialist or a neurologist, he must first learn anatomy and other basic subjects just as the plastics technician must also learn basic principles pertaining to his industry before he can continue into the area of his choice and become a specialist. The more familiar one becomes with materials and methods, the greater are the opportunities over a broader segment of the industry.

Industry's demands, in general, can be met by acquiring certain bodies of knowledge. The following is set forth as guidelines, not to be construed as either complete or required by all types of positions or all processors.

MATERIALS

1 Ability to recognize differences among types and within grades of specific types

2 Knowledge of solvents, adhesives, and coatings that apply to each material

3 Appreciation for hazards prevalent in handling a few liquid components of resin systems

4 Understanding the folly of mixing or overheating materials which could prove disastrous

5 Recognizing defective raw materials, either in process or manufactured products, and performing effective remedies

6 Knowledge of general applications (end use) of materials

MACHINERY

Processing equipment

1 Know how to safely set up, operate, and maintain specific types of machines

2 Be able to install and remove molds properly to prevent damage or injury

3 Understand the importance of, and be able to calculate hydraulic and pneumatic pressure

4 Know what constitutes good safety practices for specific operations

Auxiliary equipment

Be able to set up, operate, and maintain equipment and machines used as accessories to the major endeavor

Testing equipment

1 Be able to interpret specifications, drawings, and test methods

2   Be able to operate test equipment safely and accurately

3   Be able to extract and report data from performance tests

Tools and instruments

1   Be skilled in the use of micrometers, vernier and other calipers, dial gauges, gram scales, analytical balances, and hand tools

2   Be able to use circuit testers and ohm and volt meters

3   Be able to use power equipment, such as drill press, circular saw, band saw, router, lathe, grinding wheel, sander, and polishing equipment

VERBAL SKILLS

1   Be able to read, understand, and carry out directives in the area of responsibility

2   Be able to read and assimilate new information presented by materials suppliers, equipment manufacturers, technical journals, and other sources

3   Be able to prepare job-record cards, production orders, procedures, lab reports, and communications

4   Be able to prepare or interpret charts and graphs

5   Be able to organize filing and reference systems

6   Be able to develop good telephone technique

SOCIAL AND CHARACTER TRAITS

1   Be able to accept and exhibit responsibility

2   Be able to demonstrate dependability and honesty

3   Be able to learn new operational procedures

4   Be able to supervise other people

5   Be able to train new personnel properly

6   Be able to meet and deal properly with all people from different socio-economic levels

What are some of the job titles that apply to positions in the industry? A partial list can provide an insight into the diversity of opportunities at varying degrees of preparation.

MOLDING PLANT

Injection, compression, transfer, extrusion, blow molding, and rotational molding

Materials handling

1   Receiving and shipping
2   In-plant: dispersing, blending, coloring, granulating
3   Warehouse and inventory control

Production

Die-setter, press or equipment operator, lead man, foreman, machinery maintenance, department supervisor, and plant manager

Finishing department

Machinery operations, deflashing operations, painting and decorating, and assembly

Testing and quality control

Test technician, inspector, and supervisor

Sales

Inside sales and orders, outside sales and customer liaison, and sales manager

Purchasing

Materials, supplies, and tooling

LAMINATING PLANT

Tool builder, pattern maker, mold preparation and maintenance, laminator, preformer operator, chopper-gun operator, filament or tape winder, batch mixer, inspector, and quality control

COATING AND IMPREGNATING PLANT

Batch mixer, tower operator, roll handlers, inspectors, research technician, and test technician

CALENDERING PLANT

Batch mixer, roll-mill operator, compounder operator, calender operator, laminator, and printing or embossing operator

FOAM PRODUCTION PLANT

Batch mixer, foam-equipment operator, mold builder, and mold strippers

The foregoing list of employment opportunities points out the wide variety of skills and talents that can be utilized. Thus the role of a plastics technician is difficult to define in precise terms. Just as the term doctor can describe many kinds of doctors in medicine, each eminently qualified in his particular field, the term plastics technician describes many kinds of functions.

TABLES
OF TRADE NAMES
AND MANUFACTURERS

B

Name	Manufacturer
	Molding compounds
Phenol-formaldehyde:	
Durez	Durez Division, Hooker Chemical Corp.
Bakelite	Union Carbide Corp.
Polyplenco	Plastics Engineering Co.
Resinox	Monsanto Co.
Phenol-furfural:	
Durite	Borden Chemical Corp.
Durez	Durez Division, Hooker Chemical Corp.
Plaspreg	Furane Plastics, Inc.

Chapter 2 (Cont'd.)

Name	Manufacturer
Laminating resins	
Phenol formaldehyde:	
Resinox	Monsanto Co.
Plyophen	Reichhold Chemicals, Inc.
CTL 91 LD	Cincinnati Testing Labs, Inc.
Ironsides	Ironsides Resins, Inc.
Phenyl-silane:	
Resinox SC 1013	Monsanto Co.
CTL 37-9x	Cincinnati Testing Labs, Inc.
Casting resins	
Catalin	Catalin Corp.
Rezolin	Rezolin Division, Hexcel Corp.
Marblette	Marblette Corp.

Chapter 3

Name	Manufacturer
Molding compounds	
Phenolic:	
Fiberite	Fiberite Corp.
Urea:	
Plaskon	Allied Chemical Corp.
Beetle	American Cyanamid Co.
Fiberite	Fiberite Corp.
Melamine:	
Plaskon	Allied Chemical Corp.
Cymel	American Cyanamid Co.
Fiberite	Fiberite Corp.
Laminating resins	
Melamine:	
Resimene	Monsanto Co.
Catalin	Catalin Corp.
Diaron	Reichhold Chemicals, Inc.
Adhesives	
Urea:	
Urac	American Cyanamid Co.
Catalin	Catalin Corp.
Casco	Borden Chemical Co.
Plaskon	Allied Chemical Corp.
Melurac	American Cyanamid Co.
Plyamine	Reichhold Chemicals, Inc.

Chapter 3 (Cont'd.)

Name	Manufacturer

Adhesives (*Cont'd.*)

Name	Manufacturer
Urea (*Cont'd.*) :	
Weldwood	U.S. Plywood Corp.
Lauxite	Monsanto Co.
Melamine :	
Cymel 401	American Cyanamid Co.
Catalin 852	Catalin Corp.
Diaron	Reichhold Chemicals, Inc.
Lauxite	Monsanto Co.

Surface coatings

Name	Manufacturer
Urea :	
Uformite	Rohm and Haas Co.
Beckamine	Reichhold Chemicals, Inc.
Beetle	American Cyanamid Co.
Resimene U	Monsanto Co.

Chapter 4

Name	Manufacturer
Polyester resins :	
Atlac	Sealzit Division, Flintkote Co.
Chevron	California Chemical Co.
Hetron	Durez Division, Hooker Chemical Corp.
Koplac	Koppers Co., Inc.
Laminac	American Cyanamid Co.
Paraplex	Rohm and Haas Co.
Polylite	Reichhold Chemicals, Inc.
Selectron	Pittsburgh Plate Glass Co.
Stycast	Emerson and Cuming, Inc.
Vistron	Silmar Chemical Corp.
Diallyl phthalate resins :	
Dapon	FMC Corp.
Diallyl isophthalate resins :	
Dapon M	FMC Corp.
Peroxide catalysts :	
Aposet	Apogee Chemicals, Inc.
Luperco ⎫	
Luperox ⎬	Lucidol Division, Pennwalt Corp.
Lupersol ⎭	
Cadet ⎫	
Cadox ⎬	Chemetron-Noury, Inc.
Garalyst ⎫	
Garox ⎭	Ram Chemicals
RCI	Reichhold Chemicals, Inc.
Chlorowax	Diamond Shamrock Corp.

Chapter 4 (Cont'd.)

Name	Manufacturer
Molding compounds:	
Polyester:	
Cyglas	American Cyanamid Co.
Premix	Premix, Inc.
Alkyd:	
Glaskyd	American Cyanamid Co.
Plenco	Plastic Engineering Co.
Plaskon	Allied Chemical Corp.
Premix	Premix, Inc.
Diallyl phthalate:	
Diall	Allied Chemical Corp.

Chapter 5

Name	Manufacturer
Epi-Rez	Resins Division, Celanese Corp.
Araldite	CIBA Products Co.
DER and DEN series	Dow Chemical Co.
Epotuf	Reichhold Chemicals, Inc.
Epon	Shell Chemical Co.
Bakelite ERL series	Union Carbide Corp.

Epoxy formulators (a partial list) and the type of epoxy

Name	Manufacturer	Type*
Armstrong	Armstrong Products, Inc.	A, C, T
Devcon	Devcon Corp.	A, T
Epiall	Allied Plastics Corp.	M
Epocast	Furane Plastics, Inc.	A, C, E, F, M, T, Ct
Epolite	Rezolin Division, Hexcel Corp.	A, C, E, M, T
Epoxylite	Epoxylite Corp.	A, C, E, M, T, Ct
Helix	Carl H. Biggs Co.	A, C, E
Hyflo	Hysol Division, Dexter Corp.	M
Hysol	Hysol Division, Dexter Corp.	A, C, E, F, T, Ct
Job Ready	Job Ready Plastics	F
Kish	Kish Industries, Inc.	T
Maraset	Marblette Corp.	C, T
Plenco	Plastics Engineering Co.	M
Ren	Ren Plastics, Inc.	A, C, T

*Code: *A* = adhesives; *C* = castings; *Ct* = coatings; *E* = encapsulating resins; *F* = frozen epoxies; *M* = molding compounds; and *T* = tooling resins.

Chapter 6

Name	Manufacturer
Silicone molding compounds:	
500 series	Dow-Corning Corp.

Chapter 6 (Cont'd.)

Name	Manufacturer
Silicone laminating resins:	
DC 2104, 2105, 2106, 7141	Dow-Corning Corp.
Silicone elastomers:	
Silastic RTV	Dow-Corning Corp.
Silicone RTV	General Electric Co.

Chapter 11

Name	Manufacturer
Epoxy tooling resins:	
Epocast	Furane Plastics, Inc.
Hysol	Hysol Division, Dexter Corp.
Kish	Kish Resins
Ren-ite	Ren Plastics, Inc.
Tool-Plastik	Rezolin Division, Hexcel Corp.
Epoxical	U.S. Gypsum
Plaster:	
Hydrocal	U.S. Gypsum

Chapter 15

Name	Manufacturer
Polyethylene	
Alathon	E. I. du Pont de Nemours & Co.
Ameripol	Goodrich-Gulf Chemicals, Inc.
Chemplex	Chemplex Co.
Fortiflex	Celanese Plastics Co.
Hi-Fax	Hercules Powder Co.
Marlex	Phillips Petroleum Co.
Petrothene	Division of National Distillers Corp., U.S. Industrial Chemicals
Polyethylene	Dow Chemical Co.
	Eastman Chemical Products, Inc.
	Shell Chemical Co.
	Union Carbide Corp.
	W. R. Grace
Polypropylene:	
Avisun	Avisun Corp.
Chevron	California Chemical Co.
Escon	Enjay Chemical Co.
Marlex	Phillips Petroleum Co.
Pro-Fax	Hercules Powder Co.
Polypropylene	Shell Chemical Co.
Polyallomer:	
Tenite	Eastman Chemical Products, Inc.

Chapter 16

Name	Manufacturer
Polystyrene :	
Bakelite	Union Carbide Corp.
Dylene	Sinclair-Koppers Co.
El Rexene	Rexene Polymers Co.
Fostarene	Foster-Grant
Kralon	Uniroyal, Inc.
Lustrex	Monsanto Co.
Styron	Dow Chemical Co.
Polystyrene expandable beads or sheets :	
Dylite	Sinclair-Koppers Co.
Fostafoam	Foster-Grant Co., Inc.
Pelaspan ⎱	
Styrofoam ⎰	Dow Chemical Co.
ABS molding resins :	
Abson	B. F. Goodrich Chemical Co.
Cycolac (ABS) ⎫	
Cycoloy (ABS- ⎬	
polycarbonate)	Marbon Division, Borg-Warner, Inc.
Cycovin (ABS-PVC) ⎭	
Dylel	Sinclair-Koppers Co.
El Rexene	Rexene Polymers Co.
Kralastic	Uniroyal, Inc.
Tybrene	Dow Chemical Co.
ABS sheets :	
Boltaron	General Tire and Rubber Co.
Royalite	Uniroyal, Inc.

Chapter 17

Name	Manufacturer
Polyvinyl chloride and copolymer resins :	
Bakelite PVC ⎱	
Vinylite ⎰	Union Carbide Corp.
Exon	Firestone Tire and Rubber Co.
Geon	B. F. Goodrich Chemical Co.
Marvinol	Uniroyal, Inc.
Opalon ⎱	
Vyram ⎰	Monsanto Co.
Pliovic	Goodyear Tire and Rubber Co.
Vygen	General Tire and Rubber Co.
Polyvinylidene chloride :	
Saran	Dow Chemical Co.
Polyvinyl acetate :	
Gelva	Shawinigan Resins Div., Monsanto Co.
Lauxite	Monsanto Co.

Chapter 17 (Cont'd.)

Name	Manufacturer
Polyvinyl alcohol :	
Elvanol	E. I. du Pont de Nemours & Co.
Gelvatol	Shawinigan Resins Div., Monsanto Co.
Monosol (film)	Monosol Corp.
Reynolon (film)	Reynolds Metal, Inc.
Polyvinyl acetal :	
Alvar	Shawinigan Chemicals, Ltd.
Polyvinyl formal :	
Formvar	Shawinigan Resins Div., Monsanto Co.
Polyvinyl butyral :	
Butvar	
Saflex	Shawinigan Resins Div., Monsanto Co.
Vinylite XYHL	Union Carbide Corp.
PVC plastisols :	
Chem-O-Sol	Chemical Products Corp.
Miccrosol	Michigan Chrome and Chemical Co.
PVC film and sheet :	
Koroseal	B. F. Goodrich Chemical Co.
Krene	Union Carbide Corp.
Ultron	Monsanto Co.
Velon	Firestone Tire and Rubber Co.
PVC coated fabrics :	
Cordo	Cordo Chemical Division, Ferro Corp.
Duraleather	Masland Duraleather Co.
Naugahyde	Uniroyal, Inc.

Chapter 18

Name	Manufacturer
Teflon (TFE and FEP)	E. I. du Pont de Nemours & Co.
Halon (TFE)	Allied Chemical Corp.
Tetran (TFE)	Pennsalt Chemicals Corp.
Kel-F (CTFE)	3M Company
Tedlar (PVF)	E. I. du Pont de Nemours & Co.
Dalvor (PVF)	Diamond Alkali
Kynar (PVF_2)	Pennsalt Chemicals Corp.

Chapter 19

Name	Manufacturer
Nylon 6 :	
LNP PF 1006	Liquid Nitrogen Processing Corp.*
Nylafil G 13/40	Fiberfil, Inc.*
Nylatron	Polymer Corp.
Nylon	Foster-Grant Co., Inc.
Nypel	Nypel, Inc.

Chapter 19 (Cont'd.)

Name	Manufacturer
Nylon 6,6 :	
Celanese	Celanese Plastics Co.
LNP RF	Liquid Nitrogen Processing Corp.
Nylafil G 1/30	Fiberfil, Inc.
Plaskon	
Zytel 101-105, 109, 121-122	E. I. du Pont de Nemours & Co.
Nylon 6,10 :	
LNP QF	Liquid Nitrogen Processing Corp.
Nylafil G 2/30	Fiberfil, Inc.
Polycarbonate :	
Lexan	General Electric Co.
Merlon	Mobay Chemical Co.
LNP DF	Liquid Nitrogen Processing Corp.
Polycarbafil	Fiberfil, Inc.
Polyacetal :	
Celcon	Celanese Plastics Co.
Delrin	E. I. du Pont de Nemours & Co.
LNP KF	Liquid Nitrogen Processing Corp.
Formaldafil	Fiberfil, Inc.

*Fiberfil, Inc., and Liquid Nitrogen Processing Corp. do not manufacture base resins; they supply glass-reinforced resins to processors.

Chapter 20

Name	Manufacturer
Cast sheet :	
Acrylite	American Cyanamid Co.
Evr-Kleer	Cast Optics Corp.
Perspex	Imperial Chemicals, Ltd.
Plexiglas	Rohm and Haas Co.
Shinko Lite	Mitsubishi Rayon
Swedcast	Swedlow Plastics
Molding resins :	
Implex (high impact)	Rohm and Haas Co.
Lucite	E. I. du Pont de Nemours & Co.
Plexiglas	Rohm and Haas Co.
Film :	
Korad	Rohm and Haas Co.
Acrylic PVC sheet :	
Kydex	Rohm and Haas Co.

Chapter 21

Name	Manufacturer
Arolast (elastomer)⎱	
Arothane (foam) ⎰	Archer-Daniels-Midland Corp.
Castomer (casting rubber)	Isocyanate Products, Inc.
CPR (foam, elastomer)	CPR Division, Upjohn Co.
Estane (resins, compounds, sheets)	B. F. Goodrich Chemical Co.
Flexane (casting rubber)	Devcon Corp.
Foamthane (resins)	Pittsburgh Corning Corp.
Genthane (rubber products)	General Tire Co.
Multrathane (basic chemicals)	Mobay Chemical Co.
Niax (basic chemicals)	Union Carbide Corp.
Nopcofoam (basic chemicals)	Nopco Chemical Co.
Polylite (basic chemicals)	Reichhold Chemicals, Inc.
Selectrofoam (polyester foams)	Pittsburgh Plate Glass Co.
Texin (molding grades)	Mobay Chemical Co.
Uralane (casting rubber)	Furane Plastics, Inc.
Uralite (casting rubber)	Rezolin, Inc.

Chapter 22

Name	Manufacturer
Cellulose nitrate :	
Nixonite	Nixon Nitration Corp.
Cellulose acetate :	
Molding resin :	
Ampol C/A	American Polymers, Inc.
Lumarith	Celanese Plastics, Co.
Tenite C/A	Eastman Chemical Products, Inc.
Sheet or film :	
Kodacel	Eastman Chemical Products, Inc.
Mirro-Brite (metallized)	Coatings Products, Inc.
Cellulose acetate butyrate :	
Molding resin :	
Ampol CAB	American Polymers, Inc.
Tenite CAB	Eastman Chemical Products, Inc.
Sheet or film :	
Uvex	Eastman Chemical Products, Inc.
Cellulose propionate :	
Forticel	Celanese Plastics, Co.
Tenite C/P	Eastman Chemical Products, Inc.
Ethyl cellulose :	
Ethocel	American Polymers, Inc.
Methyl cellulose :	
Methocel	Dow Chemical Co.
Carboxy methyl cellulose :	
CMC	E. I. du Pont de Nemours & Co.
	Hercules Powder Co.
	Wyandotte Chemical Co.

PROFESSIONAL GROUPS AND ORGANIZATIONS

C

To promote the industry, disseminate information, and to effect favorable legislation, professional groups and organizations exist both locally and nationally. Some of these organizations are:

The Society of the Plastics Industry, Inc. (SPI)
250 Park Avenue
New York, N.Y. 10017

The Society of Plastic Engineers, Inc. (SPE)
656 West Putnam Avenue
Greenwich, Conn. 06830

Plastics Education Foundation (PEF)
4 Lorna Lane
Loudonville, N.Y. 12211

Society of Aerospace Materials and Process Engineers (SAMPE)
P.O. Box 613
Azusa, Calif. 91702

National Electrical Manufacturers Association (NEMA)
155 E. 44th St.
New York, N.Y. 10017

American Society for Testing and Materials (ASTM)
1916 Race Street
Philadelphia, Penna. 19103

SUGGESTED READINGS

Chapter 1

Couzens, E. G. and V. E. Yarsley: "Plastics in the Modern World," Pelican Books, Penguin Books, Inc., Baltimore, 1968.

DuBois, J. H. and F. W. John: "Plastics," Van Nostrand Reinhold Company, New York, 1967.

Kaufman, M.: "The First Century of Plastics; Celluloid and its Sequel," London Iliffe Books, Great Britain, 1964.

Miles, D. C. and J. H. Briston: "Polymer Technology," Chemical Publishing Co., New York, 1965.

Simonds, H. R.: "Source Book of the New Plastics," Van Nostrand Reinhold Company, New York, 1961.

———— and J. Church: "Concise Guide to Plastics," Van Nostrand Reinhold Company, New York, 1963.

Smith, H. W. (ed.): "Manufacture of Plastics," vol. 1, Van Nostrand Reinhold Company, New York, 1964.

Stille, J. K.: "Introduction to Polymer Chemistry," John Wiley & Sons, Inc., New York, 1962.

Whittington, L. R.: "Whittington's Dictionary of Plastics," Technomic Publishing Company, New York, 1968.

Wordingham, J. and P. Reboul: "Dictionary of Plastics," Littlefield, Adams, and Company, Totowa, N.J., 1964.

Chapter 2

Gould, D. F.: "Phenolic Resins," Van Nostrand Reinhold Company, New York, 1959.

Rosato, D. V.: "Asbestos, Its Industrial Applications," Van Nostrand Reinhold Company, New York, 1959.

Chapter 3

Blais, J. F.: "Amino Resins," Van Nostrand Reinhold Company, New York, 1959.

Kinney, G. F.: "Engineering Properties and Applications of Plastics," John Wiley & Sons, Inc., New York, 1957.

Skeist, I. (ed.): "Handbook of Adhesives," Van Nostrand Reinhold Company, New York, 1968.

Chapter 4

Doyle, E. H.: "The Development and Use of Polyester Products," McGraw-Hill Book Company, New York, 1967.

Lawrence, J. R.: "Polyester Resins," Van Nostrand Reinhold Company, New York, 1960.

Martens, C. R.: "Alkyd Resins," Van Nostrand Reinhold Company, New York, 1961.

Raech, H.: "Allylic Resins and Monomers," Van Nostrand Reinhold Company, New York, 1965.

Chapter 5

Cagle, C. V.: "Adhesive Bonding, Techniques and Applications," McGraw-Hill Book Company, New York, 1968.

"Guide for Classifying and Labelling Epoxy Products According to their Hazardous Potentialities," Epoxy Resin Formulators Division, Society of the Plastics Industry, Inc., New York, 1970.

Guttman, W. H.: "Concise Guide to Structural Adhesives," Van Nostrand Reinhold Company, New York, 1961.

Katz, I.: "Adhesive Materials, Their Properties and Uses," Foster Publishing Company, Long Beach, Ca., 1964.

Lee, H. and K. Neville: "Epoxy Resins," McGraw-Hill Book Company, New York, 1957.

"Plastics Safety Handbook," Society of the Plastics Industry, Inc., in cooperation with the National Safety Council, New York, 1959.

Skeist, I.: "Epoxy Resins," Van Nostrand Reinhold Company, New York, 1958.

Chapter 6

Meals, R. N. and T. M. Lewis, "Silicones," Van Nostrand Reinhold Company, New York, 1959.

Chapter 7

Simonds, H. R. (ed.): "Encyclopedia of Plastics Equipment," Van Nostrand Reinhold Company, New York, 1964.

"Thermoset Molding," Society of the Plastics Industry, Inc., New York, 1970.

Chapter 8

J. Butler: "Compression and Transfer Molding of Plastics," Interscience Publishers, Inc., New York, 1959.

DuBois, J. H. and W. I. Pribble (eds.): "Plastics Mold Engineering," Van Nostrand Reinhold Company, New York, 1965.

Randolph, Alan F. (ed.): "Plastics Engineering Handbook," 3d ed., Society of the Plastics Industry, Inc., Van Nostrand Reinhold Company, New York, 1960.

Sors, Laszlo: "Plastic Mold Engineering," Pergamon Press, New York, 1967.

"Compression and Transfer Molding In-plant Training Guide," Society of the Plastics Industry, Inc., New York, 1970

Chapter 9

White, R. B.: "Premix Molding," Van Nostrand Reinhold Company, New York, 1964.

Chapter 10

Berger, H.: "Asbestos with Plastics and Rubber," Chemical Publishing Company, Inc., New York, 1966.

———— and R. Oesper: "Processing of Asbestos with Plastics," Chemical Publishing Company, Inc., New York, 1966.

Brenner, W., D. Lum, and M. W. Riley: "High Temperature Plastics," Van Nostrand Reinhold Company, New York, 1962.

Dietz, A. G. H.: "Materials for Construction: Wood, Plastics and Fabrics," Van Nostrand Reinhold Company, New York, 1960.

Oleesky, S. and G. Mohr: "Handbook of Reinforced Plastics," Society of the Plastics Industry, Inc., Van Nostrand Reinhold Company, New York, 1963.

Sonneborn, R.: "Fiberglass Reinforced Plastics," Van Nostrand Reinhold Company, New York, 1954.

Chapter 11

DeDani, A.: "Glass Fiber Reinforced Plastics," John Wiley & Sons, Inc., New York, 1961.

Riley, M. W.: "Plastics Tooling," Van Nostrand Reinhold Company, New York, 1961.

Sonneborn, Ralph: "Fiberglass Reinforced Plastics," Van Nostrand Reinhold Company, New York, 1954.

Chapter 12

Gaylord, M. W.: "Reinforced Plastics, Theory and Practice," Koppers Company, Inc., Pittsburgh, 1969.

Gibbs and Cox, Inc.: "Marine Design Manual, Fiberglass Reinforced Plastics," McGraw-Hill Book Company, New York, 1960.

Rosato, D. and C. S. Grove, Jr.: "Filament Winding, Its Development, Manufacture, Application and Design," Interscience Publishers, a division of John Wiley & Sons, Inc., New York, 1964.

Steele, G. L.: "Fiberglass: Projects and Procedures," McKnight & McKnight Publishing Company, Bloomington, Ill., 1962.

Swanson, R. S.: "Plastics Technology," McKnight & McKnight Publishing Company, Bloomington, Ill., 1965.

Chapter 13

Groves, W. R.: "Plastics Molding Plant," vol. 1, London Iliffe Books, Great Britain, 1965.

Chapter 14

Holland, L.: "Vacuum Deposition of Thin Films," John Wiley & Sons, Inc., New York, 1956.

Kobayashi, A.: "Machining of Plastics," McGraw-Hill Book Company, New York, 1967.

Marcus, H.: "Metallizing of Plastics," Van Nostrand Reinhold Company, New York, 1960.

Chapter 15

Frank, H. P.: "Polypropylene," Gordon and Breach, Science Publishers, Inc., New York, 1968.

Kresser, T. O.: "Polyethylene," Van Nostrand Reinhold Company, New York, 1957.

———: "Polypropylene," Van Nostrand Reinhold Company, New York, 1960.

Palin, G. R.: "Plastics for Engineers," Pergamon Press, New York, 1967.

Chapter 16

Basdekis, C. H.: "ABS Plastics," Van Nostrand Reinhold Company, New York, 1964.

Huke, D. W.: "Introduction to Natural and Synthetic Rubbers," Chemical Publishing Company, Inc., New York, 1961.

Teach, W. C. and G. C. Kieseling: "Polystyrene," Van Nostrand Reinhold Company, New York, 1960.

Thompson, S. M.: "Gum Plastics," Van Nostrand Reinhold Company, New York, 1968.

Chapter 17

Bruins, P.: "Plasticizer Technology," Van Nostrand Reinhold Company, New York, 1965.

Schildknecht, C. E.: "Vinyl and Related Polymers," John Wiley & Sons, Inc., New York, 1960.
Smith, W. M.: "Vinyl Resins," Van Nostrand Reinhold Company, New York, 1958.

Chapter 18
Rudner, M. A.: "Fluorocarbons," Van Nostrand Reinhold Company, New York, 1958.

Chapter 19
Akin, R. B.: "Acetal Resins," Van Nostrand Reinhold Company, New York, 1962.
Carroll-Porcysynski: "Manual of Man-made Fibers," Chemical Publishing Company, Inc., New York, 1961.
Christopher, W. and D. Fox: "Polycarbonates," Van Nostrand Reinhold Company, New York, 1962.
Floyd, D. E.: "Polyamide Resins," Van Nostrand Reinhold Company, New York, 1958.
Mark, H. F., S. M. Atlas, and E. Cernia (eds.): "Man-made Fibers: Science and Technology," vol. 1, Interscience Publishers, a division of John Wiley & Sons, Inc., New York, 1967.

Chapter 20
Cope, A.: "Cope's Plastic Book," Goodheart-Wilcox, 1960.
Horn, M. B.: "Acrylic Resins," Van Nostrand Reinhold Company, New York, 1960.
Lappin, A. R.: "Plastics Projects and Techniques," McKnight & McKnight Publishing Company, Bloomington, Ill., 1965.
"Acrylite Fabrication Manual," American Cyanamid Company, Building Products Division, Wakefield, Mass., 1967.
"How to Work with Plexiglas," Cadillac Plastics and Chemical Company, Detroit, 1960.
"Fabrication of Plexiglas," Rohm and Haas Company, Philadelphia, 1966.
"Plexiglass Design, Fabrication, and Molding Data," Rohm and Haas Company, Philadelphia, 1964.
"Plexiglas Handbook for Sign Shops," Rohm and Haas Company, Philadelphia, 1965.

Chapter 21
Bender, R. J. (ed.): "Handbook of Foamed Plastics," Lake Publishing Company, Libertyville, Ill., 1965.
Donbrow, B.: "Polyurethanes," Van Nostrand Reinhold Company, New York, 1965.
Ferrigno, T. H.: "Rigid Plastic Foams," Van Nostrand Reinhold Company, New York, 1963.
Parker, D. V.: "Polyurethanes: Chemistry, Technology and Properties," Gordon and Breach, Science Publishers, Inc., New York, 1964.
Saunders, J. H.: "Polyurethanes: Chemistry and Technology," Van Nostrand Reinhold Company, New York, 1963.

Chapter 22
Haynes, W.: "Cellulose: The Chemical That Grows," Doubleday & Company, Inc., Garden City, N.Y., 1953.
Paist, W. D.: "Cellulosics," Van Nostrand Reinhold Company, New York, 1958.
Yarsley, V. E., W. Flavell, P. S. Adamson, and N. G. Perkins: "Cellulosic Plastics," London Iliffe Books, Great Britain, 1964.

Chapters 23 and 24

Mink, W.: "Practical Injection Molding of Plastics," London Iliffe Books, Great Britain, 1964.

Moslo, E. P.: "Runnerless Molding," Van Nostrand Reinhold Company, New York, 1960.

Chapter 25

Butzko, R. C.: "Plastics Sheet Forming," Van Nostrand Reinhold Company, New York, 1958.

Dow Chemical Company: "Fundamentals of Sheet Forming," technical brochure, Midland, Mich., 1962.

Chapter 26

Glyde, B. E. (ed.): "Screw Extrusion of Plastics," Gordon and Breach, Science Publishers, Inc., New York, 1963.

Griff, A. L.: "Plastics Extrusion Technology," Van Nostrand Reinhold Company, New York, 1965.

Chapter 27

Jones, D. A. and T. W. Mullen: "Blow Molding," Van Nostrand Reinhold Company, New York, 1961.

Chapter 28

Volk, M. C., J. W. Lefforge, and R. Stetson: "Electrical Encapsulation," Van Nostrand Reinhold Company, New York, 1962.

Chapter 29

Farkas, R.: "Heat Sealing," Van Nostrand Reinhold Company, New York, 1964.

Haim, G.: "Manual for Plastic Welding," vol. 3, Chemical Publishing Company, Inc., New York, 1960.

Neumann, J. A. and F. J. Bockhoff: "Welding of Plastics," Van Nostrand Reinhold Company, New York, 1959.

Zade, H. P.: "Heat Sealing and High Frequency Welding of Plastics," John Wiley & Sons, Inc., New York, 1960.

Chapter 30

"ASTM Standard for Plastics," parts 26 and 27, American Society for Testing and Materials, Philadelphia, 1968.

Lever, A. E. and J. Rhys: "The Properties and Testing of Plastics Materials," Chemical Publishing Company, Inc., New York, 1958.

Perlman, Philip: "General Laboratory Techniques," Franklin Publishing Company, Englewood, N.J., 1964.

Chapters 31 to 35

"ASTM Standards for Plastics," parts 26 and 27, American Society for Testing and Materials, Philadelphia, 1968.

INDEX